W0043619

Issues and Reviews in Teratology

Volume 5

Editorial Board

PATRICIA A. BAIRD
Vancouver, British Columbia, Canada

JOËLLE G. BOUÉ
Paris, France

F. CLARKE FRASER
Montreal, Quebec, Canada

ANDREW G. HENDRICKX
Davis, California

ANTHONY R. SCIALLI
Washington, D.C.

WILLIAM J. SCOTT, Jr.
Cincinnati, Ohio

FRANK M. SULLIVAN
London, England

MINEO YASUDA
Hiroshima, Japan

A Continuation Order Plan is available for this series. A continuation order will bring delivery of each new volume immediately upon publication. Volumes are billed only upon actual shipment. For further information please contact the publisher.

Issues and Reviews in Teratology

Volume 5

Edited by

Harold Kalter

Children's Hospital Research Foundation and
Department of Pediatrics
University of Cincinnati College of Medicine
Cincinnati, Ohio

Plenum Press • New York and London

The Library of Congress cataloged the first volume of this work as follows:

Main entry under title:

Issues and reviews in teratology.

Includes bibliographical references and index.
1. Teratogenesis. 2. Abnormalities, Human. 3. Abnormalities (Animals) I. Kalter, Harold.
QM691.I67 1983 616′.043 83-6323

ISBN-13: 978-1-4612-7847-4 e-ISBN-13: 978-1-4613-0521-7
DOI: 10.1007/978-1-4613-0521-7

© 1990 Plenum Press, New York

Softcover reprint of the hardcover 1st edition 1990

A Division of Plenum Publishing Corporation
233 Spring Street, New York, N.Y. 10013

All rights reserved

No part of this book may be reproduced, stored in a retrieval system, or transmitted in any form or by any means, electronic, mechanical, photocopying, microfilming, recording, or otherwise, without written permission from the Publisher

Contributors

Clarke Fraser • Centre for Human Genetics, McGill University, Montreal, Quebec H3A 1B1, Canada

Carl A. Huether • Department of Biological Sciences, University of Cincinnati, Cincinnati, Ohio 45221-0006

James R. Miller • Central Research Division, Takeda Chemical Industries, Juso Honmachi, Osaka, Japan; *present address*: 3744 West 12th Avenue, Vancouver, British Columbia V6R 2N6, Canada

Paul B. Selby • Biology Division, Oak Ridge National Laboratory, Oak Ridge, Tennessee 37831-8077

Irene A. Uchida • Departments of Pediatrics and Pathology, McMaster University, Hamilton, Ontario L8N 3Z5, Canada

William S. Webster • Department of Anatomy, University of Sydney, Sydney, New South Wales 2006, Australia

Frank Welsch • Department of Experimental Pathology and Toxicology, Chemical Industry Institute of Toxicology, Research Triangle Park, North Carolina 27709

Preface

Why Efforts to Expand the Meaning of "Teratogen" Are Unacceptable

Disagreement about nomenclature in teratology is not new. Dissent even about the very fabric of the discipline—what congenital malformations consist of—has often been voiced. Time, instead of resolving such difficulties, has sometimes worsened them.

For example, in the past it was agreed that congenital malformations are abnormalities of structure present at birth, but differences of opinion concerning where the line between normal and abnormal was to be drawn prevailed. It was obvious that, in order to discover the causes of congenital malformations and cast strategies for their prevention, it would be necessary to have knowledge of the baseline of their frequency, and that this required uniformity of definition of terms. Since malformations of primary social concern are those having grave outcomes (and are, paradoxically, also the commonest ones), it is logical that such conditions were the first consideration of investigators and were the defects whose frequency was considered to comprise the required baseline.

Thus, the earliest dividing line was drawn, not between the normal and abnormal—a nearly impossible task—but, in a pragmatic fashion, between what came to be called major malformations, defects with serious life and health consequences, whose totality would form the basic statistic, and minor malformations, those with little or no threat to well-being, which would be enumerated separately and so not distort the baseline frequency data. This worked well, since for the most part the distinction was obvious and was adhered to, by both experimental teratologists and those concerned with the human problems.

As long as teratological studies were confined to the epidemiological, clinical, and experimental spheres this arrangement was fairly well honored and answered the need. Even 25 and more years ago, however, restriction of the term *malformation* to the morphological apparently was not acceptable to all. For example, Potter (1964) noted that hereditary metabolic disturbances had come to be included in this category, a tendency she deplored as it would lead to the term losing all significance and specificity, because in a broad sense, as she put it, "a large share of all diseases might be considered malformations."

A more forceful drive to widen the inclusiveness of the term came from another direction. The redefined obligation to test chemicals to assess their teratogenic potential before projected marketing, stemming from the thalidomide episode, soon eventuated in the realization that concepts and terminology adopted for clinical and investigational purposes were not sufficient to serve all the new needs and concerns.

The procedures mandated by governmental regulatory agencies for performing teratological testing were such—unexpectedly, it seems—that they led to circumstances that had not ordinarily been encountered in previous teratological pursuits, i.e., the production of embryonic and fetal damage that was possibly mediated by harmful effects of the agent upon the pregnant organism itself. Such situations, created by the regulatory injunction to produce adverse maternal effects, engendered the problem of distinguishing between such fetal harm and that due to the direct prenatal action of agents.

It may be asked why, since fetal damage occurs in either case and it may even be possible for damage from both sources to happen during a given pregnancy, it is important to try to differentiate between these two avenues of deleterious action. The answer is twofold. First, so far as the practical question of testing is concerned, the distinction must be made to avoid misinterpretation, because it is often only following the application of dosages of test substances usually so large as to have little or no human relevance that maternal toxicity, and whatever fetal consequences may ensue, occurs.

But the importance of making this distinction transcends the industrial setting, since problems with regard to the possible connection between maternal and fetal toxicity are also entailed in human experience. Thus, many of the human teratogens, alleged or otherwise "discovered" AT (anno thalidomidi)—anticonvulsants, anticoagulants, synthetic progestins, synthetic estrogens, as well as environmental additives, pollutants, and contaminants, common medications, social addictives, vaginal spermicides, spray adhesives, occupational agents, and so on—raise these problems. Compounding the difficulty (aside from questions of specificity, reversibility, innocuousness, and the like of their putative developmental outcomes), exposure to such agents is usually combined with numerous troublesome and seemingly inextricable confounding factors that muddy efforts to determine causal relationships—human behavior at least as much as prenatal development consisting of intertwined features, making analysis of individual ones virtually impossible and perhaps misguided.

But this excursion into the troubled teratological byway of the interconnectedness of maternal and fetal toxicity was only a prelude to further efforts to expand the concepts and definition of congenital malformations. These perhaps had their beginning as an offshoot, in fact, of the "toxicity" puzzle, with the carefully worded inclusion by Wilson (1973) of "functional deficiency" as a type of deviant development that could theoretically be capable of resulting from an adverse influence acting in later gestation. From this diminutive tot, it seems, has sprung forth a motley assortment of lineal descendants, whose efflorescence has recently yielded some especially egregious pronunciamentos.

A by-product of the mandate noted above, which requires testing of chemicals for teratogenic potential, was that this obligation soon became transformed into testing for embryotoxic potential—quite a different kettle of fish; and the search for expressions of such phenomena soon reached into the area of early postnatal behavior. It is gratifying that such problems as disagreement about testing methodology, inconsistencies in the findings, and difficulties of interpretation, which have beset studies of such endpoints, are being ironed out, and it is hoped that this area will find a useful niche in studies of developmental damage of prenatal origin.

However, a warning must be voiced. The legitimacy of this extension within the testing process for the detection of potential fetopathogens or teratogens is not to be challenged so long as it conforms to the accepted definition of a teratogen as, e.g., Shepard (1982) has defined it, "an agent that, when applied during prenatal life, produces a permanent change in morphology or function." But when, as has been claimed (Vorhees, 1983), the territory staked out for behavioral teratology includes not only "abnormal behavioral development which results from damage to the embryo or fetus" but "also encompasses insults occurring postnatally, pregestationally and spermatogenically," the grasp on the nature of teratology of such claimants seems clouded.

This tendency to scientific aggrandizement (another example of which will be noted below) is objectionable, because its consequences are to blur, if not obliterate, distinctions between areas of research that can be profitably pursued only if their separate natures and problems are understood and respected. Behaviorists are finding it daunting enough to disentangle prenatal and postnatal effects on development (Hutchings and Fifer, 1987), in children and even in experimental situations, where etiology is not in question; widening the purview of their field is certain only to compound many of their problems.

Another recent indefensible attempt at stretching the boundaries of teratology included a mishmash of spontaneous abortion, microcephaly, growth retardation, metabolic dysfunction, cognitive dysfunction, mental deficiency, malignancy, and altered social behavior (Holmes, 1988), as well as garden variety major and minor malformations, among the potential effects of intrauterine exposure to a teratogen. If the author of this rash assertion has thought the implication of his words through, it can only be concluded that he believes all these items are teratogenic effects, as they may be due to a teratogen, and thus are equivalents of one another and are to be called malformations. Is this not reminiscent of Potter's anxious prediction of 25 years ago?

The first objection to be leveled against this sophistical approach is that it represents a drastic loosening of the bounds of widely accepted terminology, will engender a babel of mutual misunderstandings, and will lead to erosion of the infrastructure of our discipline. Such an indiscriminate linking of proven, suspect, and highly conjectural outcomes of environmental influences will also tend to permit uncritical and often unjustified acceptance of every form of developmental problem as being of prenatal origin and environmental causation, to the detriment of serious study of phenomena credibly of such derivation.

Perhaps it is because these terminological tinkerings have led some to feel that the meaning of standard terms has grown obscure, there has come about of late, unrestrained by editorial judgment, a coining of redundant neologisms. Such lexical chimeras, e.g., as *dysmorphology*, apparently considered more to fit the bill than plain *teratology*, or perhaps to be better suited to certain regional needs, are themselves in turn destined for transmogrification, and are already budding off such teratical hydras as *dysmorphogenic* and *dysmorphogenicity*.

We are not obscurantist, but we do not feel that enlightenment proceeds strictly from word juggling. Neither are we for splitting, nor for lumping; only for a clear vision of the need and the reason for keeping separate things separate—so that the focus on their individual problems will not be confused, weakened, or sidetracked.

Harold Kalter

Cincinnati, Ohio

REFERENCES

Holmes, L. B. 1988. Human teratogens: delineating the phenotypic effects, the period of greatest sensitivity, the dose–response relationship and mechanisms of action, in: *Transplacental Effects on Fetal Health*, D. G. Scarpelli and G. Migaki, eds. Liss, New York, pp. 177–192.

Hutchings, D. E., and Fifer, W. P. 1987. Neurobehavioral effects in human and animal offspring following prenatal exposure to methadone, in: *Handbook of Behavioral Teratology*, E. P. Riley and C. V. Vorhees, eds. Plenum Press, New York, pp. 141–160.
Potter, E. L. 1964. Classification and pathology of congenital anomalies. *Am. J. Obstet. Gynecol.* **90**:985–993.
Shepard, T. H. 1982. Detection of human teratogenic agents. *J. Pediatr.* **101**:810–815.
Vorhees, C. V. 1983. Behavioral teratogenicity testing as a method of screening for hazards to human health: a methodological proposal. *Neurobehav. Toxicol. Teratol.* **5**:469–474.
Wilson, J. G. 1973. *Environment and Birth Defects*, Academic, New York.

Contents

Chapter 1

Of Mice and Children: Reminiscences of a Teratogeneticist

Clarke Fraser

Chapter 2

The Concept of Homology in Comparative Mammalian Teratology

James R. Miller

Chapter 3

Short-Term Methods of Assessing Developmental Toxicity Hazard: Status and Critical Evaluation

Frank Welsch

Chapter 4

Twinning in Spontaneous Abortions and Developmental Abnormalities

Irene A. Uchida

Issues and Reviews in Teratology **5**:1–75
Plenum Press, New York, 1990, 978-1-4612-7847-4

Of Mice and Children 1

Reminiscences of a Teratogeneticist

CLARKE FRASER

1. PRENATAL AND FAMILY HISTORY

I was conceived in a little log cabin on the shores of Lake Kejimakujik, which may be why I love the lakes and woods of Nova Scotia so much. My mother neither drank nor smoked (then) and the pregnancy was uneventful, except that my parents moved to Norwich, Connecticut, where in 1920 I was born and lived for 9 months. My zygodactylous toes were hidden at first from my mother, for fear she would think her firstborn son defective, but since my father had them too (as well as my grandfather and Aunt Eva) she was not surprised. Perhaps my web toes account for my love of genetics and teratology (and swimming?), but I do not recall in my youth a particular interest in either heredity or abnormal development.

2. CHILDHOOD

We returned to Canada in 1921 and lived in Montreal until my father became a Canadian Trade Commissioner, a government representative charged with promoting trade between Canada and wherever he was posted. Thus, my sister and I spent much of our childhood outside Canada—2 years in Dublin and 10 (from age 7 to 17) in Jamaica. On leaves, or between postings, we came back to my grandfather Clarke's home in Bear River, Nova Scotia, at the western end of the Annapolis Valley, a picturesque village on a tidal river that alternated be-

CLARKE FRASER • Centre for Human Genetics, McGill University, Montreal, Quebec H3A 1B1, Canada.

tween gleaming lake and glistening mud flats twice a day. My grandfather's house has always felt like home. A big white frame house, behind a row of maples, with a large veranda sweeping around front and side, a graceful lawn sloping down to a green barn, and an apple orchard. My grandfather Clarke and his two brothers were the business magnates of the town—shipbuilding, lumber and pulpwood, a large general store, and fruit growing.

I have idyllic childhood memories of Bear River, all soaked in golden sunshine. Of picking baskets of cherries and apples, which were taken to the barn and poured out on an enormous table for sorting and packing into boxes and barrels to be shipped to who knows where. Wonderful varieties of apples—gravenstein, cox orange, yellow transparent, russet, bishop pippin, northern spy, sweet bough, and MacIntosh, each with its characteristic fragrance and texture.

I remember sitting on the veranda watching the cows come ambling down from the pasture on the other side of the road, down to the green barn to be milked (I could never quite get the hang of it myself), and pails of warm foamy milk carried up from the barn and down to the cellar where a mysterious machine called a separator made whirring noises and squirted out cream from one spout and skim milk from the other. Pasteurization? never heard of it. And ox-drawn hay carts plodding past—they seemed to take half an hour to traverse our stretch of road, but what was the hurry? The house always seemed full of people—various aunts, uncles, and cousins, and two maids who sang in the kitchen. It was a warm, loving, and gracious family, remembered with affection and gratitude.

Later my sister and I came back to Bear River when we were sent from Jamaica to go to college—Acadia University, 80 miles down the

Figure 1. The Clarke house, Bear River, Nova Scotia.

Figure 2. FCF, grandfather Clarke, and sister Mary, 1933.

road. My aunt Edie was like a second mother to us. She and my uncle How had stayed in the Bear River house almost all their lives, taking care of my grandfather after my grandmother died quite young, and maintaining the house to which various other family members retreated when a haven was needed. She was a loving martinet. No cards on Sunday, no liquor in the house (except for a small bottle of brandy in the buffet for "medicinal purposes"), no messing around with girls, particularly before marriage. Everyone (except grandfather) made his own bed and mopped his room every day. Quite a change from the relaxed atmosphere of life in Jamaica, with rum-punch parties on Sunday morning and servants to look after us. I remember one New Year's Eve when we were home from Acadia, there had just been a beautiful snowfall, the moon was full, and we wanted to go skiing down the schoolhouse hill. But alas, we were forbidden, as it was Sunday, and I guess skiing was entertainment, so we had to wait till midnight before joining our friends. Dear Aunt Edie, she lived to be 97, died in the house, and remained bright as a button to the end. She probably influenced my character formation more than she knew.

Figure 3. FCF and Aunt Edie.

3. THE NUCLEAR FAMILY

I remember my mother as an attractive, vivacious woman with a beautiful alto voice—she had taken a music degree at University, and I can remember impromptu concerts in the parlor with my father playing the piano, and her singing, or sometimes there were duets. She expressed her love liberally—in fact, I remember being rather embarrassed when she and my sister would get a bit too "gushy" with each other. On the other hand, she was a strong disciplinarian. She made sure we knew what was right and what was wrong, and that we knew when she was displeased and disappointed, which was usually enough to keep us in line. We were taught by example that one does not raise one's voice in anger (though loud, exasperated sighs were permitted), which is probably why I still dislike confrontation, and am regarded as having a very even temper. Various friends and children (who keep having psychological revelations) tell me it is bad for me to keep it all pent up (all what, dears?) and that I should stamp and scream more. But I wouldn't enjoy it, and doubt if I'd feel any better.

Because most of my father's postings were to places that had no Canadian consulate, he would find himself pinch-hitting as Consul, which meant quite an active social life. This my mother was very good at, and she was known by many, including the Canadian Navy during the war, as a gracious and charming hostess with an extraordinary ability to initiate and maintain vivacious conversations without neglecting her hostess duties. This ability she somehow failed to transmit to her son, who turned out to be more like his reticent father. Not such a bad thing, though.

My father was a quiet man, much less demonstrative about his love for her and us. He had a set of good friends, and was admired and loved by many, and he liked to sit with them, smoke his pipe, and chuckle now and then, but not dominate the conversation. He loved poetry and could recite it by the hour, but seldom did. Because of his diplomatic status we never discussed politics, which may explain my general lack of enthusiasm for political activity. There seemed also to be a taboo on discussion of money, and particularly of incomes. Could this explain my aversion for financial matters, particularly budgets, and my reluctance to take on administrative responsibilities? I have never been a departmental chairman, much less dean. Since I have no identical twin, separated at birth, we will never know. My father, too, had strong ideas about right and wrong, but was much less articulate about them. His one attempt to talk to me about sex occurred when he was driving me to my boarding

Figure 4. FCF and sister Mary, 1944.

school, Munro College, in the Jamaican hills, and he managed to say did I know it was not a good idea to play with myself. I don't know which of us was the more embarrassed. I got to know him much better when he retired to Bear River, and I was raising a family of my own, and the better I knew him the more I loved him. His death, from a sudden stroke at 75, was one of the two most grievous experiences of my life.

The third member of my nuclear family was my sister Mary, 2 years my junior. We were very close as children and I remember the pleasure I took in teaching her things—for example, when I came back from Munro for the holidays and found her struggling with elementary algebra, and the exultation of the moment when I had made it suddenly come clear. Was this a source of my love for teaching?

She grew up into a tall, slender, handsome woman who became a very successful reporter for the *London Sunday Express*. She spent some time working for the *Toronto Globe and Mail*, and would come to Montreal to see me from time to time at McGill. I remember having lunch with her at the Faculty Club in 1948 and hearing about this guy she had interviewed who said he could tell the sex of cats just by looking at slides of their tissues. Don't be ridiculous, I said, people have been looking at tissues under the microscope for ages and if you could tell sex that way it would have been recognized long since. "Don't print it," I said. Well the "guy" was Murray Barr who had just discovered the sex chromatin, so I might have done Mary out of a scoop, but she printed it anyway.

We had a memorable meeting in New York City in 1954. I had gone to talk to the Rockefeller Foundation about support for the epilepsy project of one of my colleagues, Julius Metrakos, and after that discussion had been asked if I was interested in a traveling fellowship to Europe to visit various human genetics centers and find out what was going on. Was I ever, and I did have such a tour the following year. Mary, by a strange coincidence, was on her way from London to Australia, via New York, to cover the Royal Tour to Australia and points east for the *Sunday Express*. So we arranged to meet in New York. Her plane was very late and it was about 2:00 A.M. when she finally emerged from Immigration. We went to a hotel—we'd like a room, this is my sister, ha ha yes of course—and it must have been the first time we had had a chance to really talk for many years. She had a bottle of cognac which smoothly disappeared over the next 6 hours of heart-to-heart, and then back to the airport en route to Australia and Montreal, respectively. I remember that interlude as the time when she became much more than just my kid sister: a special and wonderful person—I could talk more intimately with her than with almost anyone. This feeling was reinforced when I stayed with her and her very successful TV producer husband in London the following year, on the Rockefeller-sponsored tour. Alas, it was not too many years later than she developed signs of "nervous exhaustion" and was sent home to Bear River to recuperate, but did not do so and was diagnosed in Montreal as having Alzheimer's disease; that was in the days when it was strictly *pre*senile dementia. She went back to England in the care of her husband. After she died I had several episodes of disorientation and confusion, which led me to consult a neuropsychiatrist as to whether I might be getting it too. After an EEG and normal physical, plus a conversation with the neuropsychiatrist, these episodes ceased. I think of this whenever I counsel persons at risk for Huntington's disease or other late-onset disorders.

4. SCHOOL

In Dublin I had a governess and in Kingston went to a small private school, until my father decided I was in danger of becoming a sissy. There was probably some truth to this. I was quite fat, and not athletic, and when I wrestled with my male playmates I would always end up flat on my back with my forehead being knuckled and refusing to "cry uncle." Then, when they were sent off to boarding school I was left with my sister's female friends as playmates, so I was sent off to Munro, a boarding school in the hills (healthy climate) run on the lines of the English public (i.e., private) school system. The school was on a bluff with a marvelous view of the south coastal plain and Caribbean. The dormitories had long rows of red-blanketed cots, with washbasins in the aisles and pots under the cots. No privacy. The showers (cold) were in another building, as were the outhouses. I remember running to the showers in the midst of a hurricane, risking decapitation by sheets of corrugated iron ripped off the roofs. Probably the mildly rigorous regime of Munro reinforced the "grin and bear it" attitude I had learned from my father (who said he learned it in the army) while, for example, we trudged unendingly in the rain, through boggy heather, to picnic by the side of a foggy Irish lake. This probably helped me tolerate discomfort less complainingly later on in life.

There was a main building, built of local limestone cemented with red clay, and various smaller units—the headmaster's house, the infirmary, the masters' rooms, the chapel, built around the sides of a cement square called the barbecue because it had formerly been used to roast coffee beans in the sun. On this we used to line up to pick up teams for the games we had to play every afternoon, like it or not. Because I was fat I always seemed to be picked last, along with the other fat boy, with whom I once got into a terrible fight which must have been very funny to everyone but us—these two tubby blobs swinging wildly and for the most part harmlessly at each other.

I was very homesick, and used to suffer agonies on the few days before and after the beginning of each term. I would wake up before the 6:00 A.M. rising bell, thinking I could never live up to expectation. I was put in form 3B to begin with but, unknown to me, someone decided I should be in 3A, the next higher form. So, in the middle of a class, two boys walked in, picked up my desk, and carried it to the other classroom, where I was deposited without further explanation. At that point the master said "Kirkham—decline *hic.*" Kirkham got up and said "*Hic haec hoc hoc hac hoc huius huius huius. . . .*" I was devastated. I thought how am I ever going to survive when they don't even speak my language? But, somehow, I did.

Few of the boys at Munro were "white," so I experienced, quite early in life, what it was like to belong to a minority. Well, not really. It was a minority, but not a disadvantaged one. I learned by experience that skin color did not signify social class, intelligence, personality, or athletic prowess. The only racist views I heard at school were those of mulattoes, who considered themselves "white" and looked down on the "blacks," and I took pride in telling my Canadian friends that there was no racism in Jamaica. Snobbery among my parents' friends was based on occupation and money, not skin color. When I revisited some of these friends of my parents many years later I was surprised to hear remarks that were unmistakably racist, so probably the racism was there but I was too idealistic to see it.

Munro was a good school. The masters were mostly from England. "Fuzzy" Carter taught physics, "Maddy" Harrison mathematics, "Bulldog" Dunleavy was the sports master, and English literature was taught by, believe it or not, Claude Balls. He taught me to "murder my darlings"—i.e., strike out the phrases that you thought especially good. I am not sure that I have fully learned the lesson. The French master was Mr. Wien, a portly old Austrian gentleman known, for some inscrutable reason, as "Bolt," who also played the organ in chapel. I have a photo of the pipe organ entitled "Hic sedet Bolt." Chapel services were supposed to be nondenominational, but followed the Church of England prayer book, so the Anglican service still evokes strong (nostalgic?) religious feelings in me. I read voraciously in the school library, and came across the theory of evolution in a book by Lancelot Hogben (or was it Julian Huxley?) which seemed so eminently logical that I had no trouble accepting it in favor of divine creation.

In Kingston we lived in a gray stucco bungalow with a wide black marble veranda around two sides, and green and white awnings. The garden had lots of ferns, kala lilies, a banyan tree, and tall bougainvillea hedges. We had a butleress, a laundress, a garden boy, a chauffeur, and a lawn tennis court. Did this pampering make me lazy? Obviously not, but it may have made me a little less attentive to the demands of domestic duties when there were other outlets for my energies.

I did well in my studies at Munro, missing the top of the class in sixth form (final year) by a tenth of a point, for which I remember shedding a few tears after adding the marks up while sitting, sheltered from view, behind the whitewashed wall of the sixth form school room. The final year specialized in math, chemistry, and physics (no biology), with English grammar and composition and French thrown in, and when I returned to Canada to enter university I found I had a good grounding compared to most of my classmates.

Living in Jamaica meant that I could play tennis and golf and swim all year round. The only trouble was being fat from the age of 10 to 15 or so, and it was not until my last term at school that I discovered I could run, and made the track team. We were being made to run the mile and I was plugging along near the tail of the pack until about halfway through the fourth and final lap when I suddenly realized I wasn't tired, and put on a sprint that brought me in second. But being fat made me learn to anticipate in soccer and cricket, and did not deter me from cross-country runs which always seemed to end up at the girls' boarding school 5 miles away, where we would hide in the bushes to watch them play hockey. Being forced to participate in sports, though resented at the time, set a pattern that led to many years of activity in soccer, rugger, and tennis that have enriched my life.

5. UNIVERSITY

At 17 I returned to Nova Scotia to enroll at Acadia, a small Baptist university that had a remarkable record of turning out students who later became outstanding biologists—Nobel Laureate Charles Huggins, and Keith Porter, for example. Those were happy days. Acadia was so small that the students all knew one another. I was active in athletics (track, water polo, soccer—I was too light to make the rugger team), dramatics, and romantics. My Munro education gained me entry as a Freshie-Soph in the premed program, with a view to becoming a doctor. I don't remember why I wanted to become a doctor, but it probably had something to do with my admiration for my great-uncle Lew, literally a horse-and-buggy (or sleigh) doctor, whom I admired very much. Tall, spare, stooped, always amiable, and never in a hurry, he took care of the ills of the people of Bear River—and sometimes of their animals too—in a way that was almost parental, being loved, respected, and admired in a way that would be almost impossible today. As well, there was my romantic and idealistic state at the time, and a desire to please—particularly to please my mother, let the Freudians make of that what they will. I was also curious, not just about sex, and I liked mathematics.

Biology 1 had two lectures in genetics, given by Dr. Muriel Roscoe, who later went on to be chairman of the Botany Department and dean of women at McGill University. She noticed me because I took very brief notes, whether out of laziness or stupidity she could not make out, but since my marks turned out to be good she decided it was probably not stupidity. Anyway, I was hooked on genetics from then on. I thought I was good at mathematics (though that seems to have disappeared somewhere along the line) and I was entranced by the beauty of Mendel's laws and the rigor with which they applied to so many characters in so many organisms. I was also beginning to become aware of the beauty of biological structure, from the anatomical to the microscopic. And the extraordinary variety and ebullience of living species. I learned without complaint the names of 100 plant species in the greenhouse. I learned to observe, by drawing, in comparative anatomy. Professor Perry probably changed my life by separating me, in anatomy lab, from my cousin and long-time dear friend, Lew, and putting me with someone who could draw. He also took us on field trips up a treasure-filled ravine from which he would return with newts and other treasures under his hat. I learned to use taxonomic keys (and not to eat unidentified mushrooms) in Dr. Roscoe's mycology class. I had my first taste of research, and of cytogenetics, when I did an honors thesis in which I established the chromosome number of an obscure water plant, Najas. Big deal, but to make an original observation, however insignificant, was exciting and how many undergraduate biology students have such a chance? I became increasingly conscious of the extraordinary beauty, variety, and richness of living things. This, coupled with the intellectual rigor of genetics, was an irresistible combination.

We were very impressed by a visit from Frederick Banting, who told us of how he had had his seminal idea, of tying off the pancreatic duct, in the middle of the night, and wrote it down in a notebook he always kept by his bedside for such ideas, since he knew he would have forgotten it by morning. I resolved to do that, but never got sufficiently organized.

6. GRADUATE SCHOOL

In 1940 I applied to graduate school under Leonard Huskins at McGill University, who had just set up the first department of genetics in Canada. Apparently Botany and Zoology could not agree on who should teach genetics (I'm not sure whether both wanted it or didn't want it), so Huskins got a grant from the Rockefeller Foundation to set up a genetics department. Huskins said I could come if I got a National Research Council bursary and, to our mutual surprise, I did. Since I hadn't had a course in genetics, he required me to read Sinnott and Dunn's textbook before I came, but never tested me on it and to this day I have never taken or given an elementary genetics course. I did take, and was inspired by, Arthur Steinberg's course in developmental genetics, on which I modeled the course I later gave. I was also inspired by John Berrill and his course in embryology, who taught me the Socratic method and about the beauty of development. George Scarth's course in plant physiology taught me about cells and membranes, and the elegance of the physicochemical approach to biology. There were also lots of interesting discussions with fellow graduate students, including Roger Boothroyd (still at McGill) and Herbert Stern (subsequently chairman of biology at the University of California at San Diego). It was a young department, and closely knit—we met around a big wooden table for coffee in the morning and tea in the afternoon, and the discussions were vigorous and stimulating. They taught me to argue with passionate rigor without rancor. These days we seem to be too busy for such time-outs, but they were valuable mind-stretchers, and it's too bad they're gone.

Every paper that came out of the department had been read by several members and there were long arguments about the precise wording, often around the coffee table, to the benefit of the students. There was less pressure to publish in those years, and more aspiration to perfection, to the point where a manuscript even when thought to be ready would be put away for 6 months to see how it looked after a period of "maturation," before it was submitted for publication.

Sheldon Reed was there, working on the genetics of cleft lip in the mouse, and the genetics of coat color as revealed by skin transplants. He went on to head the Dight Institute of Human Genetics in Minneapolis. Alma Howard (Rolleston), a tall, handsome, elegant, kind young woman, had recently finished a postdoc and was working on the relation of chiasmata to mammary cancer susceptibility. She went on to a distinguished career at the Medical Research Council Radiotherapeutic Research Unit at Hammersmith where she pioneered the use of radioisotopes to reveal the stages of the cell cycle. Arthur Steinberg (who has recently retired from Western Reserve University after a distinguished career) had just arrived to take his first job after graduating from Columbia University under T. H. Morgan and the great but short-lived C. B. Bridges. There were visits from various "big names," and I remember the acute anxiety state when asked to explain my work to Curt Stern or Herman Muller, in spite of their kindly and entirely nonintimidating interest. It was an exciting place to be, but I don't remember realizing at the time how lucky I was to be there.

7. GRADUATE STUDIES

When I got to McGill I was told to look around the department and decide what to work on. There were corn, Trillium (nice chromosomes), sawflies (spruce bud-worm parasites), Drosophila, and mice. I must still have had some affinity for medicine, because the minute I walked into the mouse room I knew that that was where I wanted to work. But there was no room in the mouse room that year, so I did my master's with Arthur Steinberg, working on the interchromosomal effects of inversions on crossing-over in Drosophila. I was disappointed at first, but the exposure to rigorous genetic analysis did me good. Steinberg was a hard taskmaster, but once he decided I was worth taking seriously he worked hard to whip me into shape. While I sat in the front row during his developmental genetics lectures, my furious concentration would result in a sort of glassy-eyed stare, and I still took only brief notes. One day he said, "Fraser, if you must sleep in my class, please do so in the back, not the front row!" I was devastated, but meekly moved to the back row and tried to look more wakeful. Now I take notes mainly to keep awake; perhaps that's where the habit started.

We looked at the effects of X-chromosome inversions on crossing-over in chromosome 3, and I was doing a 12-point crossover test that involved a lot of very tedious scoring of the flies. The first results didn't make sense and Steinberg finally decided we should run duplicate cultures to see where I was goofing up. But his results didn't make sense either, and we finally tracked the trouble down to an inversion in the control stock! Our personalities did not hit it off at first. Perhaps we were both too arrogant (yes, I too was arrogant at that age) but we soon overcame the initial abrasiveness and he accepted me as a friend; I valued this greatly and remember my pleasure when Edith and he would invite me and my current girlfriend over for spaghetti suppers.

For my Ph.D. I finally got to work on mice, also under Steinberg's guidance. We investigated the epigenetics of various hair and skin mutants by histological studies and skin graft experiments. Two experiences again impressed me about the importance of good controls. A wavy hair mutation had occurred in our laboratory, and, since the study of mutant autonomy was au courant in developmental biology, I grafted wavy skin from newborns to nonwavy sibs of different coat colors, thereby labeling graft and host tissues. The *host* hair around the graft grew in wavy, suggesting an influence of the graft on the host follicles. We thought we had evidence of a morphogenetic substance. But the control grafts, nonwaved to nonwaved, showed the same thing, so it was presumably an effect of the scar tissue at the edges of the graft. Second, we called the new mutation "marcelled," since we had crossed it with the two previously known nonallelic waved mutations, *wa-1* and *wa-2* (received from another lab, which shall be nameless), and shown the F_1 to be normal. An abstract accepted for the upcoming Genetics Society of America meeting records this exciting news. But just to be on the safe side we intercrossed wa-1 and wa-2 mice, and their F_1 were also waved! We finally sorted out that the "wa-2" stock was really wa-1 that had been mislabeled somewhere along the line, and marcelled was a recurrence of wa-2.

I also worked on hr^{rh} (rhino), a hyperkeratotic cystic skin mutation. Rhino skin grafts on nonrhino hosts showed less hyperkeratosis around the edges, suggesting that rhino was an inborn error of metabolism, with nonrhino skin providing a corrective factor. Since vitamin A deficiency causes hyperkeratosis, we wondered about a block in vitamin A metabolism. Treatment with large doses of vitamin A reduced the hyperkeratosis quite dramatically. In retrospect, I have wondered whether this was a nonspecific toxic effect of the vitamin A rather than evidence that the mutant gene was blocking a step in vitamin A metabolism. Anyway, it was enough to satisfy the examiners, and I got my Ph.D. It would have made a good story to say that some of my treated animals got pregnant and produced malformed young, thus turning me into a teratologist, but that is not the way it happened.

8. ROYAL CANADIAN AIR FORCE

During this period a war was going on. General McNaughton, president of the National Research Council of Canada, had called for science graduate students to stay with their studies, since maintaining a vigorous science manpower was important to the war effort. But when I had finished my thesis there didn't seem to be much to do in genetics that would help the war effort (though I visited the NRC once to discuss whether hairless mice would be good for testing mustard gases—they wouldn't), so I joined the Royal Canadian Air Force. Herb Stern and I joined up and spent the first few months together. His philosophical good humor helped to ease the culture shock. We spent some frustrating months in a holding center, called the Manning Pool, which was in what had been the bull pen of the Canadian National Exhibition buildings in Toronto, waiting to be posted to a course, which led to this exasperated effusion:

Tell me Daddy what did you
Do to help win World War II?
What great & noble deeds of fame
Were done by you to quench the flame
Of fascist hate, and fully tame
The dastard Nazi crew?
My son, I was a bold P.P.
(Alternately named G.D.*).
While others swept the German skies
I swept Toronto Manning's sties,
(With whirling broom and flashing eyes)
To keep my country free.

While others dumped upon the Hun
Explosive missiles by the ton
I dumped garbage cans galore.
They polished Japs off by the score
I polished windows—this was war!
Could I do less, my son?

I finally started pilot training and got in 12 hours of flying time before they decided they had enough pilots and remustered me to bombardier training. Those 12 hours were among the most exhilarating of my life. The sense of power as we blasted down the runway (in a little single-engine Cornell trainer), the feeling of Olympian remoteness sitting high above the world, the apprehension while starting a spin and the exultation of recovering gave a feeling of excitement to life that I have seldom experienced since.

Learning to be a bombardier was not as exciting, not after I had dropped my first practice bomb anyway. Sitting in the nose of an Anson, you have to guide the pilot to fly a course that keeps the target moving toward you in the long slot of the bomb sight (on which you have set air speed, wind speed, and altitude). The plane is bouncing around in a

*General duties.

crosswind, but you manage to get the target in the cross wires and press the button. It was a direct hit and I thought, "This is a piece of cake." But I never came anywhere near the target again!

By the time I got my wings the war was almost over, and I was seconded back to McGill where I worked with an RAF type, Henry Browning, on the biological effects of the recently invented DDT (so there *was* war-related research for a geneticist after all). We used Drosophila to develop a bioassay, but by the time we got it worked out a chemical assay had been invented. We also demonstrated differences in susceptibility and were able to select resistant and susceptible strains of Drosophila, predicting what later happened in the field, but because we could not agree on the format of the paper it never got published. We also demonstrated its toxicity in dogs. It gave them "the jitters." DDT was a great insecticide, since it stayed on surfaces for a long time and killed insects on contact. It saved countless lives from death by typhus, to say nothing of lesser credits like keeping our mouse room free of flies for months at a time. What a pity it was not used in a rational way, rather than the indiscriminate drenching of the countryside that led to its prohibition.

9. MEDICAL SCHOOL

Demobilization (1945) made me eligible for a veteran's education allowance that allowed me to go back to school, so I thought I would go into medicine after all, and combine that with genetics. Getting into medicine was a bit simpler than it is today. I went to the dean of medicine at McGill, told him my plan (doing medical genetics), and he said "OK Clarke, you're in." I had no idea how dramatically medical genetics was going to take off, but knew I wanted to do it, and I thought a medical degree would give me better access to patients and records, as well as credibility.

I did not do well in medical school, partly because of my intention to take from it only what would be useful to me in genetics. Also because I was not good at things that depended on rote memory rather than logic, like anatomy. And partly because I took on the teaching of a course in developmental genetics, and running a small research program in mouse genetics. Frustration with my first-year medical studies led to this parody of an A. A. Milne verse about a shipwrecked sailor his grandfather knew, who couldn't decide what to do first, and "did nothing at all until he was saved."

The Case of the Conscientious Med. Student

There once was a Med. student (could it be you?)
With so many things he was called on to do
That whenever he thought it was time to begin
He couldn't because of the state he was in.

In first year the Prof. of Anatomy said,
"The whole human frame from its foot to its head,
"(That's caput to pedal extreme) must be learned
"E'er to other Med. matters your efforts are turned."

So our student began with an Atlas and Gray,
Dissectors and scalpel and skull to essay
To commit to his cortex each muscle and bone
And tendon and artery and vein and neurone.

But he found that each structure he studied this way
Was related to six or more others, and they
Intertwined with some more, so that when he had thought
The last one was learned, the first was forgot!

The axillary art'ry is simple I know
But where does it come from and where does it go?
Poor fellow! Some weeks had gone by e'er it dawned,
'Twas subclavian before and brachial beyond!

Then up spoke Professor of Physiology
"Learning structure is fine but it still seems to me
"It's useless to study Man's structural quirks
"Unless you know how each one of them works."

So our student put Gray on the shelf and began
To study the basis of function in man.
Nerve impulse, digestion, the shift of chlorides
And such things that go on in a person's insides.

But he found that such subjects to be understood
Required that he know other things that he should
have learned in those long Undergraduate days
That he diddled away, if you'll pardon the phrase.

So he started instead to acquire the essentials
Of thermodynamics, and ergs, and potentials
And such Mathematic and Physical laws
That he needed to answer the 'Whys' with 'Because.'

Biochemistry Prof. then came to the fore
And said (quote), "Before you do anything more
"Remember that when the exams come at last
"Biochem. will be one of the ones to be passed."

"And success in this course will depend on how well
"You know pH, mass action, and logs, and can tell
"From a knowledge of calories, dynes and R.Q.
"How much what is required, when, why and by who."

So our student continued to cram his poor head
With the facts that he needed to understand 'Med.'
But it took him so long this aim to fulfill,
For all that I know he's Pre-clinical still!

The summer following my first year of medical school, which I spent at the Carnegie Laboratory in Cold Spring Harbor, was somewhat marred by the need to prepare for a supplemental exam in anatomy. It was a good summer though. Bacterial genetics was just beginning to emerge and Cold Spring Harbor was at the center of things. Max Delbruck was there (a fine Ping-Pong player), Joshua Lederberg, Leo Szilard, and of course M. Demerec. They were just discovering that bacteria have sex. Also Barbara McClintock in her corn patch, beginning to realize that genes might jump and suffering much skepticism from her colleagues. The physicists were turning to biology because they were disillusioned with the military use of nuclear physics, and the application of their particulate approach to bacterial and viral biology led to the exponential growth of bacterial genetics. I learned so much that summer that I was able to include two whole lectures in bacterial genetics in my developmental genetics course the next year! It was very exciting but did not divert me from my path to medical genetics.

I managed to flunk two courses in the third year, which was very discouraging, though it was my own fault, partly the result of a certain flippancy that may have been interpreted by some as insolence. In my obstetrics oral I was quizzed about pernicious vomiting of pregnancy. I suggested several approaches and each time was asked what would I do if that didn't work; after the third suggestion the examiner said "and she's *still* vomiting, Doctor, in your *face*—what would you do? To which I replied "wipe if off, sir," which did not appear to be the right answer. I thought seriously of quitting medical school, and I remember writing a long self-pitying letter to Dr. J. S. L. Browne (the great endocrinologist, who had taken an interest in my plans), saying I was not properly appreciated and maybe I should give up medicine and get back to genetics. He wrote a very kind letter, which I still have, telling me, in essence, not to be silly. I also wrote James V. Neel (one of the few M.D./Ph.D. geneticists in those days) to ask if he thought it was worth going on to get the M.D. in view of all these difficulties. Yes it was. So I decided to take a year off to get myself reorganized, and then go back for fourth year.

I was hired by the McGill Genetics Department to teach the course in biometry (what gall!), and continued to teach developmental genetics as well; my excitement that year was in seeing if I could keep one lecture ahead of the biometry class. Part of my reorganization involved getting married to Beryl, who finished her B.Comm. while I worked, and then went to work to support me during my final year in medicine, but not for long, since she attended my graduation in the late stages of pregnancy. The obstetricians had apparently forgiven me by fourth year, since the first question I was asked was about the inheritance of achondroplasia and the second was about the management of a primipara, with the head still not descended at 40 weeks—a situation that I was well aware of at the moment! Beryl's family origins in Prince Edward Island resulted in strong ties with The Island, and happy memories of summer holidays on the red sand beaches of Keppoch.

I had fun with the developmental genetics course. The class was small enough for the Socratic method, and we used to have some fine arguments, using specific experiments as the basis for discussion. These included Huxley's "rate" experiments with the crayfish, Gammarus, the heat shock phenocopies of Goldschmidt (shades of modern heat shock protein studies), the Drosophila imaginal eye disc transplantation experiments of Beadle and Tatum and, later, their Neurospora experiments leading to the one-gene–one-enzyme hypothesis, the classical Landauer studies of how gene effects depended on rates of development, and Grüneberg's studies on quasi-continuous variation and "pedigrees of causes." The Socratic approach did not appeal to some students who didn't have a clue as to what I was trying to do, since I wasn't going systematically through an information base, but others did respond and this was rewarding.

I would sometimes surprise the students at the beginning of a class by asking them to write a one-sentence definition of something (penetrance, dominance, allele, whatever), and then we would go over the answers the next day—I think it helped some of them to learn about the precise use of words, a subject dear to my heart. Years of badgering my students, children, and yes, even colleagues, about some of their looser phraseology eventually led to this constructive advice about how (not) to fill in the pause while searching for just the right word:

Are you ever afraid of appearing absurd
For not being able to think of a word?
There's a ruse you can use when your speech is abortive
Fill in with basically, you know, like, sort of.

Do the gaps in your speech make you seem like a nerd
And basically feeling like, you know, a turd?
Like—the patient's expired and you search in your head
For a word that means you know like sort of—dead?

Do you fear you'll forget all the words that you do know?
Not to worry, there's basically like sort of you know
So you'll never be struck like—you know—sort of—dumb
With these useful non-words to like sort of choose from!

The following year I went back to fourth-year medicine and did a lot better by concentrating on medicine as a whole, not just the part relevant to genetics. In fact, if there had been a prize for the greatest improvement between third and fourth year, I would have won it with ease. But there wasn't.

10. GETTING STARTED—MEDICAL GENETICS

In the late 1940s interest in human genetics was growing, and there was a move on to start a separate society for that discipline. It is interesting to look back on how intense the debate was between those who thought human genetics had a body of knowledge sufficiently different from that of general genetics that it needed a separate base from which to grow, and those who thought separation would only weaken both disciplines. Anyway, a separate society *was* formed, and subsequent developments seem to have affirmed the first view. It certainly aided my efforts to start a human genetics unit at McGill.

The logical place was The Montreal Children's Hospital. I had been discussing the ways and means with J. Wallace Boyes, chairman of the Genetics Department, and with the professor of Paediatrics, Alton Goldbloom, and his second-in-command and successor, Alan Ross, during my "time-out" year in 1949, and during my final year in medicine. It was their vision of how genetics would fit into pediatrics that made the whole thing possible. Nevertheless, the directorship almost went to a human geneticist visiting from another country, and I had to make a bit of a fuss, pointing out how unfair it would be to give the position to an import when I had been working toward it for so long to say nothing of my superior qualifications for the position. This was successful, and I got the job. That was one of the few times I had to fight for something rather than having it handed to me on a silver platter.

Just in case I didn't get the job, I had been looking around at other possibilities. There were several offers from departments of anatomy, none of which seemed to care whether I could teach anatomy or that I had once flunked it. There was an opening in Japan, in the Atomic Energy Commission study of the genetic effects of the A-bomb explosion, under Jim Neel's direction; William J. Schull (whom I had not yet met) wrote me a long, encouraging letter about the prospects of their group. It would have been interesting. Bruce Chown, the renowned blood-grouper, wanted me to come to Winnipeg and extend medical genetics there beyond the blood groups. Norma Ford Walker invited me to look at and be looked at by Toronto. There were some conversations with the noted neurologist Wilder Penfield about studying the genetics of epilepsy, but we could not agree about the importance of genes versus the mother's pelvis, or how to go about studying it. And even some correspondence with Macfarlane Burnett, in Melbourne, where my parents were at the time. An opportunity to work in Australia was hard to resist.

Why did I stay at McGill? As a child I remember being anxious about new ventures—travel, boarding school, even visiting a friend's home for the weekend. So perhaps there was a bit of fear of the unknown. But apart from this it made good sense to stay where there was an outstanding medical school, a good department of general genetics, enthusiasm for developing medical genetics, and ties, both interpersonal and interdepartmental, being formed. Why go somewhere else when you can be what you want where you are?

I graduated from medical school in 1950. At that time it was possible to substitute a year of clinical research for an internship. My "clinical research" involved getting the medical genetics department set up at the Montreal Children's Hospital, beginning to collect some family data, and beginning to learn about counseling—by trial and error, since there was no one to teach me. I am still trying and erring. I sometimes regretted

Figure 5. FCF, 1950.

not doing an internship, as I never felt like a "real doctor" who did things for (to) people, and I still avoid taking blood when possible. My attraction to medicine did not extend to cutting into people, and I felt that even a physical examination was an invasion of privacy with which I was not comfortable. Nevertheless, for a study of congenital heart disease, I found myself visiting families in their homes, stethoscope in hand, and checking women for murmurs in an array of informal circumstances that I blush to contemplate. It's a wonder that I was not shot by some irate husband, and it was, if you'll pardon the phrase, heart-warming to see how trustingly women would bare their bosoms to this callow young researcher.

How much difference would it have made to my career if I had delayed getting to work for yet another year? We will never know. Today an internship/residency would be virtually obligatory. Then it may have been better to strike while the genetical iron was hot.

The Montreal Children's Hospital was a good place to begin. It was physically small and had a close-knit staff with a wonderful esprit de corps. It was easy to wander around the wards, spotting genetically interesting patients and quizzing the interns and residents about the family histories. The staff seemed eager to consult and there was no evidence of resentment, or fear that the geneticists would usurp any prerogatives of the pediatricians. There was no lack of referrals as the word spread. The simple presence of a geneticist on the wards and at rounds was a great advantage in spreading awareness of the "genetics connection."

Links between the hospital Medical Genetics Department and the McGill University Genetics Department always remained strong; I resisted the pressure to move into the medical school, feeling that it was healthy to maintain the roots in basic genetics. The McGill department benefited by being able to draw on "real" examples for teaching, and having access to the patients and resources of the hospital for graduate research, and the hospital department benefited because this exposure provided a good supply of excellent graduate students. Julius Metrakos was my first Ph.D. student, in human genetics, and working with twins ascertained at the hospital; he and his electroencephalographer wife, Katherine, went on to do their classical study on the genetics of epilepsy with hospital probands.

It was not always easy, though. The vigorous expansion of studies in human genetics had to be done partly at the expense of other areas of genetics, and not all members of the genetics department were as convinced as I was that medical genetics was taking off. Why should there be courses in human genetics? Why couldn't human genetics be taught as part of general genetics, simply by using more human examples (as I had indeed been doing in my developmental genetics course)? The recognition that human/medical genetics was acquiring a specific body of knowledge that allowed it to stand on its own came slowly, but each dramatic advance made it more difficult to ignore. After Jérôme Lejeune announced the discovery of trisomy 21 (at a seminar in the McGill Genetics Department, in 1958) the department did not leap at the opportunity to get into human genetics on the ground floor, and there were all sorts of logistical objections to working with the chromosomes of human subjects in a university laboratory. Dorothy Warburton, then a graduate student working on the genetics of abortion (and now a cytogeneticist at Columbia University), finally did a human karyotype (mine) just to show it could be done. But it was not till 3 years later, in 1961, that McGill became formally involved in human cytogenetics when Louis Dallaire started his Ph.D. research in a small room in the Medical School building (since there was no room in the genetics department). Dallaire, after a pediatric residency at the Children's Hospital, saw the opportunities for cytogenetics in pediatrics, and chose that to work on for his Ph.D. My cytogenetics training was limited to the chromosomes of Najas and inversions in Drosphila salivary glands, so I felt rather temerarious in undertaking to supervise a Ph.D. project in cytogenetics. But we managed to pull it off by choosing a clinically oriented problem—looking for translocations in the parents of sibs with multiple malformations—a new approach in those days, which revealed five new translocations. Dallaire went on to set up a service laboratory for The Montreal Children's Hospital in 1964 at the Douglas Psychiatric Hospital (since there was no room at MCH).

From the time the MCH Medical Genetics Department was formed, the need for a biochemical genetics arm was obvious. I found it increasingly frustrating to pronounce at rounds that such and such a disease was caused by an autosomal recessive gene without being able to say anything about what the gene was doing—or, more likely, not doing. Charles Scriver, who was a resident at MCH, foresaw the need and went for training to Harvard University and then for 2 years to work on inborn errors of metabolism with Charles Dent at University College, London. When he returned as chief resident at MCH, with the idea of setting up a biochemical genetics laboratory, I welcomed it with the greatest enthusiasm. When the medical board approved the idea, there was one dissenting vote—the chief of Pathology, who said that two such ambitious empire-builders would never be able to work together. How wrong he was. The two groups, biochemical and clinical, coexisted happily as a medical genetics team, which produced four appointments to the Order of Canada, and their symbiotic nature was recognized by the formation of the Medical Research Council Medical Genetics Group in 1972 (roughly equivalent to a Center of Excellence, with only one in Canada for each discipline), now in its fourth support renewal. The dissenter's mistake was that we were not *both* empire-builders. Nor was I prepared to preside over the dissolution of clinical genetics. Scriver regarded me as something of an academic father figure (an honor, as both his biological parents were McGill gold medalists, as was he himself) and seemed to value my support, and I found his enthusiasm and unfailingly optimistic view of even the bleakest landscape inspiring. And I admired the determination, energy, and eloquence with which he fought for his aspirations—to apply the advances in genetics to the practice of medicine. The only problem for me was that, because I referred all biochemical problems to his end of the floor, I remained lamentably ignorant about the biochemical aspects of genetics. And I could not match his ability to convince the political powers that be. Hence, we have the Quebec Network of Genetic Medicine and the Inborn Errors Food Bank (a repository of special foods for children with inborn errors of metabolism), but not a Birth Defects Registry or an Alphafetoprotein Maternal Serum screening program.

My efforts to give human genetics a presence in the undergraduate scene, rather than setting up shop in the Medical School faculty, resulted in the Human Genetics Sector, which remained part of the Genetics Department as far as teaching was concerned, but had its own budget for its research activities. (That was in the days when there *was* some hard money for research.) This arrangement worked quite well, but disappeared when genetics was merged with botany and zoology into a Biology Department. Later, human genetics regained something of an identity as the Centre for Human Genetics, a tricephalic monster (clearly the invention of a teratologist) which answered to three deans (science, medicine, and graduate studies) and provided a focus for those working in human genetics in departments throughout the university. So far, under Leonard Pinsky's adept direction, this has worked out to the mutual benefit of biology and human genetics.

11. MEDICAL GENETICS COMES OF AGE

On looking back to 1950 it is hard to see why we were so enthusiastic about the future of human genetics. The literature consisted mainly of more or less misguided efforts to fit human traits, both normal and abnormal, to the Mendelian laws. There were a handful of inborn errors of metabolism, none treatable. Human chromosomes were small, numerous (too numerous by two as a matter of fact), and unsuitable for cytogenetic analysis. DNA had only recently been recognized as the genetic material, but how it worked was a mystery. There were no chromosomal diseases, no molecular diseases, no genetic screening, no prenatal diagnosis. Genetic counseling consisted of providing recurrence risks, mostly for Mendelian disorders, and discussing the options of sterilization or abortion. Articles on genetics for physicians preached the importance of taking (and paying attention to) the family history and contradicted the prevalent idea that genetic diseases were untreatable. With a few outstanding exceptions, psychologists and psychiatrists held that mental traits were largely nongenetic. Victor McKusick was beginning to develop lists of Mendelian diseases, though some objected that these should not be made available to physicians, who might underestimate the complexity of deriving from them recurrence risks for individual cases. The first attempt at a list of recurrence risks for pediatric diseases had all of 206 entries (Fraser, 1954).

One of the first catalogues of human genetic diseases was R. Ruggles Gates's two-volume treatise *Human Genetics,* which was a valuable source book in its time, for all its faults. Gates was trained as a botanist, but had an interest in anthropology, including human genetics. He had taught at McGill, before my time, and used to visit the Genetics Department occasionally. I don't know why it took me so long to realize he was a racist. For example, one of the "horror stories" used by anti-Negro racists was the myth of "black babies"—which might crop out in "white" families where there was Negro ancestry on one side. Gates told me he knew of several cases where white women with black ancestry had had black babies by white husbands. I traced one of them down, since it had been told to him by his cousin, a dear old lady who was the best friend of my aunt Edie, in Bear River. It turned out that, in this case, both parents were mulatto and the baby not all that dark, so that example certainly didn't qualify. Gates was an editor of an anthropological journal called the *Mankind Quarterly.* He asked me if I would be on the editorial board and I, being young and naive enough to be flattered, agreed without having a careful look at it. It was only after Arthur Steinberg (still *in loco parentus academici*) wrote to ask if I realized what I was doing that I actually read some of the papers. It included regular contributions from people like Putnam of the nefariously racist *Putnam Papers.* With a red face, I resigned.

If things looked dull in the 1940s, they certainly did not in the 1950s. I remember the exhilaration at the meeting where the treatment of phenylketonuria was announced. First there was the dramatic response to the diet; the children were described as seeming to wake up, recognize their parents, and stop having seizures and temper tantrums. Then the disappointment as they began to regress. And then the realization that, just as in the Drosophila eye-pigment story of Beadle and Tatum, they were responding to a total lack of phenylalanine by breaking down their own proteins which provided endogenous phenylalanine, and thus the need for strict control of phenylalanine intake. This discovery gave medical genetics a big boost in the eyes of physicians, as it was dramatic evidence that genetic diseases were treatable. (For some reason hemophilia and diabetes did not count.)

A still more dramatic boost came from the recognition of sickle-cell anemia as the first molecular disease and its role in showing how DNA worked by controlling the sequence of amino acids. It seems that Jim Neel, who had been working out the genetics of sickle-cell anemia in the Detroit black population, was on a committee that also included Linus Pauling. When he told Pauling about this genetic disease that caused hypoxic red blood cells to curl up into odd shapes, Pauling suggested that there might be an abnormality in the hemoglobin molecule, and he gave the problem to a graduate student, who soon showed that sickle and normal hemoglobin did, indeed, have different charges. Then Vernon Ingram produced his famous fingerprints (two-dimensional chromatograms using paper electrophoresis followed by partition chromatography) showing that the difference lay in a single polypeptide. And finally, the seminal, one-column note to *Nature* reporting that the polypeptide difference lay in one amino acid, the first evidence that genes act by controlling amino acid sequences. Surely this paper must take the prize for the most significant information published in the fewest words. To watch this story unfold was a thrilling experience, just as exciting then as current developments in DNA technology and genetic engineering are now. Well, almost.

The next exciting episode was the (re)discovery of human chromosomes. Human cytogenetics was a mess, as there were so many chromosomes and they were so small. But interest in mammalian chromosomes revived with the idea of treating radiation sickness by marrow transplantation. A marker for transplanted cells was needed to tag cells so they could be positively identified as graft or host. This was provided by a mouse translocation, easily spotted even in such unfavorable material. Then one of T. C. Hsu's technicians accidentally made up a culture medium that was hypotonic—and the swollen cells had beautifully separated chromosomes. Mammalian cytogenetics was in business! A Swedish group reported that fetal fibroblasts had 46, not 48, chromosomes. This was greeted with skepticism—maybe fetal cells were different (though where postnatal cells could have found two extra chromosomes was not explained). Then Jérôme Lejeune, working in a garret in Paris, with a 100-year-old microscope, and two cocks to provide serum for his cultures, found that fibroblasts from children with Down syndrome had an extra chromosome. He first reported this, rather diffidently, in a seminar in the Genetics Department at McGill, right after the 1958 International Genetics Congress. I well remember the excitement, though some were skeptical, since (they said) Down syndrome was more likely to be a dominant mutation. Skepticism seems to be an almost automatic reaction to any new finding, no matter how plausible—I remember similar doubts being expressed about the Lyon hypothesis. It is difficult to strike a balance between skepticism and gullibility—an excess of either can be harmful.

I was on the National Institutes of Health Genetics Study Section when the chromosome story first broke, and the committee labored hard trying to find a basis for evaluating the flood of grant applications that appeared. Many of them proposed looking at the chromosomes of patients with every imaginable disease. Other applicants wanted to study large numbers of consecutive live births, or other groups, to estimate the frequencies of aberrations. Others wished to test hypotheses of varying degrees of imagination with varying degrees of rigor. What proportion of the available funds to put into "just counting," or "fishing expeditions" rather than hypothesis testing? Counting is dull, but can be important. How much should inexperience count against an applicant when almost no one had experience? No easy answers. But my experience with the Study Section and later with the NIH Genetics Training Grants Committee, as well as the Canadian MRC Genetics Grants Committee has left me with a high regard for peer review as about as fair and responsible a process for distributing research funds as one could hope for.

Human cytogenetics had another boost when David Hungerford, in Philadelphia, noticed that lymphocytes, treated with phytohemagglutinin (a bean extract used to lyse red cells), underwent mitosis. This opened the way to chromosome studies on lymphocytes, much easier than fibroblasts to separate from patients. One (possibly apocryphal) story goes that at one point the suppliers of phytohemagglutinin tried to purify it, and managed to purify out whatever the active agent is. Labs all over the country were desperate because their cultures stopped mitosing. But John H. Edwards (then at St. Christophers Hospital in Philadelphia, where he discovered the Edwards syndrome) solved this problem by getting some beans from the local supermarket and dropping them in his cultures. I remember the first time I met Edwards (now Professor of Medical Genetics at Oxford) in a London hotel room, where he demonstrated how to take skin biopsies by removing slices of skin from Joe Warkany and I. Pick a fold of skin up in a forceps so it shows just above the forceps surface and slice it off with a safety razor blade. The forceps pinch hurts enough so you hardly feel the cut and inhibits bleeding. Much easier than a punch biopsy. A tall, thin man with an unruly forelock, Edwards has a wonderful sense of humor, and can give very amusing talks, but he is so brilliant that he is sometimes difficult to understand—he would leave out obvious (to him) steps in the reasoning. But he was always willing to go back and fill them in. One of my M.Sc. students, Gilbert Coté (now a medical geneticist in Athens, Greece), who was far too brilliant, mathematically, for me to supervise for his Ph.D., went to Edwards instead. Every few months he would send me some of Edwards's reprints, with a neat note at the end of each one that said "This means that. . . ." Would that he were still there; now I have to try and figure them out for myself.

The discovery of chromosomal diseases was a major boost for medical genetics in the eyes of physicians. Genes and DNA were interesting hypotheses but here was a cause of genetic disease that they could actually *see*. In fact, there was something of an overreaction; genetics was regarded by some as synonymous with cytogenetics, genetic diseases with chromosomal diseases. Medical genetics units had to decide how to handle the numerous referrals for karyotyping of patients with clearly Mendelian diseases. Recent improvements in cytogenetic techniques have attained such a high degree of resolution that such requests are now seeming more reasonable. And so it goes.

12. GENETIC COUNSELING

As the Department of Medical Genetics took root at The Montreal Children's Hospital, and I began to see families in connection with the budding research program, I got drawn more and more into counseling. Nothing learned in medical school prepared me for this, and I knew nothing of interviewing techniques, the dynamics of decision-making, the marital dance, coping with grief, and so on. I thought counseling would be fairly simple, since genetics was so logical, and providing recurrence risks was pretty straightforward, so the only complications would be when the question arose, in the occasional case, of whether abortion would be justified. But it soon became clear that it was not all that easy—so much so that I wrote a paper on the darker side of counseling (Fraser, 1956). I wanted to say "seamy" side but was advised against it. "Rough" would have been better. It described some problems that had beset me, and presumably were also bothering other counselors. Rereading this is a vivid reminder of how much things have changed in the counseling field, and how much they remain the same. One category of the "unsatisfactory aspects of counseling" was the genetic heterogeneity of clinical entities. I think I coined, in this respect, the term *genetic heterogeneity*, though not the concept, of course. I cited as examples: a dominant pedigree of diabetes mellitus (subsequently recognized as "MODY"); cases of cleft lip and of retrolental fibroplasia in offspring of first cousins, suggesting that there might be autosomal recessive forms of these disorders—still relevant examples of ambiguity in provision of genetic risks. In the latter case I actually stated that, since the parents were practicing Roman Catholics and would undoubtedly go on having children (no oral contraceptives then), it might be better not to apprise them of the small chance of a one-in-four risk, even though "it does stick the counselor's neck out a bit." But I was still very young!

Other examples highlighted the psychological and emotional pressures that may impinge on the counseling process, and which I felt very incompetent to deal with. I still do. These included maternal guilt from an attempted abortion followed by a malformation in the child; the use of the family history as a weapon in the marital cold-war; ways in which counselees may try to use information supplied by the counselor to forward their ulterior motives (e.g., an albino who wanted me to say the risk of his having an albino child was high, because he did not want to marry his fiancée), and how differently people can perceive the same information [the two fathers of first cousins contemplating marriage, given the same information; one reported to the family that the counselor (me) said the risk was low and there was no objection to the marriage and the other that the risk was high and the marriage would be ill-advised]. There was also the pressure that might be put on the counselor—parents who said that if I did not say the risk was high, and support abortion of the pregnancy, the mother would have a nervous breakdown and it would be my fault.

When I started my counseling career, the few writings there were on it took a very directive approach. Madge Macklin, for example, writing in the 1940s of a woman with Friedreich's ataxia who wanted to have children, said "there was only one answer to her question, namely an unqualified NO." And speaking of devastating diseases of early onset ". . . [parents] have no moral right to bring children into the world who may suffer from such diseases. . . ."

Dr. Macklin was a pioneer in medical genetics who fought tenaciously to get genetics into the medical curriculum and the purview of physicians, and her writings in this area make interesting reading. I knew her as a plump, short lady with her hair piled up in an imposing coiffure, a wonderful person, kindly except when involved in academic discussion. I can still feel the withering scorn with which she would say "What can you expect from a mere man!" The pendulum has since swung far toward nondirectiveness, though with the advent of prenatal diagnosis for so many disorders, there are now those who would agree with her view on the morality of bringing defective children into the world.

Looking back through my own writings I am surprised to see some early signs of directive tendencies to which I would no longer subscribe, but I still do not go so far as to refuse flatly ever to indicate, after repeated caveats, what I *think* I *might* do in a given situation. Having said that in public, I was immediately labeled a directive counselor. But it is probably impossible to be completely nondirective. Abby Lippman, in her follow-up study of counselees, asked a woman if I had been directive, and she said no, but if Dr. Fraser had thought I shouldn't have children he would have said so, wouldn't he? More than one counselee has said in retrospect that she wished I *had* been a bit more directive. And sometimes the facts are so complex that parents may need some guidance. For example, a woman with a 1% risk of a neural-tube-defect fetus—should she have a serum AFP or go straight to amniocentesis? Some counselors present all the facts and let the couple decide, but the couple may not be able to grasp the facts adequately. So I try to be nondirective, but there are times when I let my intuition decide how to play it, and hope I am doing it for the best. There will probably never be any data to provide an answer. After listening to a panel on directive counseling at a meeting of the American Society of Human Genetics I was moved to pen the following would-be humorous lines.

Lines Inspired by a Panel Discussion on Directive Counseling*

Whenever you, a counselor, feel tempted to advise,
Remember that it's frowned upon, in many people's eyes,
To make suggestions re directions they should take if they were you
So never be directive, just tell them what to do.

If you feel the urge to try and purge consultands' points of views,
Exert restraint, correction ain't what a counselor ought to use.
If you should sense any evidence of thinking that's askew,
Never be directive, just tell them what to do.

These lines were accepted for publication and actually got into galley proof, but the publisher objected—he thought it might offend some

*(Without permission of Liss, Inc.)

readers, and people wouldn't know whether I was for or against directive counseling! I hope you, dear reader, are not offended.

Since those early days a great deal has been learned about the techniques, psychodynamics, and effectiveness of genetic counseling. Sue Wright did one of the early follow-up studies for her M.Sc. but because of the exigencies of medical school it was published only in abstract. Like several previous and subsequent studies it showed that parents have fewer subsequent offspring (are more deterred?) when the risk is high, and when the proband's disorder results in chronic debilitating disease rather than death in infancy. Abby Lippman, in her Ph.D. thesis, pointed out the fallacy of the implication that deterrence is a "correct" or "reasonable" response, and is therefore a measure of the effectiveness of counseling. She had trouble getting her thesis project accepted by the Biology Department graduate training committee, which said it was not "real biology"; fortunately, they were persuaded otherwise. Lippman's status as a mature student, with a family and some experience with psychiatric/psychological research, uniquely prepared her for this project. Rather than the usual questionnaire, she used tapes of counseling sessions, from which she developed insights into parents' perceptions of the problems, and how they make decisions, which have been referred to as a milestone in the field (Lippman-Hand and Fraser, 1979).

Since then much has been written about the methodology of counseling, but there is still much to learn. We now have programs to train much-needed genetic counselors at the master's level, which provide courses on how to counsel. No doubt these represent important progress, but I confess that I still sometimes feel the same inadequacies and frustrations that I did in 1956, when I know that the parents I am seeing have a lot on their minds that they are not saying, and I seem unable to reach them, or I see signs of psychopathology that I know I am incompetent to handle but the couple will not accept referral to a more qualified expert. And I wonder if the teaching now provided in genetic counseling programs may suffer the dangers of a little knowledge. Are we doing more harm than good by recognizing and suddenly bringing to a boil areas of conflict that couples have over the years succeeded in reducing to a simmer? How much good do we do by telling people they should not feel guilty about their bad genes, when it might take years of psychotherapy to resolve these feelings? Or to attempt some psychotherapy ourselves without the appropriate years of training? Such questions seem to be left to the "wisdom" of the counselor, who (like me) may not feel very wise. They are unlikely to be answered by any research techniques at our disposal. So the final words of my "darker side" paper may still apply. Until we know more, we can at least tread carefully, think sympathetically, speak tactfully.

On the other hand, there is a "brighter side." Counselees often seem pleased and grateful when they leave and sometimes put this in writing. If I could get them all together, I would have quite a touching array of letters thanking me for my sympathy, concern, and guidance. No doubt other counselors share the experience that counseling, in spite of its frustrations, is a rich and rewarding experience.

13. THE TERATOLOGY CONNECTION

Before I got the McGill appointment there was another job opportunity I had sought. On reading about Josef Warkany's early teratological experiments with riboflavin deficiency, and feeling drawn to that kind of work, I wrote and asked for a chance to come and work for him. Unfortunately, he had come down with tuberculosis, so that didn't work out (Warkany, 1988); my regret was confirmed later when we met and I could appreciate what a wonderful person, as well as great scientist, he is. He told me that my letter about a job had been a big boost to his morale, which was very low at the time, and maybe that was why he accepted me as a friend even though I was a geneticist and used words like *penetrance* and *multifactorial*. His view of geneticists had been influenced by seeing some of them adopt the Nazi view of eugenics. Also because when he got his first riboflavin-induced malformations, about 25% of the pups were affected and a geneticist colleague said it must be autosomal recessive! I'm sorry about *multifactorial*, Joe. I recognize that everything is multifactorial, but it's just that geneticists insist on speaking of Mendelian conditions being caused by single factors—mutant genes—and there wasn't another term that meant "not unifactorial."

Figure 6. FCF and Joe Warkany, Gainesville, 1962.

During this time the pendulum of opinion was swinging away from the idea that malformations are genetic in origin (by exclusion, since the uterus was thought to protect the embryo from environmental insults) to the other extreme—that malformations are mostly caused by environmental factors. There was the rubella story, Warkany's work, and Theodore Ingalls's demonstration that maternal hypoxia could be teratogenic in rats and mice. In fact, a woman who had been in an automobile accident on the 52nd day of pregnancy, and had had some bleeding, per vagina, successfully sued on the basis that the Down syndrome of her child had been caused by hypoxia resulting from the blood loss. Ingalls's evidence that all the anomalies characteristic of Down syndrome originated around the 52nd day of pregnancy and that hypoxia was teratogenic were strong arguments for the plaintiff. This was years before the discovery of trisomy 21.

I felt it was time to come to the defense of genetics, by showing that different strains of mice exposed to maternal hypoxia would have different frequencies—and possibly different types—of malformation. Theodore Fainstat (now Professor of Obstetrics and Gynecology at the University of California in San Diego) took this on as his master's project.

14. CLEFT PALATE AND CORTISONE

While we were building the hypoxia apparatus, a plastic surgeon, Dr. Hamilton ("Happy") Baxter, who was making his cleft lip and cleft palate patients available to me for genetic studies, happened to get hold of some cortisone, at that time a newly discovered "wonder drug" that nobody knew much about except that it was good for arthritis. He thought that since cortisone was a steroid, and the embryonic "organizer" was a steroid, maybe treatment of pregnant mice with cortisone would cause neural tube defects. We didn't think much of the argument (the organizer was thought probably not to be a steroid), but what did we have to lose? Making a series of wild guesses about dosage we treated some pregnant females we happened to have around, and in one of the first few litters we found, not CNS defects, but cleft palates. Fainstat had injected the female, but I was the one who found her litter, working alone one evening in the mouse room. I remember vividly the excitement when I opened the newborn's mouth with a dental probe that we kept around for looking for vaginal plugs, and saw this cavernous opening which I finally realized must be a cleft palate. We published a short note in the *McGill Medical Journal* to establish priority. Baxter was given senior authorship for his essential contribution—the cortisone (Baxter and Fraser, 1950). We never got back to hypoxia.

We then started testing several strains that we happened to have in the mouse room, A/J and C57BL in particular, and were lucky that they showed striking differences in cortisone-induced cleft palate frequency, so our hypothesis was supported (Fraser and Fainstat, 1951). This led to a long series of studies by my graduate students and me, in which we looked both at the genetics of the system and at how the palate closed and what the cortisone was doing to prevent it. I was blessed with very good students who worked together almost as a family. We would meet for beer and teratological conversation every other week or so at the home of some member of the group, and these sessions were very productive—many of our experiments were planned therein and then chosen (or not) by one of the group. The experiments were usually simple and did not require elaborate statistics; I appreciated J. B. S. Haldane's (wasn't it?) maxim that if the data require statistical analysis, it isn't a good experiment. We worked as a team, and when I say "we" did this or that, I really mean we, since we all had a part in it. Furthermore, most of these students became lifelong friends, an enrichment of my life that I cherish dearly. Later generations of students missed these sessions, as it seemed harder, particularly for women with families, to free up evenings, and the meetings gradually petered out, to my regret.

Ted Fainstat made a start on establishing critical periods and dose ranges and the existence of strain differences. Harold Kalter did a series of reciprocal crosses and backcrosses that showed for the first time that maternal as well as fetal genes influenced susceptibility, that the maternal differences were not cytoplasmic, and that there must be several genes involved (Fraser *et al.*, 1953; Kalter, 1954). Kalter, after his Ph.D. and a postdoc with me, went on to work for Joe Warkany in Cincinnati, which stimulated this poetic parody:

TO BELLA AND HAROLD

To say

AU REVOIR and GODSPEED

(apologies to Robert W. Service)

There are strange things done 'neath the Montreal sun
By the men who moil with mice
Where the DBA's moan for cortisone
And the A/Jax scratch for lice.
Oh Daphne pricks them and Karin sticks them,
But what really makes them falter
Is to see coming in with a fiendish grin
The phenomenal doctor Kalter.

The C57's raise to the heavens
Their pleading eyes, and pray.
Oh the DBA's shake, and the A/Jax quake
In confusion and dismay.

What devilish way will it be today
That the fiend will find to alter
The pattern of stress, make the foetus a mess.
What next, oh redoubtable Kalter?

But tearful are we, for to Cincinnati
Shifts the scene of this fearful story,
And Joe Warkany is the one who will see
Its dénouement, in all its glory.
But before you go, we want you to know
The monsters from your murine harem
Are produced not by starving, or freezing or carving
Not by physical stress, you just scare 'em!

At that point we put the genetic analysis on hold and concentrated more on palate closure and how it goes wrong. We could have focused either on how the cortisone works or on how the palate closes.

We rejected the strategy of pursuing the physiological and biochemical effects of cortisone treatment (of which there are a great many) on the mother/embryo to find which one was hitting the palate. Instead, we chose to focus on the palate. The reason for this choice illustrates how one's personality affects one's approach to research. First, it is very difficult to design incisive experiments to test the question of which effect of cortisone causes cleft palate. Second, we did not have the biochemical knowledge, techniques, or equipment to pursue this very far. Some biologists I know would not let this stop them—they would learn the techniques, acquire the equipment by persistent application for research funds and/or by borrowing, and carry on. We felt that there are so many interesting things to do that we *can* do; we would choose some of them rather than get radically reprogrammed. Furthermore, I am by nature distrustful of machines and untalented in their use and repair, and put more faith in counting what I see, like open or closed palates, than in making inferences from a series of curves produced from a "black box." This is an illusion, of course, because the palates we count are probably no closer to what we are trying to measure than a Cot curve or Southern blot band is from what the molecular biologists is trying to measure. This is not to defend one attitude or the other, just to observe that personality differences make people approach problems in different ways, which is, no doubt, a good thing. Several groups have taken the biochemical approach, but it is still not clear which effect of cortisone is the one (if it *is* only one) that causes the cleft.

So we chose to observe the morphology of how the palate closes and where it goes wrong in treated embryos. The approach had some merit; the biochemists might spend a lot of time looking in the wrong place unless they had some idea of where the disturbance of development began. There would be no use in looking for relevant changes in (say) acid mucopolysaccharide synthesis in the palate shelves if the cortisone prevents their closure by altering growth of the cranial base or diminishing amniotic fluid volume. Looking at homogenized embryos would be even less likely to render a solution to the problem. And, from the structural changes we saw, there developed the multifactorial/threshold model for cleft palate.

15. THE MULTIFACTORIAL/THRESHOLD MODEL

When it was suggested to Bruce Walker that he might look at the process of palate closure and how cortisone affects it, he asked why, and was referred to John Berrill's adage "If you sit and watch an embryo long enough it will talk to you." So he did and it did. His paper on how the palate closes, and the existence of an intrinsic shelf force (Walker and Fraser, 1956), has become a Citations Classic (*Current Contents*, 1988). Walker established that cortisone delays palate closure relative to differentiation of the embryo as a whole, and also that *untreated* A/J embryos close their palates later than C57BL/6J embryos (Walker and Fraser, 1957). Daphne Trasler showed that early closure correlated with resistance to cortisone in six different crosses, establishing that stage of normal palate closure was an indication of liability to cleft palate (Trasler and Fraser, 1957; Trasler, 1965). Daphne stayed at McGill University after her Ph.D., and I am thankful for our long and fruitful interaction. From histological studies of the closing palate we conceived a model in which the shelves develop an intrinsic force that eventually allows them to push the intervening tongue out of the way and extend above it to meet in the midline. A certain degree of delay in shelf movement, with continuing enlargement of the head, would mean that the shelves could not meet each other and a cleft would result; this degree of delay constitutes a threshold, separating normal from abnormal.

I remember fooling around with some fetuses from a discarded pregnant female mouse one night, and wondering what unfixed shelves looked like. They were quite jellylike, and when I pulled the tongue down from between the shelves they flipped up and moved toward each other in a matter of seconds, as if they had been all ready to go but for the resistance of the intervening tongue. This confirmed our impression that there was indeed an intrinsic shelf force, though we spent some time proving (with mirrors) that the movement was not just due to gravity or (with immersed embryos) to surface tension. We still do not know the nature of the shelf force, though acid mucopolysaccharides are probably involved, since Walker showed that they build up in the shelves at the right time and Karin Heiburg showed that cortisone treatment diminishes their concentration. Carl Verrusio invented an ingenious model, with Pliofilm and string, to show that changes in tension along the base of the shelf (such as those caused by extension of the cranial base) would make the (Pliofilm) shelves rise appropriately.

So the concept of a multifactorial/threshold (MFT) model gradually emerged—a continuously distributed variable (stage of shelf-becoming-horizontal) and a threshold (maximum tolerable delay) separating embryos whose shelves came up soon enough to close from those whose shelves came up too late and had clefts. The position of the distribution relative to the threshold determined the frequency of clefts. The important thing was that the amount of delay imposed by the cortisone could be the same in both strains. Difference in cleft palate frequency between strains lay in differences in their *normal developmental pattern*. The pattern of shelf closure was influenced by many things—the shelf force, the resistance of the intervening tongue, the width of the shelves, the width of the head—and all these could be influenced in various ways by genetic and environmental factors. The position of an embryo on the distribution curve was a measure of its liability to cleft palate induced by anything that delayed closure. Obviously the system was multifactorial.

The first diagrammatic depiction of the model (Fraser *et al.*, 1957) was worked out on a train, traveling to Cincinnati to attend the first meeting, organized by Warkany, of a group that evolved into the Teratology Society (Warkany, 1988). Somewhat more sophisticated (but still quite naive) versions of the diagram evolved as more elements were recognized, including depiction of the threshold, as well as the liability distribution, as continuously distributed (Fraser, 1980; Vekemans and Biddle, 1984).

We collected examples of specific factors that would cause cleft palate by affecting one of these variables. Trasler showed that oligohydramnios could cause cleft palate by increasing flexion of the head, thus compressing the lower jaw, and jamming the tongue up between the shelves. This was a serendipitous discovery; the amniotic punctures that caused the oligohydramnios were made to see if injecting cortisone directly into the embryo would cause cleft palate. We never did solve that one. She also showed how the large prolabium of an embryo with a cleft lip would prevent the tongue from sliding forward under the primary palate, and out of the space between the shelves, causing it to arch up between the shelves and delay their closure (Trasler and Fraser, 1963). This provided a reasonable explanation for why babies with cleft lip often have cleft palate as well, whereas isolated cleft palate is an embryologically (and therefore genetically) different entity. Bob Seegmiller showed how the mouse mutant gene *cho* (chondrodystrophy) causes cleft palate by inhibiting forward growth of the jaw, so that the tongue is prevented from sliding forward and downward from between the shelves—a possible model for the Robin syndrome. Other mouse mutants cause cleft palate in still other ways; *ur* (urogenital) diminished shelf width and *bm* brachymorphic increased head width. And of course the shelves may reach each other but fail to fuse, as in the case of clefts due to the dioxin TCDD. Finally, I was able by selection and inbreeding to develop a strain, SW/Fr, that had about a 10% incidence of spontaneous cleft palate. It had the greatest susceptibility to cortisone of any strain so far tested, and the latest closing shelves. Thus, this genotype had enough "slow closure" genes to bring the tail of the liability distribution well beyond the threshold even without delay by treatment.

Some 20 years later Fred Biddle and Michel Vekemans picked up the genetics again. Biddle came to us as a postdoctoral student from James R. Miller's lab, and Vekemans was an M.D. from Belgium who wanted training in genetics and took his Ph.D. with me. Biddle used dose responses, converted to probits to gain linearity, and an elegant system of reciprocal backcrosses, to parcel out the relative contributions of maternal and fetal genes to susceptibility (Biddle and Fraser, 1976) and to estimate the number (2–3) and dominance relationships of the fetal genes influencing the difference in susceptibility between the A/J and C57BL/6 strains (Biddle and Fraser, 1977). He also drew attention to the slope of the curve, which was different for teratogens that (presumably) acted by different mechanisms (Biddle, 1978). Vekemans developed a method of converting binomial statistics to a frequency distribution and showed that those normal curves we have been postulating all along were really true (Vekemans and Fraser, 1979)! He also used recombinant inbreds to locate one of the susceptibility genes on chromosome 5 and to show that H-2, which influenced the A/C57BL difference, did not influence the A/DBA difference (Vekemans *et al.*, 1981). Thus, progress is being made in identifying specific genes in this multifactorial system. Other laboratories have identified other differences. Note that any two strains may differ by only a few genes for liability, but that other pairs of strains will differ by other genes, so the whole array of genes affecting liability is quite large.

Thus, we have a process involving many factors that have to act together in a coordinated and synchronized way to achieve palate closure. Various megagenes and teratogens may also interfere with the process at various points. But it is reasonable to suppose that many palates fail to close because of a combination of factors, each contributing a small amount to shelf delay relative to threshold, none of them abnormal per se, but in aggregate delaying closure beyond the threshold.

What use was all this? Some say an MFT model has no heuristic value, but others think it does. Cortisone-induced cleft palate was the first example of a drug-induced malformation in mammals (excepting nitrogen mustard and trypan blue, not usually considered drugs), and led to the first serious studies bringing genetics into teratology. More important, perhaps, is the demonstration of how an embryo's *normal* developmental pattern may influence its liability to teratogenic insult. Thus, differences in metabolism or placental transport of cortisone are not necessary to explain strain differences in liability (though they undoubtedly exist)—the difference may lie simply in the inborn developmental pattern. Also, the concept can be useful in genetic counseling. Parents of children with a cleft palate or other MFT defects seem to appreciate this concept of how the problem arose, and why it has a chance of happening again, since it doesn't imply that they carry bad genes or did something wrong during the pregnancy. Finally, it is intellectually gratifying (to some) to have an example of an MFT defect in which the biological nature of the liability distribution and threshold can be demonstrated and many of the possible multifactors identified. Sewall Wright (1934) had postulated a threshold many years previously to explain his data on polydactyly in guinea pigs, and Hans Grüneberg (1952) had postulated that his "quasi-continuous variants" were threshold characters, but our model had the advantage that we knew what the distribution of liability and threshold actually were.

16. THE HUMAN CONNECTION

While we were developing the MFT model in the mouse, we had been collecting family data on human cleft lip and cleft palate, mainly to improve recurrence risk estimates, but had not thought much about the genetics except to conclude that the data did not fit any simple Mendelian model and were probably multifactorial. But Cedric Carter, in London, had been exploiting a similar MFT model (see Fraser, 1980, for its history) to explain some apparently paradoxical results he was getting with the recurrence risks for hypertrophic pyloric stenosis in human babies. Although pyloric stenosis occurs five times as often in males as in females, offspring of affected females are far more often affected than are those of affected males. This made sense if females had a distribution of liability with a mean farther from a threshold of abnormality than that of males. Insofar as genes contribute to liability, affected females would have more liability genes to transmit than would males, since they have to be nearer the tail of their distribution to be over the threshold. And so their children would be more likely to be affected (Carter, 1961). It should be emphasized that when Carter referred to a "polygenic" genetic component he used the term very loosely, meaning "not monofactorial," and involving several genes. He did not think in strict terms of a great many genes, each with an equal and additive effect.

It was very exciting to hear Carter present these data at the Second International Conference on Human Genetics in Rome, in 1961, and to realize that our MFT model had a human counterpart. In fact, his diagram, with two distributions and a threshold, was practically a carbon copy of ours. We had an exciting talk about the predictions from the model that could be tested with human data on recurrence risks. In fact, some of the predicted results were there in our cleft lip data, but we had not caught on. I remember telling Carter that our sib recurrence rate was higher after two affected than after one affected child, and Cedric saying, "But of course! Those parents would have more liability genes" and I had one of those "stupid me, why didn't I see that" reactions.

Beside the variation in recurrence risk with sex of proband and number of affected relatives, the MFT model predicted that risk would increase with severity of proband's defect (which it does) and (relatively) with frequency of the defect in the population. Finally, the drop-off in frequency from first- to second- to third-degree relatives would not be by half, as for a dominant with reduced penetrance, but by an amount determined by the shape of the tail of the normal curve—a much larger drop from first- to second- than from second- to third-degree relatives. It was another 9 years before there was a serious attempt to show that these predictions applied to cleft lip; since no one had enough data to do it alone, it required a collaborative pooling, achieved in a workshop sponsored by the National Institute of Dental Research (Fraser, 1970, 1980).

Of course there were those who pointed out that the observed recurrence rates did not fit exactly the predictions of the model based on a polygenic (in the strict sense) basis for liability. A "mixed model" in which a "major" gene results in three distributions (for the heterozygote and two homozygotes), each with some polygenic and/or environmental variation as well, would predict rather similar recurrence rates, which in some cases fitted the data somewhat better. I think this is a specious controversy. The strictly polygenic model is unrealistic, since liability genes are unlikely to have small equal and additive effects (i.e., no dominance or epistasis). Genes just aren't like that! On the other hand, palate closure or any other morphogenetic process is likely to be so complex that attributing cleft palate susceptibility in general to the presence of a single "major" gene is also unrealistic. Furthermore, any such major gene must have a low penetrance (or else recurrence risks would be higher), and as penetrance decreases, the distinction between major and minor ("poly-") genes becomes blurred. One alternative hypothesis, invoking allelic restriction as a basis for reduced penetrance of a postulated major gene, evoked the following quotation from the famous poet, Anon:

> Models tetrachorial
> For traits multifactorial
> Are no less idealical
> Than restrictions allelical

[It was gratifying, some years later, to be able to show that nonsyndromic macrocephaly, for which autosomal dominant inheritance had been claimed, fitted the expectation for a polygenic genetic basis (Arbour *et al.*, 1989). The pedigree pattern of dominant inheritance could be created by imposing an arbitrary level of abnormality on a continuously distributed variable.]

The important thing is to have two or more distributions with tails extending beyond the threshold. This is the aspect that leads to the MFT predictions about recurrence rates. In effect, one tries to infer the shape of the underlying distribution of liability by how much of its tail is beyond the threshold in various categories of relatives (Fraser, 1980). This is quite an insensitive approach, and it is small wonder that even with the sophisticated computer programs of today the results are still ambiguous. The attraction of the major-gene model is that major genes can be mapped and perhaps their mode of action identified, opening the way for prenatal diagnosis and preventive regimes. For every common malformation there are occasional families where a major gene does seem to be segregating. And even for the rest there are now striking examples of single genes contributing to susceptibility in a detectable way—the HLA loci to insulin-dependent diabetes mellitus, and the various genes for peptic ulcer, for example, and to ferret these out is very important. But they all have low penetrance and the distinction between these and "polygenes" is tenuous. I once proposed "plurigenic" for situations where there were several detectable genes with effects not big enough to be major but not small enough to be poly—but it never caught on. Another problem is that in many cases the predisposing genes seem to be different in different families, making the task of mapping difficult.

The attraction of the MFT model is that it offers the possibility of identifying the *biological,* epigenetic, basis of liability (stage of palate closure for cleft palate, face shape for cleft lip, stage of neural tube closure for neural tube defects, and so on), and perhaps of finding means of altering this characteristic in a way that would diminish liability to any environmental or genetic insult. The use of periconceptional vitamin supplements to reduce recurrence rates for NTDs—and perhaps cleft lip—may be an example of this (Smithells, 1984). In any case it does not help to argue the relative merits of the models. The important thing is to find out more about the biology.

17. 6-AMINONICOTINAMIDE

To get back to teratology, we had some fun with 6-aminonicotinamide (6-AN). We hit on the idea that we could control the stage of embryonic exposure to a nicotinamide deficiency induced by 6-AN more precisely (and reduce the resorption frequency) by terminating the deficiency at a chosen time with a flooding dose of nicotinamide. Len Pinsky did the preliminary experiments that defined the system. Marc Goldstein (now an ophthalmologist) and Merrille Feiner Pinsky (now a radiologist) looked at reciprocal cross differences between the A/J and C57BL mouse strains and showed how differences in maternal and fetal susceptibility genes were organ-specific and could in some combinations result in *patroclinous* reciprocal cross differences. Merrille and Len also fell in love and become engaged during their summer work in the mouse room, one of the most significant nonteratological achievements of our teratology program! Carl Verrusio showed how one could, in effect, titrate the amount of nicotinamide that would correct the 6-AN-induced deficiency, and showed it to be less in C57BL than A/J. There was also evidence of a slower turnover of nicotinamide in C57BLs.

Carl Verrusio and Russell Pollard demonstrated some matroclinous reciprocal cross differences and did backcrosses that indicated a cytoplasmically transmitted resistance factor—one of the few cytoplasmically transmitted effects known in mammals at the time. Curiously it disappeared in the second backcross, and was not present when a different diet was fed—one that had less nicotinamide. I felt that there were many interesting questions left unanswered here, but no one was inspired to pick up on them, and since my philosophy was for students to choose something that appealed to them, rather than to do something else, that *I* wanted, that trail petered out.

18. SPIN-OFFS

Throughout these years I had the habit of checking the newborn mice each morning, and setting up the new matings myself, which allowed me to do some long-range projects that didn't require much time, but spread over long periods of time. This resulted in the CL/Fr strain, with a high cleft lip frequency, still being studied in various parts of the world. The SW/Fr strain, which has a high cleft palate frequency, began when I started inbreeding a commercial "mass-inbred" strain just to see how inbred it really was, and saw some cleft palates in the F_2. I also found two cataract mutations, Cat^{Fr} and lens rupture (*lr*). And there was the inbred Nn strain that had a high frequency of prolapsus uteri, and an eyeless strain that got prematurely gray (old?). Interesting opportunities that no one seized.

19. CLEFT LIP

Another trail led toward the biological nature of cleft lip. Jeff Davidson, a Ph.D. student from Trinidad, did reciprocal crosses of A/J (high) and C57BL (low) and a series of reciprocal backcrosses to A/J. These showed that the spontaneous cleft lip occurring in that strain was genetically not simple, and that, as with cortisone-induced cleft palate, there was a matroclinous reciprocal cross difference. He also showed, quite strikingly, that the resorption rate in these crosses was determined entirely by maternal rather than fetal genes, something we never followed up. Diana Juriloff (now at the University of British Columbia), much later, showed that one fetal locus may determine the difference in fetal cleft lip susceptibility, plus genetically determined maternal effects. She is still working to map and characterize this gene.

Arguing by analogy from the cortisone-induced cleft palate model, we wondered what it was about the forming lip that might constitute liability to clefts. One likely guess seemed to be the shape of the face and how it would influence the relationships of the facial processes. So Daphne Trasler looked at face shapes of A/J and C57BL embryos (high and low spontaneous cleft lip frequency, respectively), and sure enough the morphology of their facial processes differed in plausible ways (Trasler, 1968). She and her students undertook a series of studies that elaborated this, second, example of how the embryo's normal developmental pattern influences its susceptibility to malformations. The A/J embryonic medial nasal processes diverge much less sharply than the C57BL's, presumably making it less easy for them to meet the lateral processes and fuse. Treatment of A/J mice with aspirin led to a lateral cleft between the lateral and (less divergent) medial nasal processes, but C57BL mice had medial clefts between the widely divergent medial processes. I developed a strain (CL/Fr) by selection and inbreeding, as mentioned above, that had a high frequency of cleft lip, and median processes that diverged even less sharply than in A/J. So here was another example of how the *normal developmental pattern influenced liability* to a malformation. Trasler also showed that, as with cleft palate, the defect can arise in several ways, e.g., the mutant gene *dc* (dancer) reduces the medial nasal processes, whereas 6-AN reduces both the medial and lateral processes.

Other examples of genes influencing susceptibility via developmental patterns include slower closure of the atrial septum, making the A/J strain more susceptible to *d*-amphetamine-induced atrial septal defects (Nora *et al.*, 1968), and later closure of the neural tube increasing susceptibility to retinoic acid-induced spina bifida (Kapron-Bras and Trasler, 1984).

The face-shape hypothesis as it developed from the mouse data encouraged us to see if the same thing might be demonstrated in people. If susceptibility to cleft lip was influenced by embryonic face shape, and if the vestiges of the relevant differences in shape persisted postnatally (which was shown to be so in the mouse), and if the differences were at least in part genetic (and not recessive), then parents of children with cleft lip should have face shapes that differed from controls. On the basis of external features (which we thought might be more reflective of embryonic shape than cephalograms), the late Hermine Pashayan did find some differences in face shape, which were substantiated by others, though not consistently so. They were probably real, but not large and clear enough to be useful for predicting liability in a useful way. It would be interesting to try this in Japanese families, where the cleft lip frequency is higher and the differences may be more clear-cut. In fact, in the light of the causal heterogeneity found experimentally, one might expect to find different types of difference in different families, so it would take very large numbers to get significant results. The same problem will apply, perhaps even more so, to attempts to find genetic markers for susceptibility—as in mice, they will probably turn out to be different in different families.

20. STUDIES IN MEDICAL GENETICS

Concurrent with our studies in experimental teratology and attempts to relate the MFT model to facial clefts, there were a series of clinical genetics projects. These were chosen on a very ad hoc basis, mainly depending on what sort of interesting conditions turned up on the wards, and what struck my fancy and that of the graduate students, summer students, undergraduates, and even high school students who worked with me. Melodie Williams (now Buxman) and I revisited the families with hydrotic ectodermal dystrophy described by Clouston; her paper won a national prize in a U.S. High School Science Fair and she went on to become a dermatologist. Julius Metrakos and I published early evidence for the autosomal recessive nature of the Ellis van Creveld syndrome, and Dr. Jessie Boyd Scriver (Charles's mother) and I did the same for chondrodystrophia calcificans congenita. Alice Lytwin and I looked at the array of anomalies in the sibs of probands with indubitable Meckel syndrome to get a more unbiased view of the spectrum of anomalies resulting from homozygosity for this gene—an underutilized approach. Cystic dysplasia of the kidneys was the only constant feature, but since I don't believe in obligate features in dysmorphic syndromes (Jack Rubinstein take note) I will not be surprised by reports to the contrary. Tony Glanz and I derived some figures for counseling for congenital myotonic dystrophy. Jack Naiman and I reported the first case of mental retardation and corpus callosum absence in sibs—later shown by Eva and Fred Andermann to result from a gene endemic to a French-Canadian deme. Charles Pender and I reported a large pedigree of dominant diabetes insipidus tracing back to an ancestor named Waterman. Adele Sadovnick and I showed that the IQs of Down syndrome children raised at home had the usual correlation of 0.5 with those of their parents and sibs, with the mean shifted downwards—but the IQs of those raised in institutions clustered at the low end of the distribution. And so on.

I was also blessed with a procession of Ph.D. students, and a few fellows and postdocs, who took on weightier problems and helped to advance my career as well as (I hope) their own. They attended those evening bull sessions and contributed to the ideas that would emerge from the arguments that developed. Some of them managed to combine medical and teratological themes, particularly Jim Miller who studied neural tube defects in both mice (effects of maternal fasting) and man (genetics and prenatal factors). He maintained this double interest for many years, after moving to the University of British Columbia. Dorothy Warburton, in addition to her now-classical study of recurrence rates in spontaneous abortions, got involved in studying the interactions between cortisone, genotype, and diet in determining cleft palate frequency. Julius Metrakos collected twins at The Montreal Children's Hospital for his Ph.D. thesis, and helped with the counseling in the days when the

Medical Genetics Department was contained in one small office with a secretary's desk in the hall. I helped him and his wife Kay launch their monumental study of the genetics of epilepsy, which is still accumulating data. Louis Dallaire was a pioneer in cytogenetics at McGill, working on translocations ascertained by screening sibs with multiple malformations. Renny Gold investigated the nature of dominance by studying proteins of dominant mouse mutations involving hair and lens proteins. Denise Theodosis and I showed that vitamin A, in causing exencephaly, acts first on the neuroepithelium, rather than the mesenchyme. Marilyn Preus brought some objectivity into diagnostic dysmorphology by her seminal use of numerical taxonomy in the classification of syndromes (Preus, 1985). Abby Lippman and I delved into the psychodynamics of decision-making by genetic counselees. And Maya Thangavelu (whom I codirected, with Penny Allderdice, at Memorial University in Newfoundland) studied the effects of paracentric inversions on fertility in mice, including interchromosomal effects, thus coming almost full circle to the subject of my M.Sc. thesis. My apologies also to the numerous (27 or so) M.Sc. students I have not mentioned by name, though they too added to the fun I've had. Thus, I got involved in a lot of subjects I knew very little about, relying on the good sense of colleagues and graduate students to keep me from making too many mistakes. If I were to take any credit, it would be for the knack of being able to separate out from a tangled network of possible projects a problem of suitable size for a particular student, whether a Ph.D. or undergraduate summer student.

I owe a lot to the MRC of Canada (and previously to the NIH) for their liberal attitude toward my grant applications, which were for some years entitled simply "Studies in Medical Genetics" and reported what I had been doing, what I was doing on various topics, and the sort of thing I might do in the future depending on what opportunities presented themselves. The council accepted the principle that I could not plan very far ahead, as what I would do depended in part on what would turn up on the wards. So I benefited from their willingness to bet on the investigator's track record and take some things on faith. Later on, when we were awarded an MRC Group Grant, I also benefited greatly from the 5-year grant periods, which allowed us to take on projects we could not have gambled with on the usual 3-year basis.

21. THE TERATO-CLINICAL GENETICS DILEMMA

And so my scientific life journey was more of a happy gambol than a planned itinerary. It rambled off on false trails here, dashed after clues to new ideas there, usually within the dysmorphogenic landscape, but following new signposts as they turned up, rather than steadfastly pursuing a foreseen goal. I seemed to be more attracted by genetically complex (i.e., "messy") systems and how to ask meaningful questions at the epigenetic level than by the more rigorous classical approach of counting and mapping genes. The trails of experimental teratology and clinical genetics kept interweaving. I always felt that either was a full-time job, and that I could not do both of them well, but I could never bring myself to give up one of them. Medical genetics had the attraction of being human-oriented, but lacked the rigor of experimental teratology. Teratology was more intellectually satisfying, but lacked the human interest. I continued to be frustrated by my less than adequate efforts to keep up with both, but I benefited greatly from being involved with both. The mouse room gave me the opportunity to set up critical tests of ideas that could not be done with people. The hospital gave me the opportunity to gather data on people to see how they fit the teratological models we made in the mouse room, and also provided the richness of the contacts I had with parents and children in counseling sessions. I even sometimes got the feeling I was helping people—one of the attractions that made me choose medicine in the first place.

One of my major faults was my tendency to have my fingers in too many pies, due largely to an inability to say no to any offer of an interesting project or request for help. The result was that if there was a characteristic affect on my life, it would be the feeling of being driven by too many responsibilities—lying awake at night thinking of unfinished manuscripts, grant applications to prepare or review, graduate students to take care of, lectures to prepare, wife and children to be with, and, of course, the growing clientele of counselees over the years. I am sure that many of you who read this will recognize the symptoms. I could never bring myself to take the time off to go fishing for more than a day, even though that was one of the reasons that made going to Newfoundland (see below) so attractive. And I could never manage to take 6 weeks off for an immersion course in French, though I badly wanted to become bilingual. And I never had a proper sabbatical, because I couldn't divest myself of my hospital responsibilities for that long. I envy my colleagues who find time to become expert in a hobby—Warkany with his etching, Wilson with his sculpture. I find a few hours a week for tennis but can't seem to fit in even a little serious photography. But of course there were also rewards—the satisfaction of a well-written paper or a lecture you know has gone over, the feeling of surprise at progress report time when you see how much you've actually managed to do, almost without noticing, and the occasional formal recognition of achievement. When I was chosen president of the American Society of Human Genetics (ASHG) or the Teratology Society, given a D.Sc. by Acadia, the Allen Award of the ASHG, the March of Dimes Award, or the Order of Canada, the initial surprise, followed by a warm glow of pleasure, was tempered by the secret thought—if they only knew how incompetent I really am! But the greatest rewards have been happy times with family—both biological and academic—and the host of good friends I've made over the years.

It is interesting to reflect upon how much one's successes and failures are governed by chance. I was certainly lucky in beginning my career just at the time that both medical genetics and teratology were about to take off, so I was able to get in on the ground floor. This may be why I was the youngest president of both the ASHG and the Teratology Society, in successive years. I was also the only president of the ASHG to compose and sing a song dedicated to the ASHG as part of the presidential address. I had fun writing that talk, mostly in the hammock on the side veranda of the Bear River house, and it is still on the reading list for genetic counseling students. Not only did I coin the term NID (natural insemination donor) but predicted the use of genetic engineering to transform genes, and suggested that the ASHG should patent the process, using the following jingle for TV commercials advertising its "superior DNA":

(sung to the tune of "Smiles")

There are genes that make you happy
There are genes that make you blue*
There are genes that tell you who's your father†
And how you'll rate on your I.Q.
There are genes that make your blood clot quickly
And genes that tell how much you'll weigh‡
But if you don't like the genes you're born with
TRY A.S.H.G. D.N.A.§

Quite some years later, when the ASHG presented me with the Allen Award, they asked me, with no forewarning, to sing the song again, thinking to take me aback. But they did not know I had been singing it at the end of the last lecture to my class in human genetics all during the intervening years!

*Congenital methemoglobinemia, for instance.
†Poetic license; actually, of course, they can usually only tell who's *not* your father.
‡If you don't make a pig of yourself!
§Copyright pending.

22. THE BOOKS

Over the years I had been approached many times by publishers who thought it would be a good idea for me to write a textbook of medical genetics, but I had always been too busy. Then I met Jim Nora. He was a pediatric cardiologist at Baylor University, who had noted what appeared to be an excessive number of children with heart malformations whose mothers had taken dexamphetamine during their pregnancy. He decided he wanted to learn something about teratology so, with the aid of an NIH fellowship, he came to McGill for a couple of years (1964–1965), and succeeded in showing that dexamphetamine did, indeed, cause heart malformations in mice. Furthermore, the type of defect depended on the normal developmental pattern of septum formation in the mouse strains used, so he learned some genetics at McGill, as well as teratology. So much so that, a few years later, at the March of Dimes Third International Conference on Congenital Malformations at The Hague, in 1969, he suggested that we write a medical genetics text together. He would do all the clinical stuff and I would just fill in a few genetic details. It wasn't quite as simple as that, but how could I resist? Thus, Nora and Fraser's *Medical Genetics* appeared in 1974. I had drawn heavily on the material that I used in my undergraduate lectures, and I thought it would be nice to have a shorter version, stripped of most of the clinical material, for my class, and so Fraser and Nora's *Genetics of Man* was born. Jointly, they have gone through five editions, so we must have done something right. Jim was very easy to work with, and I hope I was too. Our cooperation was perhaps promoted by the fact that many of our conferences were held in gourmet restaurants. But it may be time to stop cutting and pasting for the next edition and write a new version from scratch.

The impetus for Wilson and Fraser's *Handbook of Teratology* came from Jim Wilson, who first raised the idea on a flight from Vienna to Stockholm after the Vienna International Conference on Congenital Malformations, and it fell on receptive ground, as we had a fairly complete outline by the end of the flight. Getting manuscripts from our numerous collaborators was a much harder job, but the result was worth it. Alas, there will be no second edition, as Jim is no longer with us. He was a wonderful person to work with, always good-humored and considerate. He was a sculptor in his spare time, and the only time I saw him angry was in the Museum of Modern Art in Stockholm; he didn't think a dog's head with an old tire around its neck was modern art!

23. TRAVELS

One of the most gratifying things about being a scientist in general, and a teratogeneticist in particular, is the privilege of going to meetings in various parts of the world. I'm not sure that we appreciate how much they enrich our lives. For one thing, we make friends in almost all parts of the world, whom we meet again at intervals, and often correspond with in between. I think there are few parts of the world where I could go and not feel I would be welcomed as a friend by someone, and I value this greatly. Second, we get to see many parts of the world that we would not otherwise visit. I have, in my memory, a long series of vivid images around the world, often supported by slides, of experiences that have enriched my life. I wish I was a Paul Theroux, who could make such travels a delight to read about, but I am not. Arthur Koestler pokes fun at us "Call-Girls" in a very amusing novel, but although we may sometimes prostitute ourselves intellectually for the lure of prestige and travel to exotic places, we work very hard for our rewards and not without altruism. And I would always come back from meetings feeling that everyone else was making so much more progress than I was, which I guess was a stimulus to work even harder.

My most chastening travel experience happened in the men's washroom of the Prince Edward Island ferry. A little girl came in with her even smaller brother. They were speaking French, so I thought I would take advantage of the opportunity to practice mine. I said "Ce chambre est pour les hommes seulement. Ce n'est pas pour toi." And she said "Je ne parle pas l'Anglais."

24. INTERNATIONAL CONFERENCES ON CONGENITAL MALFORMATIONS

One notable series of meetings was the International Conferences on Congenital Malformations series sponsored by the March of Dimes. The first one was in London, in Church House with a reception in St. James Palace. I remember visiting the Tower of London with Joe Warkany and how angry he got with the morbid emphasis of the guide on the details of various decapitations of crowned heads, and the automatonic march of the sentry, which he considered demeaning. I also recall asking him to step back a bit so I could take his picture on the river boat en route to Kew Gardens, not realizing that that would have deposited him in the Thames. I was on the program committee for the second conference in New York (1963), after thalidomide had put teratology in the limelight, and where Joe was somewhat displeased with me for saying that animal testing would not necessarily predict human teratogenicity. Not that he disagreed, but he thought the drug companies should not be let off the hook too easily. I was chairman of the program committee for the third (1969) in The Hague (lovely early morning swims in the North Sea), and general chairman of the fourth in Vienna (1973). There was a wonderful evening in Grinzing, in a group of cafés serving cold cuts and jugfuls of spätlese, the new white wine, and Dave Smith, with his little mouth organ and stentorian voice, roaming from table to table starting singsongs. And Paul Polani, running up the down escalator in the subway, until moved on by a polite policeman. Finally, the Montreal conference (1977), of which I was the Honorary President, which gave me the privilege of being the Gouverneur at the final banquet, and the choice of whether to banish to prison or pardon the Rascal, Dave Smith. Naturally, I banished him. There was also the annual East versus West soccer match, a tradition started by Dave Smith and Eberhardt Passarge

Figure 7. Eberhardt Passarge, FCF, Dave Smith, Montreal, 1977. (Courtesy of Dr. Judith Hall.)

at the informal Malformations and Morphogenesis meetings that splintered off the McKusick Syndrome Delineation meetings when they got too large and diffuse. Anyone could play, the division between East and West was very pragmatic, the rules were very casual, even permitting Jim Hanson's ear-splitting yells behind whoever was about to kick the ball, and a good time was had by all. Guess who scored the winning goal!

The international conferences were then discontinued, partly because of some sentiment against their sumptuous entertainments, which were thought to be a poor way to spend those hard-won dimes. But they served an important purpose, by improving international communication and producing their state-of-the-art reviews of the many biological and medical areas relevant to malformations.

25. WORLD HEALTH ORGANIZATION COMMITTEE

The series of technical reports produced by the WHO on various topics relevant to the health of the world were produced at Expert Advisory Committee meetings with a rather unusual format. Experts were chosen with national representation in mind. The committee members brought with them position papers on assigned topics and were charged with producing a report within a fixed period of time, usually 3 or 4 days. The report had to be complete and agreed upon by the end of the meeting. I accepted an invitation to be a rapporteur at my first one, without knowing what I was getting into. The rapporteur's task was to take notes of the day's discussion and, overnight, to convert them into a draft for discussion the next day, so there was little time for sleep. Recommendations had to be compatible with the philosophies of the nations represented (we could not refer to social class effects, for example) and with the technical capabilities of third world countries, so they tended to be a bit bland, but the information in the reports was often useful. It is an exhausting experience, especially for the rapporteur, but a satisfying one. For one thing it is an opportunity to meet and get to know colleagues from various countries, some of them for the first time, in a very informal atmosphere. I remember J. A. Fraser Roberts (a tall, white-haired man with a handsome aristocratic face, who took snuff from a silver box) arriving fresh from another meeting where the discovery of the Xg blood group had been announced, and Marco Siniscalco literally pirouetting around the room saying "Now we can test the Lyon hypothesis!"

26. THALIDOMIDE

Thalidomide was responsible for quite a few of my trips. I first saw the news in a headline in *La Stampa* on a newsstand in Geneva, and my first reaction, like that of Joe Warkany, was incredulity—just another scare headline, I thought. When I got back to Montreal a surgeon friend said, "Come up to the ward for a minute, I've got something to show you." The shock was like a slap in the face. I was not only shocked, but embarrassed, since it was only a few years earlier that I had written that there was little reason to fear such a tragedy, since all teratogenic agents up to that point had been so only in doses harmful to the mother, not in the therapeutic range (Fraser, 1959). I had declared:

> If you suffocate a pregnant animal to the point of prostration, or give her amounts of cortisone that would kill her if she were not pregnant, or starve her until she loses 20% of her body weight, or batter her with sublethal doses of roentgen rays, or create in her an acute vitamin deficiency, the offspring are likely to be malformed. These are what Landauer calls "sledgehammer blows" to the embryo. They are useful tools in the analysis of abnormal development, but I doubt that they have very much to do with the bulk of malformations where no such violent maternal insults are evident.

True, but the implication was that drugs in the therapeutic range are not likely to be teratogens. That was 2 years before thalidomide.

Thalidomide really put the limelight on teratology, including the fact of genetic differences in susceptibility, which meant that findings in experimental animals could not automatically be applied to man. Teratologists spent a lot of time at committee meetings, conferences, workshops, and symposia talking about why one could not design a screening procedure in animals that would ensure that only "safe" drugs got on the market. I spelled out, in some detail, the problems of animal screening for teratogenicity at the March of Dimes Second International Conference on Congenital Malformations in New York in 1963 and concluded that ". . . screening for drug teratogenicity in experimental animals is not likely to be an effective safeguard against teratogenicity in man. . . . Extrapolation from experimental teratology to man is unwarranted unless supported by evidence in man" (Fraser, 1964). That made me popular with the pharmaceutical company representatives but not with some of my colleagues, who thought it might be letting the drug houses off too lightly. On a more positive note the teratologists also spent a lot of time organizing and running workshops to instruct biologists from pharmaceutical companies how to screen experimental animals for malformations.

And the media had to be clued up about teratology; I remember one meeting in Ann Arbor that had been set up by the March of Dimes to this end. Leading science writers from all over the country were there. My talk emphasized how genetic differences made it difficult to predict from teratogenic effects in one species what would happen in other species. I ended up by saying that no pregnant woman should take *any* drug unless she really needed it. In the discussion, one writer asked, "Would that include aspirin?" and I said, "Yes, I suppose it would." Next morning there were headlines all over the country—"Aspirin harmful to unborn children" and the like—and of course teratologists all over were getting phone calls from frightened women. It took Joe Warkany a long time to forgive me for that. Anyway, that was what got Daphne Trasler to try aspirin on her mice, so this ill wind blew some good. Yes, it was teratogenic, and no, it does not seem to be so in people except, perhaps, at very high doses.

My experience with thalidomide also involved two extraordinary tricks of fate, both illustrating how careful you have to be not to be fooled by incredible coincidences. The first was an attempt to develop a mouse model for thalidomide. We treated some pregnant C57BL mice from a line I had had in the mouse room for many years. I had the habit of inspecting every newborn litter in my own stocks, so I knew the frequency of malformations very well. About the third treated litter produced some offspring with short limbs and cleft palates, something we had never seen in thousands of newborns from that line. Eureka! So did the next couple of litters. Great excitement. I drafted a letter to the *Lancet* which was being typed when one of the *untreated* C57BL sisters produced the same thing! It turned out to be mutation (chondrodystrophic, *cho*) that had occurred in my C57BL stock two generations previously. Clearly the gods were joking with me.

A second extraordinary coincidence involved a student, Tim Heshka. We met at a student social gathering. I noticed the reduction deformities of his fingers and he was about the right age, so I asked if he was a thalidomide baby. He was, he said, and, as well as having vestigial fingers, he had no feet. Yet he skis, plays soccer, and recently completed a master's project which involved stereotactic injections into rat brains. He took my class in human genetics and we became good friends. In my lecture on prenatal diagnosis I would run through a series of conditions and ask the students whether they would have an abortion if this or that condition were found prenatally in their baby. Tim always voted against. In 1986 I was asked to organize a Teratology Society symposium to recognize the 25th anniversary of the thalidomide experience (see *Teratology* **38:**201, 1988). Widukind Lenz (who first recognized its teratogenic effects) came, and Frances Kelsey (who kept it off the U.S.

market), Joe Warkany, who told why he did not believe the first reports, and Bob Brent, who spoke on the phenotype. Finally, I asked Tim to speak for those most directly involved. He did so with eloquent simplicity, and was a great success. Afterwards, Lenz drew me aside and said, "You know, I think he may not have the thalidomide syndrome." He probably has the hypodactylia–hypoglossia syndrome, which, fortunately, is usually nonrecurrent. I pondered long and hard about when and how to tell Tim, and about the repercussions on the family who had lived so long with the anger and guilt that went with a drug-induced malformation. He was devastated, but has readjusted very well, and he and his fiancée have decided against prenatal diagnosis when the time comes. But my point here is the astronomical odds against such a rare condition occurring coincidentally in one of the comparatively few babies with reduction deformities born in Canada to mothers who had taken thalidomide!

Though I did my share of consulting with lawyers about thalidomide (they liked me because our data said you could not predict human teratogenicity from animal testing, which tended to let them off the hook), I was only once directly involved in a thalidomide trial. It was the first Canadian trial, in a little town near Quebec City. I was subpoenaed to appear in court on the same day I was to give my first paper before the Royal Society of Canada. I drove to the courthouse in due time, but there had been an injunction, so we had to wait. In the meantime I chatted with the defense lawyer. He told me they planned to argue that thalidomide did not cause the malformation, but acted as an immunosuppressant that interfered with the mother's rejection of an already abnormal embryo. After I told him what I thought of that argument he said, "I think you would not be a very good witness." So I jumped in my car and drove back to the Royal Society meeting, arriving just in time for my paper, only to find that I had left my briefcase, with notes and slides, in the courthouse. So I learned, perforce, that notes are expendable and that slides are not all that essential if there is a chalkboard. It turned out, incidentally, that the manufacturer could prove that they had not released any thalidomide at the time it was supposed to have been taken. Since the Canadian government was providing funds for the care of thalidomide victims, the mother had persuaded the doctor to say he had prescribed it so she could get the support for her baby.

27. SACCHARIN A TERATOGEN?

Another teratological experience showed how fast news travels when matters of economic importance are concerned. My pediatric colleague and friend Don Hillman and I began to get worried when we saw three, unrelated patients at the Montreal Children's Hospital with a lobster-claw defect; one of them also had a cleft lip (she later turned out to have the EEC syndrome), and one had a scar that was thought to be a forme fruste of a cleft lip. They all came from the same town and had been delivered by the same pediatrician. Questioning of the mothers did not reveal anything in common except that they were all overweight and had been using saccharin as a sweetener in fairly large amounts. We doubted that that was the cause, else why weren't there a lot more cases around, but felt we should report the cases, which we did in a letter to the editor of *Pediatrics*. But before the letter appeared I went to a meeting of the American Society of Human Genetics and asked a few friends if they had seen anything like this. They had not, but the day after I got back there was a call from the U.S. Food and Drug Administration saying what was this about me giving a paper (sic!) at the ASHG meeting claiming that saccharin caused split hand and foot, and why hadn't we notified them? No sooner had we calmed them down when there was a call from the Canadian FDA—what was this about a paper. . . ? And the next day there was a deputation from the Sugar Research Foundation, who were very interested to hear that saccharin might be teratogenic. Finally, one of the friends I talked to at the meeting got quite excited because he had heard something about saccharin fed to pigs during pregnancy causing split foot. He later phoned to say that this was a false alarm—the pigs got diarrhea from the saccharin and their hooves had split from overhydration. As far as I knew, saccharin has yet to be implicated as a teratogen. So what caused our "epidemic" of ectrodactyly? We will never know.

28. THE ANOMALAD

One of the many things I never became famous for was the anomalad. Posterity deserves to have the full story on record. I was part of a small committee on syndromological nomenclature sponsored by the NIH and comprising several well known syndromologists including Dave Smith, John Opitz, and Eberhardt Passarge. We were discussing, among other things, what to call a group of anomalies all arising from the same primary cause, e.g., micrognathia, cleft palate, and glossoptosis. Dave wanted to use *anomaly*, but there were objections to using a singular noun, already in use to mean something else, for a plural entity. More or less in jest I suggested *anomalad*—a triad is a group of three, tetrad a group of four, why not anomalad, a group of anomalies? Quite an argument developed—some liked it, others thought it an etymological terata. To my surprise I found myself defending it vigorously. The meeting adjourned for dinner without consensus. Overnight and who knows how many beers, Dave, Eberhardt, and others became converted (I wasn't even there) and the next morning declared that *anomalad* was official. They even sang a little song to the tune of funiculi–funicula which started:

Fraser, Fraser, goodness what a lad,
He invented the anomalad . . .

Alas, publication of the report drew a fusillade of criticism, mostly because it didn't roll trippingly off the tongue (nobody has trouble with *marmalade!*). I still think it's a useful term, but it got supplanted by *sequence*—a poor term since the common usage of the word does not reflect the meaning intended. Perhaps that's why it is used so uncritically.

29. AGENT ORANGE

I also devoted some time to another suspected teratogen, Agent Orange, and its dioxin contaminant, TCDD. The most memorable moments were spent in Vietnam. After TCDD was found to be teratogenic in rats, the American Association for the Advancement of Science sent a committee to Vietnam to look into the question of human teratogenicity, which made a cautiously worded report that did not rule out the possibility. The press kept referring repeatedly, and in my view irresponsibly, to defoliant spraying causing birth defects, and the National Academy of Sciences was asked to form a Committee on the Biological Effects of Herbicides in Vietnam to study this, among many other questions. I don't know how many geneticists/teratologists had declined before they asked me; I said it would be impossible to obtain valid data on birth defect frequencies and their relation to spraying in this war-torn country, riddled with malnutrition and disease. They said it would be better to say that after I had been there, and I could not resist. It was an extraordinary team, who were to look at all kinds of biological effects, not just birth defects, and contained a great range of biologists from psychiatrists and anthropologists through parasitologists, foresters, ecologists, botanists—and me. We flew over the countryside around Saigon in helicopters and light planes. We visited villages by U.S. Army motorboat—just like the ones in "Apocalypse Now." We counted malformations in the Rung Sat (a district near Saigon where pilots returning from uncompleted missions would dump their unused Agent Orange). We talked with midwives, and inspected birth records at Saigon hospitals. We saw several children with cleft lip in the Rung Sat—mostly isolated unilateral cleft lip—but no more than expected in that Oriental population. A few of our planes were shot at, and we could often hear gunfire in the distance. We were not allowed to visit the Highlands, as the spring offensive heated up the week we got there. We spent one Sunday afternoon relaxing on a beautiful beach at Vung Tau. Suddenly there were two loud explosions on the far side of the sand dunes and two fighter planes came screaming across the dunes, very low, toward us. I remember wondering whether a man would be a smaller target standing or lying, but before I could decide they had gone. We were told they had been getting rid of some obsolete explosives; it was an exciting moment.

My trip to Vietnam gave me a fresh insight into the question of how safe is safe. We were to fly over the Rung Sat in a helicopter at 3000 feet. Since the V.C. were known to have hand-held rocket launchers, we inquired if 3000 feet was a safe height. Indeed yes, they said, and off we went. Some time later we noted that the altimeter read 1500 feet. Hey, pilot, is this a safe height? Oh sure. But you said it was safe at 3000 feet, how come we're at 1500? Well, 1500 is safe too but 3000 is safer! Would that politicians and the public would appreciate the same relativity for drug "safety."

It was a fascinating trip, the upshot being that we could find no evidence relating spraying to birth defects—but that as predicted there could have been significant effects that we could not discern from the data available. This did not lead to any decrease in references to Agent Orange causing birth defects.

30. SICKLE-CELL SCREENING

Another interesting field trip was sponsored by the WHO to Jamaica, along with Arno Motulsky, to look at the feasibility of sickle-cell screening. Arno knew about hemoglobins and I knew something of the country and the dialect. We visited hospitals, health clinics, and schools, including my old school, Munro, where we soon found ourselves giving a lecture to the sixth form on sickle-cell disease. We produced a report which anticipated quite a few of the problems seen in other screening programs, and made some sensible suggestions, but as far as I know, it is still moldering in the files of the WHO.

31. NEWFOUNDLAND

My longest field trip took 3 years. I was on a search committee for a medical geneticist being sought by Memorial University in St. John's, Newfoundland. They had been looking for quite a long time. One day the chairman of the committee phoned to ask why I didn't take myself off the committee and apply for the job. Well, after 42 years at McGill I was beginning to feel in a bit of a rut. My syndromologist wife (Dr. Marilyn Preus) also felt adventuresome so we went, primarily to set up a clinical genetics service. There were several good human geneticists there, but no organized service.

We had a great time. We loved being near the sea and the spectacular walk up the precipitous cliffs from the harbor mouth to the top of Signal Hill, where blueberries could be picked in season—and just 10 minutes from our house. The Newfoundland people were wonderful, there were lots of fascinating genetical problems, and we made a lot of good friends. Of course there were some snags. For very complex reasons, I could not, as I had when I began at MCH, walk freely about the hospital wards to see what was passing through and bug the resident staff about family histories. So it was difficult to get a good view of the amazing variety of genetic problems passing by. But the main reason for leaving was that Marilyn felt moved to extend her horizons by going into law school. McGill offered me an Emeritus Professorship, which was hard to resist, so we decided to return to the fold. We didn't completely succeed in establishing the genetics service, as there was a certain amount of inertia to be overcome, but I think we made it easier for the next incumbent. I am still a visiting professor there and visit regularly for clinics and consultations—a privilege that I value greatly.

A lot of my time at Memorial was spent and, as it turned out, wasted on neural tube defects (NTDs) and periconceptional maternal vitamin supplements. A British group, headed by R. W. Smithells (1984), had some evidence suggesting that mothers who had given birth to children with NTDs had serum levels of certain vitamins lower than those of controls. This led him to organize a multicenter trial of vitamin supplementation, begun *preconception,* in subsequent pregnancies of mothers who had had an affected child. He wanted to do a randomized controlled trial, but several ethics committees withheld approval for this on the grounds that it would be unfair to deprive the women on placebos of the possible benefits of the treatment. It was decided, therefore, to derive a comparison group from women who, for one reason or another, did not get into the trial. The results showed a striking decrease in frequency of NTDs in the supplemented group. A small randomized trial organized by Michael Laurence in Wales showed a similar decrease. But there were some flaws in the randomization procedure in the Laurence trial, and the results of the Smithells trial were met with skepticism by some epidemiologists on the grounds that the comparison group might not be strictly comparable to the treated group since the women had not been randomized. In the best of all possible worlds, this would be a good argument—a randomized trial would remove all possibility of error. In the real world, some of us thought that the data were fairly convincing—they were adjusted to allow for differences in all known variables, and it seemed highly unlikely that some hitherto unrecognized variable would account for the difference. But the skeptics won out, and a large multicenter randomized controlled trial was launched by the British Medical Research Council. They managed to get approval from the various ethics committees involved. The results will eventually be in, but it is taking a long time because it is difficult to recruit women into the trial when they know that taking vitamins may reduce their risk and that, in the trial, they may not get them. An interesting example of how societal, nonscientific matters can interfere with the progress of science—for better or worse, time will tell.

Anyway, some of the Canadian geneticists thought it would be easy, and interesting, to do a Smithells-type trial but with a better control—the previous recurrence rates in the populations served by the collaborating centers. The Canadian MRC agreed to fund a workshop to work out a protocol, and we drafted what I thought was quite a good proposal. But not good enough—the epidemiologist reviewers were adamant that a randomized trial was the only way to go. By then the British MRC trial was recruiting centers, so since we couldn't have a Canadian trial I thought we at Memorial might volunteer for the British one. But the local ethics committee, in their wisdom, decided that this would be unethical! So I was stymied, whichever hole I went for.

32. PERSONAL HISTORY

I haven't said much about my personal life, which isn't very interesting. I was a romantic youth, and was usually deeply in love with and committed to someone, but only one at a time. The girl I was first engaged to, a lovely person, with artistic leanings, didn't want to wait till I was through medical school, and I did, so she passed. It might have been a great combination. I was almost engaged to another lovely girl during medical school, but was badly shaken up at flunking out in third year, and somehow that shook me loose from her, I don't understand why. Beryl and I had a pretty happy family life for quite a while, and raised a family with lots of happy memories, but eventually it went sour and we separated after a period of increasing unhappiness. This was the other most agonizing time of my life, and I did not handle it well, so my children tell me. Fortunately, my second marriage, to Marilyn, has not suffered the same stresses, and we have learned to live together in a complementary and rewarding way; I count my blessings.

I love athletics and played rugger till I was 40—a wonderful game; there are few things so satisfying as the excitement when you see a punt (and several large opponents) coming your way, the feeling of a good tackle, or the surge of a well-knit scrum, unless its coming into the net and taking a solid volley. Perhaps the appeal is controlled aggression, or that it's slightly dangerous, or maybe just the feeling of total involvement. In spite of aching knees and tennis elbow I keep up the tennis, and can hardly imagine life without it.

Neither have I mentioned my children, biological rather than academic—Norah, Noel, Alan, and Scott. They must have felt that Dad was away from home a lot, either on trips or down at the lab. But I read them bedtime stories, cooked Sunday supper, took them for walks, played music for them, sang with them, taught them to swim, fish, and play tennis, and learned from them (not very well) how to skate. We saved a lot of time for these pleasures by not having a television set. Every Christmas I would take pictures of them and make one into a Christmas card and (when they were old enough) they would come to the lab and help me print them. We had some good family times, and they do not seem to have resented my absences. We seem to be on good terms now. I have some great men's doubles with my boys, and Norah and I go to concerts and the market together. If I failed them, I ask their forgiveness. Curiously, three of the four are professional musicians—hard to fit to a Mendelian ratio. The other is a fish ecologist, so perhaps that is the recessive trait. And they all show a drive to excel (though, like me, not to make money) and, like me, they enjoy getting fully absorbed in something.

Why did I choose to do the things I have done, and why have I worked to hard at them? Some of my choices have surely been influenced by chance. If I had not gone into the air force, I would not have been able to finance medical school. If Happy Baxter had not come across some cortisone, I might never have become interested in the palate. Second, I like problems that have a practical orientation, and seem to have a knack for asking simple questions and designing projects of the right size for various kinds of students. If I wanted to ask elegant questions and get elegant and precise answers, I would have gone into

Figure 8. FCF, a recent portrait.

molecular biology. But you can't learn very much about the palate by studying organisms that don't have palates. And doing molecular biology on medical problems would mean sticking needles into people, which I do not like. I would rather just look at, listen to, and talk to them. And there is the satisfaction that you may actually be helping them sometimes.

But there are elements other than altruism. Curiosity—it is fun to solve puzzles, and to see whether your logical deductions, or intuitive hunches work out. One of my graduate students claimed that he had an orgasm when he saw the first embryo that confirmed the deduction he was testing. My responses are not that intense, but the experience is certainly gratifying. This is true whether you are trying to find out the mechanisms causing cleft palate, the factors influencing susceptibility, how often it recurs in relatives, how people feel when they have a baby with a cleft palate, or how to help them resolve their destructive feelings. Basic or applied, solving puzzles is fun. There is the esthetic appeal of patterns—when you stop to appreciate the mystery of a developing embryo, whether in a delicately molded little hand or the regular pattern of nuclei in the fibers of the growing lens, there is both the sensual response to beauty and the sense of wonder that this complex order emerges so mysteriously from the apparently simple egg. How on earth does it know what to do?

Recognition, prestige, and praise are nice, but not major elements. To interact with students, to see them develop, and to see hypotheses and concepts emerge from these interactions is rewarding. But perhaps the most valuable element is the opportunity to become entirely absorbed in something. That is why I work hard. My happiest moments are when I am totally wrapped up in a problem; whether it be designing an experiment, analyzing data, searching a microscope slide for a clue as to what happened to the palate, writing a paper, or even trying to compose a poem.

Lines composed in response to a challenge to write a sonnet:

I watched two larvae spin, through sunsplashed hours
Their living threads, one golden and one red.
Each wandered independent through the flowers
And chose to go wherever fancy led
Then, quite by chance, fate crossed their paths, and they
Their threads entwined in multicolored skein
No longer took at random each their way
But wove a fabric neither could attain
Alone, each the other gave of strength and joy.
And so the crimson cocoon that they wove
Did golden joy and scarlet love alloy
Until, unique creation, two in one
A glowing butterfly ascended toward the sun.

(And this from a geneticist!)

Getting totally absorbed takes periods of uninterrupted time, and one has to fight for these, against many competing pressures, but it is surprising what your mind can do if you let it really bear down on something for a while. Many people never do find out because their minds never get a chance to concentrate. Research is a great way to get your mind focused on something. I hope I never stop.

REFERENCES

Arbour, L., Watters, G. V., Hall, J., and Fraser, F. C. 1989. The inheritance of nonsyndromic macrocrania and its association with psychomotor impairment. *Proc. Greenwood Genetic Centre* **8:**193.

Baxter, H., and Fraser, F. C. 1950. The production of congenital defects in the offspring of female mice treated with cortisone. *McGill Med. J.* **19:**245–249.

Biddle, F. G. 1978. Use of dose–response relationships to discriminate between the mechanisms of cleft-palate induction by different teratogens: an argument for discussion. *Teratology* **18:**247–261.

Biddle, F. G., and Fraser, F. C. 1976. Genetics of cortisone-induced cleft palate in the mouse—embryonic and maternal effects. *Genetics* **84:**743–754.

Biddle, F. G., and Fraser, F. C. 1977. Cortisone-induced cleft palate in the mouse: a search for the genetic control of the embryonic response trait. *Genetics* **85:**289–302.

Carter, C. O. 1961. The inheritance of congenital pyloric stenosis. *Br. Med. Bull.* **17:**251–254.

Fraser, F. C. 1954. Medical genetics in pediatrics. *J. Pediatr.* **44:**85–103.

Fraser, F. C. 1956. Heredity counseling: the darker side. *Eugen. Q.* **3:**45–51.

Fraser, F. C. 1959. Antenatal factors in congenital defects: problems and pitfalls. *N.Y. State J. Med.* **59:**1597–1605.

Fraser, F. C. 1963. On being a medical geneticist. *Am. J. Hum. Genet.* **15:**1–10.

Fraser, F. C. 1964. Experimental teratogenesis in relation to congenital malformations in man, in: *International Conference on Congenital Malformations.* M. Fishbein, ed. Lippincott, Philadelphia, pp. 277–287.

Fraser, F. C. 1970. The genetics of cleft and cleft palate. *Am. J. Hum. Genet.* **22:**336–352.

Fraser, F. C. 1980. The William Allan Memorial Address: evolution of a palatable multifactorial threshold model. *Am. J. Hum. Genet.* **32:**796–813.

Fraser, F. C., and Fainstat, T. D. 1951. The production of congenital defects in the offspring of pregnant mice treated with cortisone: a progress report. *Pediatrics* **8:**527–533.

Fraser, F. C., Fainstat, T. D., and Kalter, H. 1953. The experimental production of congenital defects with particular reference to cleft palate. *Etud. Neo-Natales* **2:**43–58.

Fraser, F. C., Walker, B. E., and Trasler, D. G. 1957. Experimental production of congenital cleft palate: genetic and environmental factors. *Pediatrics* **19:**782–787.

Grüneberg, H. 1952. Genetical studies on the skeleton of the mouse. IV. Quasi-continuous variations. *J. Genet.* **51:**95–114.

Kalter, H. 1954. The inheritance of susceptibility to the teratogenic action of cortisone in mice. *Genetics* **39:**185–196.

Kapron-Bras, C. M., and Trasler, D. G. 1984. Gene–teratogen interaction and its morphological basis in retinoic acid induced mouse spina bifida. *Teratology* **30:**143–150.

Lippman-Hand, A., and Fraser, F. C. 1979. Genetics counseling—the postcounseling period. II. Making reproductive choices. *Am. J. Med. Genet.* **4:**73–87.

Nora, J. J., Sommerville, R. J., and Fraser, F. C. 1968. Homologies for congenital heart diseases: murine models, influenced by dextroamphetamine. *Teratology* **1:**413–416.

Preus, M. 1985. Numerical classification of syndromes. *Hosp. Pract.* **20:**111–129.

Smithells, R. W. 1984. Can vitamins prevent neural tube defects? *Can. Med. Assoc. J.* **131:** 273–274, 276.

Trasler, D. G. 1965. Strain differences in susceptibility to teratogenesis: survey of spontaneously occurring malformations in mice, in: *Teratology: Principles and Techniques*, J. G. Wilson and J. Warkany, eds. University of Chicago Press, Chicago, pp. 38–55.

Trasler, D. G. 1968. Pathogenesis of cleft lip and its relation to embryonic face shape in A/J and C57BL mice. *Teratology* **1:**33–49.

Trasler, D. G., and Fraser, F. C. 1957. A morphological basis for strain difference in frequency of cortisone-induced cleft palate in mice. *Proc. Genet. Soc. Can.* **2:**39.

Trasler, D. G., and Fraser, F. C. 1963. Role of the tongue in producing cleft palate in mice with spontaneous cleft lip. *Dev. Biol.* **6:**45–60.

Vekemans, M. J. J., and Biddle, F. G. 1984. Genetics of palate development, *Curr. Top. Dev. Biol.* **19:**165–192.

Vekemans, M., and Fraser, F. C. 1979. Stage of palate closure as one indication of "liability" to cleft palate. *Am. J. Med. Genet.* **4:**95–102.

Vekemans, M., Taylor, B. A., and Fraser, F. C. 1981. The susceptibility to cortisone-induced cleft palate of recombinant inbred strains of mice: lack of association with the *H-2* haplotype. *Genet. Res.* **38:**327–331.

Walker, B. E., and Fraser, F. C. 1956. Closure of the secondary palate in three strains of mice. *J. Embryol. Exp. Morphol.* **4:**176–189.

Walker, B. E., and Fraser, F. C. 1957. The embryology of cortisone-induced cleft palate. *J. Embryol. Exp. Morphol.* **5:**201–209.

Warkany, J. 1988. Story of a teratologist, in: *Issues and Reviews in Teratology,* Volume 4, H. Kalter, ed. Plenum Press, New York, pp. 1–79.

Wright, S. 1934. An analysis of variability in number of digits in an inbred strain of guinea pigs. *Genetics* **19:**506–536.

BIBLIOGRAPHY

Steinberg, A. G., and Fraser, F. C. 1944. Studies on the effect of X chromosomes on crossing over in the third chromosome of *Drosophila melanogaster. Genetics* **29:**83–103.

Steinberg, A. G., and Fraser, F. C. 1946. The expression and interaction of hereditary factors affecting hair growth in mice: external observations. *Can. J. Res. Sect. D* **24:**1–9.

Fraser, F. C. 1946. The expression and interaction of hereditary factors producing hypotrichosis in the mouse: histology and experimental results. *Can. J. Res. Sect. D* **24:**10–25.

Fraser, F. C. 1947. The use of genetics in medical practice. *McGill Med. J.* **16:**349–358.

Browning, H. C., Fraser, F. C., Shapiro, S. D., Glickman, I., and Dubrule, M. 1948. The biological activity of DDT and related compounds. *Can. J. Res. Sect. D* **26:**282–300.

Fraser, F. C. 1949. The use of genetics in clinical medicine. II. Dominant inheritance. *McGill Med. J.* **18:**19–24.

Fraser, F. C. 1949. The effect of vitamin A on hereditary hyperkeratosis in the mouse. *Can. J. Res. Sect. D.* **27:**179–195.

Fraser, F. C. 1949. The use of genetics in clinical medicine. III. Recessive inheritance. *McGill Med. J.* **18:**176–182.

Boyes, J. W., Fraser, F. C., Lawler, S. D., and MacKenzie, H. J. 1949. A pedigree of hereditary progressive muscular dystrophy. *Ann. Eugen.* **15:**46–51.

Fraser, F. C., and Herer, M. L. 1950. The inheritance and expression of the "lens rupture" gene in the house mouse. *J. Hered.* **41:**3–7.

Fraser, F. C. 1950. The use of genetics in clinical medicine. IV. Sex-linked inheritance. *McGill Med. J.* **19:**194–198.

Fraser, F. C. 1951. The use of genetics in clinical medicine. V. On taking the family history. *McGill Med. J.* **20:**184–190.

Fraser, F. C., and Fainstat, T. D. 1951. The production of congenital defects in the offspring of pregnant mice treated with cortisone: a progress report. *Pediatrics* **8:**527–533.

Fraser, F. C., and Fainstat, T. D. 1951. Causes of congenital defects: a review. *Am. J. Dis. Child.* **82:**593–603.

Fraser, F. C. 1952. Consanguinity and its significance in the family history. *Can. Med. Assoc. J.* **66:**258–260.

Kalter, H., and Fraser, F. C. 1952. Production of congenital defects in the offspring of pregnant mice treated with compound F. *Nature* **169:**665.

Fraser, F. C. 1953. Cleft palate induced in mice by cortisone: prematurity, congenital malformation and birth injury. Association for the Aid of Crippled Children, New York.

Pender, C. B., and Fraser, F. C. 1953. Dominant inheritance of diabetes insipidus: a family study. *Pediatrics* **11:**246–254.

Fraser, F. C., Fainstat, T. D., and Kalter, H. 1953. The experimental production of congenital defects with particular reference to cleft palate. *Etud. Neo-Natales* **2:**43–58.

Publications are listed chronologically.

Kalter, H., and Fraser, F. C. 1953. The modification of the teratogenic action of cortisone by parity. *Science* **118**:625–626.

Fraser, F. C. 1954. Medical genetics in pediatrics. *J. Pediatr.* **44**:85–103.

Fraser, F. C., and Scriver, J. B. 1954. A hereditary factor in chondrodystrophia calcificans congenita. *New Engl. J. Med.* **250**:272–277.

Fraser, F. C., and Baxter, H. 1954. The familial distribution of congenital clefts of the lip and palate. *Am. J. Surg.* **87**:656–659.

Fraser, F. C., Kalter, H., Walker, B. E., and Fainstat, T. D. 1954. The experimental production of cleft palate with cortisone and other hormones. *J. Cell. Comp. Physiol.* **43**:237–259.

Metrakos, J. D., and Fraser, F. C. 1954. Evidence for a hereditary factor in chondroectodermal dysplasia (Ellis–van Creveld syndrome). *Am. J. Hum. Genet.* **6**:260–269.

Naiman, J., and Fraser, F. C. 1955. Agenesis of the corpus callosum: a report of two cases in siblings. *Arch. Neurol. Psychiat.* **74**:182–185.

Fraser, F. C. 1955. Hérédité dominante d'un diabète insipide due à une déficience en pitressine. *J. Génet. Hum.* **4**:195–203.

Fraser, F. C. 1955. Dominant inheritance of absent nipples and breasts. Mendel Memorial Volume. Istituto Gregorio Mendel, Rome, pp. 360–362.

Fraser, F. C. 1955. Thoughts on the etiology of clefts of the palate and lip. *Acta Genet. Stat. Med.* **5**:358–369.

Fraser, F. C. 1956. Heredity counselling: the darker side. *Eugen. Q.* **3**:45–51.

Trasler, D. G., Clark, K. H., and Fraser, F. C. 1956. No cleft palates in offspring of pregnant mice given cortisone after fetal palate closure. *J. Hered.* **47**:99–100.

Warkany, J., and Fraser, F. C. 1956. The role of genetics and other prenatal factors in disorders of childhood. *Pediatrics* **18**:314–317.

Trasler, D. G., Walker, B. E., and Fraser, F. C. 1956. Congenital malformations produced by amniotic-sac puncture. *Science* **124**:439.

Walker, B. E., and Fraser, F. C. 1956. Closure of the secondary palate in three strains of mice. *J. Embryol. Exp. Morphol.* **4**:176–189.

Metrakos, J. D., Metrakos, K., and Fraser, F. C. 1956. Juvenile epilepsy: genetic and electroencephalographic studies. *Electroencephalogr. Clin. Neurophysiol.* **8**:164.

Goldbloom, R. B., Fraser, F. C., Waugh, D., Aronovitch, M., and Wigglesworth, F. W. 1957. Hereditary renal disease associated with nerve deafness and ocular lesions. *Pediatrics* **20**:241–252.

Walker, B. E., and Fraser, F. C. 1957. The embryology of cortisone-induced cleft palate. *J. Embryol. Exp. Morphol.* **8**:201–209.

Fraser, F. C., Walker, B. E., and Trasler, D. G. 1957. Experimental production of congenital cleft palate: genetic and environmental factors. *Pediatrics* **19**(Suppl.):782–787.

Fraser, F. C. 1957. Antenatal factors in congenital defects: problems and pitfalls. *Proc. Genet. Soc. Can.* **2**:37–38.

Fraser, F. C. 1957. Genetic background of congenital malformations. *Ross Pediatr. Res. Conf.* **23**:59–63.

Trasler, D. G., and Fraser, F. C. 1957. A morphological basis for strain difference in frequency of cortisone-induced cleft palate in mice. *Proc. Genet. Soc. Can.* **2**:39.

Fraser, F. C. 1957. Etiological factors in clefts of the palate and lip. *Acta Genet. Stat. Med.* **7**: 229–230.

Fraser, F. C. 1957. Review of "Heritable disorders of connective tissue" by V. McKusick. *Eugen. Q.* **4**:168–170.

Fraser, F. C. 1958. Types of problems presented to genetic counselors. *Eugen. Q.* **5**:46–47.

Fraser, F. C. 1958. Genetic counselling in some common paediatric diseases. *Pediatr. Clin. North Am.* **1958**:475–491.

Fraser, F. C. 1958. Recent advances in genetics in relation to pediatrics. *J. Pediatr.* **52:**734–757.

Fraser, F. C. 1958. Human genetics in eastern Canada. *Proc. Genet. Soc. Can.* **3:**68–69.

Trasler, D. G., and Fraser, F. C. 1958. Factors underlying strain, reciprocal cross, and maternal weight differences in embryo susceptibility to cortisone-induced cleft palate in mice. *Proc. X Int. Congr. Genet.* **2:**296–297.

Heiberg, K., Kalter, H., and Fraser, F. C. 1959. Production of cleft palates in the offspring of mice treated with ACTH during pregnancy. *Biol. Neonat.* **1:**33–37.

Warburton, D., and Fraser, F. C. 1959. Genetic aspects of abortion. *Clin. Obstet. Gynecol.* **2:** 22–35.

Fraser, F. C. 1959. Antenatal factors in congenital defects. Problems and pitfalls. *N.Y. State J. Med.* **59:**1597–1605.

Pinsky, L., and Fraser, F. C. 1959. Production of skeletal malformations in the offspring of pregnant mice treated with 6-aminonicotinamide. *Biol. Neonat.* **1:**106–112.

Warkany, J., and Fraser, F. C. 1959. Prenatal factors in diseases of children, in: *Textbook of Pediatrics*, W. E. Nelson, ed. Saunders, Philadelphia, pp. 234–250.

Fraser, F. C. 1958. Rapport le Xeme Congrès International de Génétique. *J. Génet. Hum.* **7:** 315–323.

Fraser, F. C. 1959. Causes of congenital malformations in human beings. *J. Chron. Dis.* **10:** 97–110.

Fraser, F. C. 1959. Genetic factors in deafness. *Voice* **2:**4–8.

Fraser, F. C. 1959. What is the embryological basis for cleft lip and palate? *Mod. Med.* **27:** 33–34.

Fraser, F. C. 1960. Genetics and medical practice. *Modern Medicine of Canada* (Feb. 15). pp. 69–73.

Fraser, F. C. 1960. Some experimental and clinical studies on the causes of congenital clefts of the palate and of the lip. *Arch. Pediatr.* **77:**151–156.

Pinsky, L., and Fraser, F. C. 1960. Congenital malformations after a two-hour inactivation of nicotinamide in pregnant mice. *Br. Med. J.* **2:**195–197.

Fishman, J., Fraser, F. C., Watanabe, M., Sodhi, H. S., and Beck, J. C. 1960. Familial nerve deafness and goitre. *Can. Med. Assoc. J.* **83:**889–892.

Miller, J. R., and Fraser, F. C. 1961. The First International Conference on Congenital Malformations. *Can. Med. Assoc. J.* **84:**60–63.

Fraser, F. C. 1961. Experimental induction of cleft palate, in: *Congenital Anomalies of the Face and Associated Structures,* S. Pruzansky, ed. Thomas, Springfield, Ill., pp. 188–197.

Fraser, F. C. 1961. The use of teratogens in the analysis of abnormal developmental mechanisms, in *First International Conference on Congenital Malformations,* M. Fishbein, ed. Lippincott, Philadelphia, pp. 179–186.

Fraser, F. C. 1961. Genetics and congenital malformations, in: *Progress in Medical Genetics,* A. G. Steinberg, ed. Grune & Stratton, New York, pp. 38–80.

Fraser, F. C. 1961. Congenital defects—intrinsic and extrinsic factors. *Acad. Med. Bull.* **7.**

Fraser, F. C. 1961. Congenital clefts of the face, in *De Genetica Medica,* L. Gedda, ed. Istituto Gregorio Mendel, Rome.

Warburton, D., and Fraser, F. C. 1961. On the probability that a woman who has had a spontaneous abortion will abort in subsequent pregnancies. *J. Obstet. Gynaecol. Br. Commonw.* **68:**784–788.

Curtis, E., Fraser, F. C., and Warburton, D. 1961. Congenital cleft lip and palate: risk figures for counselling. *Am. J. Dis. Child.* **102:**853–857.

Miller, J. R., Fraser, F. C., and MacEwan, D. W. 1962. The frequency of spina bifida occulata and rib anomalies in the parents of children with spina bifida aperta and meningocele. *Am. J. Hum. Genet.* **14:**245–248.

Fraser, F. C., and others. 1962. The teaching of genetics in the undergraduate medical curriculum and in post-graduate training. *W.H.O. Tech. Rep. Ser.* **238.**

Warburton, D., Trasler, D. G., Naylor, A., Miller, J. R., and Fraser, F. C. 1962. Pitfalls in tests for teratogenicity. *Lancet* **2:**1116–1117.

Fraser, F. C. 1962. Drug-induced teratogenesis. *Can. Med. Assoc. J.* **87:**683–684.

Fraser, F. C., and Schabtach, G. 1962. "Shrivelled": a hereditary degeneration of the lens in the house mouse. *Genet Res.* **3:**383–387.

Fraser, F. C. 1962. Experimental analysis of developmental disturbances, in: *Genetics and Dental Health,* C. J. Witkop, ed. McGraw–Hill, New York, pp. 129–134.

Fraser, F. C. 1963. On being a medical geneticist. *Am. J. Hum. Genet.* **15:**1–10.

Trasler, D. G., and Fraser, F. C. 1963. Role of the tongue in producing cleft palate in mice with spontaneous cleft lip. *Dev. Biol.* **6:**45–60.

Fraser, F. C. 1963. Harelip and cleft palate, in: *Birth Defects,* M. Fishbein, ed. Lippincott, Philadelphia, pp. 235–244.

Fraser, F. C. 1963. Recent knowledge in the pathogenesis of congenital anomalies. Institute on the Development of Community Health Services for Children with Congenital Anomalies, Ann Arbor, pp. 16–27.

Fraser, F. C. 1963. Taking the family history. *Am. J. Med.* **34:**585–593.

Goldstein, M., Pinsky, M. F., and Fraser, F. C. 1963. Genetically determined organ specific responses to the teratogenic action of 6-aminonicotinamide in the mouse. *Genet. Res.* **4:** 258–265.

Fraser, F. C. 1963. Methodology of experimental mammalian teratology, in: *Methodology in Mammalian Genetics,* W. J. Burdette, ed. Holden–Day, San Francisco, pp. 233–246.

Fraser, F. C. 1963. Report: Conference on Prenatal Effects of Drugs. Commission on Drug Safety, Chicago, pp. 11–12.

Fraser, F. C. 1964. Experimental teratogenesis in relation to congenital malformations in man, in: *Second International Conference on Congenital Malformations,* M. Fishbein, ed. Lippincott, Philadelphia, pp. 277–287.

Fraser, F. C. 1963. Hereditary disorders of the nose and mouth, in: *Second International Congress of Human Genetics,* Volume 3. Istituto Gregorio Mendel, Rome, pp. 1852–1855.

Fraser, F. C., and Warburton, D. 1964. No association of emotional stress or vitamin supplement during pregnancy to cleft lip or palate in man. *J. Plast. Reconstr. Surg.* **33:** 395–399.

Warburton, D., and Fraser, F. C. 1964. Spontaneous abortion risks in man: data from reproductive histories collected in a medical genetics unit. *Hum. Genet.* **16:**1–25.

Warkany, J., and Fraser, F. C. 1964. Prenatal factors in diseases of children, in: *Textbook of Pediatrics,* 8th ed., W. E. Nelson, ed. Saunders, Philadelphia, pp. 260–279.

Fraser, F. C. 1964. Teratogenesis of the central nervous system, in: *Mental Retardation: A Review of Research,* H. E. Stevens and R. Heber, ed. University of Chicago Press, Chicago, pp. 395–428.

Dallaire, L., and Fraser, F. C. 1964. Two unusual cases of familial mongolism. *Can. J. Genet. Cytol.* **6:**540–547.

Fraser, F. C. 1965. Some genetic aspects of teratology, in: *Teratology: Principles Techniques,* J. G. Wilson and J. Warkany, eds. University of Chicago Press, Chicago, pp. 21–38.

Goldstein, M., Fraser, F. C., and Roth, K. 1965. Resistance of A/Jax mouse embryos with spontaneous congenital cleft lip to the lethal effect of 6-aminonicotinamide. *J. Med. Genet.* **2:**128–130.

Tadjoedin, M. K., and Fraser, F. C. 1965. Heredity of ataxia-telangiectasia (Louis Bar syndrome). *Am. J. Dis. Child.* **110:**64–68.

Fraser, F. C. 1965. Some teratological implications of quasi-continuous variation. Proceedings of the 5th Annual Meeting of the Teratology Society.

Nora, J. J., Trasler, D. G., and Fraser, F. C. 1965. Malformations in mice induced by dexamphetamine sulphate. *Lancet* **2**:1021–1022.

Kallio, E. I. S., Bacal, H. L., Eisen, A., and Fraser, F. C. 1966. A familial tendency toward skin sensitivity to ragweed pollen. *J. Allergy* **38**:241–249.

Dallaire, L., and Fraser, F. C. 1966. The syndrome of retardation with urogenital and skeletal anomalies in siblings. *J. Pediatr.* **69**:459–460.

Fraser, F. C. 1966. Impact on the family of a child with a genetically determined disease, in: *Third International Congress of Human Genetics,* Chicago, pp. 33–34 (abstr.).

Verrusio, A. C., and Fraser, F. C. 1966. Identity of mutant genes "Shrivelled" and cataracta congenita subcapsularis in the mouse. *Genet. Res.* **8**:377–378.

Williams, M., and Fraser, F. C. 1966. Hydrotic ectodermal dysplasia—Clouston's family revisited. *Can. Med. Assoc. J.* **96**:36–38.

Fraser, F. C., Chew, D., and Verrusio, A. C. 1967. Oligohydramnios and cortisone-induced cleft palate in the mouse. *Nature* **214**:417–418.

Nora, J. J., McNamara, D. G., and Fraser, F. C. 1967. Dexamphetamine sulphate and human malformations. *Lancet* **1**:570–571.

Nora, J. J., McNamara, D. G., and Fraser, F. C. 1967. Hereditary factors in atrial septal defect. *Circulation* **35**:448–456.

Fraser, F. C. 1967. Workshop on embryology of cleft lip and cleft palate. *Science* **158**:1603–1606.

Verrusio, A. C., Pollard, D. R., and Fraser, F. C. 1968. A cytoplasmically transmitted, diet-dependent difference in response to the teratogenic effects of 6-aminonicotinamide. *Science* **160**:206–207.

Fraser, F. C., and Latour, A. 1968. Birth rates in families following birth of a child with mongolism. *Am. J. Ment. Defic.* **72**:883–886.

Fraser, F. C. 1968. Genetic counselling and the physician (the Blackader Lecture 1968). *Can. Med. Assoc. J.* **99**:927–934.

Fraser, F. C. 1968. Genetic factors in experimental teratology. *Cong. Anom.* **8**:78–79.

Pollard, D. R., and Fraser, F. C. 1968. Further studies on a cytoplasmically transmitted difference in response to the teratogen 6-aminonicotinamide. *Teratology* **1**:335–338.

Harris, M. W., and Fraser, F. C. 1968. Lid gap in newborn mice: a study of its cause and prevention. *Teratology* **1**:417–423.

Nora, J. J., Sommerville, R. J., and Fraser, F. C. 1968. Homologies for congenital heart diseases: murine models, influenced by dextroamphetamine. *Teratology* **1**:413–416.

Fraser, F. C. 1968. Are salicylates teratogenic? Proceedings of the Conference on the Effects of Chronic Salicylate Administration. U.S. D.H.E.W., Washington, D.C.

Davidson, J. G., Fraser, F. C., and Schlager, G. 1969. A maternal effect on the frequency of spontaneous cleft lip in the A/J mouse. *Teratology* **2**:371–376.

Dallaire, L., and Fraser, F. C. 1969. The Smith–Lemli–Opitz syndrome of retardation, urogenital and skeletal anomalies. *Birth Defects* **5**(2):180–182.

Pashayan, H. M., Whelan, D., Guttman, S., and Fraser, F. C. 1969. Variability of the Cornelia deLange syndrome. *J. Pediatr.* **75**:853–858.

Fraser, F. C. 1969. Gene–environment interactions in the production of cleft palate, in: *Methods for Teratological Studies in Experimental Animals and Man,* H. Nishimura and J. R. Miller, eds. Igaku Shoin, Tokyo, pp. 34–48.

Fraser, F. C., Pashayan, H. M., and Kadish, M. E. 1970. Cranio-carpo-tarsal dysplasia. *J. Am. Med. Assoc.* **211**:1374–1376.

Fraser, F. C. 1970. The genetics of cleft lip and cleft palate. *Am. J. Hum. Genet.* **2**:336–352.

Fraser, F. C. 1970. Developmental genetics: a status of research report. *Teratology* **3**:73–88.
Levy, E. P., Pashayan, H. M., Fraser, F. C., and Pinsky, L. 1970. XX and XY Turner phenotype in a family. *Am. J. Dis. Child.* **120**:36–43.
Fraser, F. C., and Pashayan, H. M. 1970. Relation of face shape to susceptibility to congenital cleft lip: a preliminary report. *J. Med. Genet.* **7**:112–117.
Fraser, F. C., and McKusick, V.A., eds. 1970. *Congenital Malformations.* Proceedings of 3rd International Conference on Congenital Malformations. Excerpta Medica, Amsterdam.
Fraser, F. C. 1970. Counselling in genetics: its intent and scope. *Birth Defects* **5**(1):7–12.
Pashayan, H. M., Pinsky, L., and Fraser, F. C. 1970. Hemifacial microsomia: oculo-auriculo-vertebral dysplasia. *J. Med. Genet.* **7**:185–188.
Hamly, C.-A., Trasler, D. G., and Fraser, F. C. 1970. Reduction of 6-aminonicotinamide teratogenicity in mice by etherization. *Teratology* **3**:293–294.
Bornstein, S., Trasler, D. G., and Fraser, F. C. 1970. Effect of the uterine environment on the frequency of spontaneous cleft lip in CL/Fr mice. *Teratology* **3**:295–298.
Pashayan, H. M., Levy, E. P., and Fraser, F. C. 1970. Can the deLange syndrome always be diagnosed at birth? *Pediatrics* **46**:940–942.
Preus, M., Fraser, F. C., and Levy, E. P. 1970. Dermatoglyphics in congenital heart malformations. *Hum. Hered.* **21**:388–402.
Fraser, F. C. 1971. Book review of An ABC of Medical Genetics, by C. O. Carter. *Teratology* **4**:112.
Seegmiller, R. E., Fraser, F. C., and Sheldon, H. 1971. A new chondrodystrophic mutant in mice: electron microscopy of normal and abnormal chondrogenesis. *J. Cell Biol.* **48**: 580–593.
Fraser, F. C. 1971. Etiology of cleft lip and palate, in: *Cleft Lip and Palate,* W. C. Grabb, S. W. Rosenstein, and K. R. Bzoch, eds. Little, Brown, Boston, pp. 54–65.
Fraser, F. C. 1971. Genetic counselling. *Hosp. Prac.* **6**:49–56.
Lowry, B., Miller, J. R., and Fraser, F. C. 1971. A new dominant gene mental retardation syndrome. *Am. J. Dis. Child.* **121**:496–500.
Pashayan, H. M., and Fraser, F. C. 1971. Nostril asymmetry not a microform of cleft lip. *Cleft Pal. J.* **8**:185–188.
Pashayan, H. M., Fraser, F. C., McIntyre, J. M., and Dunbar, J. S. 1971. Bilateral aplasia of the tibia, polydactyly and absent thumb in father and daughter. *J. Bone J. Surg.* **53B**: 459–499.
Pashayan, H. M., and Fraser, F. C. 1971. Facial features associated with predisposition to cleft lip. *Birth Defects* **7**(7):58–63.
Dallaire, L., Fraser, F. C., and Wigglesworth, F. W. 1971. Familial holoprosencephaly. *Birth Defects* **7**(7):137–142.
Fraser, F. C. 1971. Discussion of Third Conference on Clinical Delineation of Birth Defects. *Birth Defects* **7**(7):101–102.
Fraser, F. C. 1971. The epidemiology of the common major malformations as related to environmental monitoring, in: *Monitoring, Birth Defects and Environment,* E. B. Hook, Academic Press, New York, pp. 85–96.
Pashayan, H. M., Fraser, F. C., and Goldbloom, R. B. 1971. A family showing hereditary nephropathy. *Am. J. Hum. Genet.* **23**:555–567.
Preus, M., and Fraser, F. C. 1971. Genetics of hereditary nephropathy with deafness (Alport's disease). *Clin. Genet.* **2**:331–337.
Pinsky, L., and Fraser, F. C. 1972 Atypical malformation syndromes. *J. Pediatr.* **80**:141–144.
Perry, T. B., and Fraser, F. C. 1972. Paternal age and congenital cleft lip and cleft palate. *Teratology* **6**:241–246.

Cox, D., Fraser, F. C., and Sass-Kortsak, A. 1972. A genetic study of Wilson's disease: evidence for heterogeneity. *Am. J. Hum. Genet.* **24:**646–666.

Burdi, A., Feingold, M., Larsson, K. S., Leck, I., Zimmerman, E. F., and Fraser, F. C. 1972. Etiology and pathogenesis of congenital cleft lip and cleft palate, an NIDR state of the art report. *Teratology* **6:**255–270.

Preus, M., and Fraser, F. C. 1972. Dermatoglyphics and syndromes. *Am. J. Dis. Child.* **124:** 933–943.

Fraser, F. C. 1973. Survey of counselling practices, in: *Ethical Issues in Human Genetics*, B. Hilton, ed. Plenum Press, New York, pp. 7–22.

Clow, C. L., Fraser, F. C., Laberge, C., and Scriver, C. R. 1973. On the application of knowledge to the patient with genetic disease, in: *Progress in Medical Genetics*, Volume 9, A. G. Steinberg and A. G. Bearn, eds. Grune & Stratton, New York, pp. 159–213.

Spriestersbach, D. C., Dickson, D. R., Fraser, F. C., Horowitz, S. L., McWilliams, B. J., Paradise, J. L., and Randall, P. 1973. Clinical research in cleft lip and cleft palate: the state of the art. *Cleft Pal. J.* **10:**113–165.

MacLeod, P. M., and Fraser, F. C. 1973. Case report 2. *Syndr. Ident.* **1:**10–11.

Pollard, D. R., and Fraser, F. C. 1973. Induction of a cytoplasmic factor increasing resistance to the teratogenic effect of 6-aminonicotinamide in mice. *Teratology* **7:**267–270.

Levy, E. P., Cohen, A., and Fraser, F. C. 1973. Hormone treatment during pregnancy and congenital heart defects. *Lancet* **1:**611.

Marsh, L., and Fraser, F. C. 1973. Chelating agents and teratogenesis. *Lancet* **2:**846.

Preus, M., and Fraser, F. C. 1973. The lobster claw defect with ectodermal defects, cleft lip/palate, tear duct anomaly and renal anomalies. *Clin. Genet.* **4:**369–375.

MacLeod, P. M., and Fraser, F. C. 1973. Congenital contractural arachonodactyly: a heritable disorder of connective tissue distinct from Marfan syndrome. *Am. J. Dis. Child.* **126:** 810–812.

Perry, T. B., and Fraser, F. C. 1973. Variability of serum creatine phosphokinase activity in normal women and carriers of the gene for Duchenne muscular dystrophy. *Neurology* **23:**1316–1323.

Fraser, F. C. 1974. Some aspects of maternal effects on congenital malformations, in: *Congenital Defects. New Directions in Research*, D. T. Janerich, R. G. Skalko, and I. H. Porter, eds. Academic Press, New York, pp. 17–22.

Preus, M., Fraser, F. C., and Fuhrmann, W. 1974. Cleft palate lateral synechia syndrome without the lateral synechia (CP ± LS syndrome). *Teratology* **9:**135–141.

Shih, L.-Y., Trasler, D. G., and Fraser, F. C. 1974. Relation of mandible growth to palate closure in mice. *Teratology* **9:**191–201.

Berman, P., Desjardins, C., and Fraser, F. C. 1974. Inheritance of the Aarskog syndrome. *Birth Defects* **10**(7):151–159.

Fraser, F. C. 1974. Updating the genetics of cleft lip and palate. *Birth Defects* **10**(8):107–111.

Fraser, F. C., 1974. Genetic counselling. *Am. J. Hum. Genet.* **26:**636–659.

Preus, M., and Fraser, F. C. 1974. The cerebro-oculo-facio-skeletal syndrome. *Clin. Genet.* **5:**636–659.

Levy, E. P., Fletcher, B. D., and Fraser, F. C. 1974. Mohr syndrome with subclinical expression of the bifid great toe. *Am. J. Dis. Child.* **128:**531–533.

Fraser, F. C., and Rosen, J. 1975. Association of cleft lip and atrial septal defect in mice: a preliminary report. *Teratology* **11:**321–324.

Preus, M., Alexander, W. J., and Fraser, F. C. 1975. The C syndrome. *Birth Defects* **11**(2): 58–62.

Berman, P., Desjardins, C., and Fraser, F. C. 1975. The inheritance of the Aarskog facial-digital-genital-syndrome. *J. Pediatr.* **86:**885–891.

Pashayan, H. M., Fraser, F. C., and Pruzansky, S. 1975. Variable limb malformations in the Brachmann–Cornelia leLange syndrome. *Birth Defects* **11**(5):147–156.

Preus, M., Feingold, M., and Fraser, F. C. 1975. Internipple distance and hand measurements in various syndromes. *Birth Defects* **11**(5):3–6.

Fraser, F. C. 1975. Non-scientific influences on decisions concerning human chemical exposure—a personal commentary. *Mutat. Res.* **33:**93.

Fraser, F. C., and Hunter, A. G. W. 1975. Etiologic relations among categories of congenital heart malformations. *Am. J. Cardiol.* **36:**793–796.

Fraser, F. C. 1976. Letter to the editor. *Clin. Genet.* **9:**444–445.

Benirschke, K., Carpenter, G., Epstein, C., Fraser, F. C., Jackson, L., Motulsky, A., and Nyhan, W. 1976. Genetic diseases, in: *Prevention of Embryonic, Fetal, and Perinatal Disease,* Fogarty International Center Series on Preventive Medicine **12:**219–261.

Fraser, F. C. 1976. Genetics as a health-care service. *N. Engl. J. Med.* **295:**486–488.

Kaplan, P., Cummings, C., and Fraser, F. C. 1976. A "community" of face–limb malformation syndromes. *J. Pediatr.* **89:**241–247.

Fraser, F. C. 1976. The multifactorial/threshold concept—uses and misuses. *Teratology* **14:** 267–280.

Fraser, F. C., and Biddle, C. J. 1976. Estimating the risks for offspring of first-cousin matings: an approach. *Am. J. Hum. Genet.* **28:**522–526.

Fraser, F. C., and Sadovnick, A. 1976. Correlation of IQ in subjects with Down syndrome and their parents and sibs. *J. Ment. Defic. Res.* **20:**179–182.

Preus, M., and Fraser, F. C. 1976. A methodology for establishing a diagnostic index for syndromes of unknown etiology. *Clin. Genet.* **10:**249–259.

Gunn, T., Bortolussi, R., Little, J. M., Andermann, F., Fraser, F. C., and Belmonte, M. M. 1976. Juvenile diabetes mellitus, optic atrophy, sensory nerve deafness and diabetes insipidus—a syndrome. *J. Pediatr.* **89:**565–570.

Biddle, F. G., and Fraser, F. C. 1976. Genetics of cortisone-induced cleft palate in the mouse—embryonic and maternal effects. *Genetics* **84:**743–754.

Kaplan, P., Hollenberg, R. D., and Fraser, F. C. 1976. A spinal arteriovenous malformation with hereditary cutaneous hemangiomas. *Am. J. Dis. Child.* **130:**1329–1331.

Fraser, F. C. 1977. Relation of animal studies to the problem in man, in: *Handbook of Teratology,* Volume 1, J. G. Wilson and F. C. Fraser, eds. Plenum Press, New York, pp. 75–96.

Fraser, F. C. 1977. Interactions and multiple causes, in: *Handbook of Teratology,* Volume 1, J. G. Wilson and F. C. Fraser, eds. Plenum Press, New York, pp. 445–463.

Biddle, F. G., and Fraser, F. C. 1977. Cortisone-induced cleft palate in the mouse. A search for the genetic control of the embryonic response trait. *Genetics* **85:**289–302.

Trasler, D. G., and Fraser, F. C. 1977. Time–position relationships with particular reference to cleft lip and cleft palate, in: *Handbook of Teratology,* Volume 2, J. G. Wilson and F. C. Fraser, eds. Plenum Press, New York, pp. 271–292.

Fraser, F. C. 1977. The chondrodystrophies, in: *Diseases of the Newborn,* 4th ed., A. J. Schaffer and M. E. Avery, eds. Saunders, Philadelphia, pp. 881–887.

Fraser, F. C. 1977. Genetic counseling, in: *Diseases of the Newborn,* 4th ed., A. J. Schaffer and M. E. Avery, eds. Saunders, Philadelphia, pp. 920–927.

Fraser, F. C. 1977. Factors influencing the occurrence of cleft lip and cleft palate, in: *Gene–Environment Interaction in Common Diseases,* Japan Medical Research Foundation, ed. University of Tokyo Press, Tokyo, pp. 96–104.

Fraser, F. C., and Gunn, T. 1977. Diabetes mellitus, diabetes insipidus, and optic atrophy: an autosomal recessive syndrome? *J. Med. Genet.* **14:**190–193.

Fraser, F. C., and Pressor, C. 1977. Attitudes of counselors in relation to prenatal sex-determination simply for choice of sex, in: *Genetic Counselling*, H. A. Lubs and F. de la Cruz, eds. Raven Press, New York, pp. 109–112.

Seegmiller, R. E., and Fraser, F. C. 1977. Mandibular growth retardation as a cause of cleft palate in mice homozygous for the chondrodysplasia gene. *J. Embryol. Exp. Morphol.* **38:**227–238.

White, R. A., Preus, M., Watters, G. V., and Fraser, F. C. 1977. Familial occurrence of the Williams syndrome. *J. Pediatr.* **91:**614–616.

Biddle, F. G., and Fraser, F. C. 1977. Maternal and cytoplasmic effects in experimental teratology, *Handbook of Teratology*, Volume 3, J. G. Wilson and F. C. Fraser, eds. Plenum Press, New York, pp. 3–33.

Lippman-Hand, A., Fraser, F. C., and Cushman Biddle, C. J. 1978. Indications for prenatal diagnosis in relatives of patients with neural tube defects. *J. Obstet. Gynecol.* **51:**72–76.

Hunter, A. G. W., McAlpine, P. J., Rudd, N. L., and Fraser, F. C. 1977. A "new" syndrome of mental retardation with characteristic facies and brachyphalangy. *J. Med. Genet.* **14:** 430–437.

Fraser, F. C., Metrakos, J. D., and Zlatkin, M. 1978. Is the epileptic genotype teratogenic? *Lancet* **1:**884.

Fraser, F. C. 1978. Prevention of birth defects: how are we doing? *Teratology* **17:**193–202.

Scriver, C. R., Laberge, C., Clow, C. L., and Fraser, F. C. 1978. Genetics and medicine: an evolving relationship. *Science* **200:**946–952.

Fraser, F. C., and Challis, E. B. 1978. Genetic counseling, in: *Family Practice*, R. E. Rakel and H. F. Conn, eds. Saunders, Philadelphia, pp. 465–478.

Fraser, F. C. 1978. Future prospects—clinical, in: *Birth Defects*. Proceedings of the Fifth International Conference, J. W. Littlefield and J. de Grouchy, eds. Excerpta Medica, Amsterdam, pp. 396–400.

Fraser, F. C., and Skelton, J. 1978. Possible teratogenicity of maternal fever. *Lancet* **2:**634.

Kalousek, D., Cushman Biddle, C. J., Rudner, M., Arronet, G. H., and Fraser, F. C. 1978. 47,X,i(Xq),Y karyotype in Klinefelter's syndrome. *Hum. Genet.* **43:**107–110.

Fraser, F. C., Ling, D., Clogg, D., and Nogrady, B. 1978. Genetic aspects of the BOR syndrome—branchial fistulas, ear pits, hearing loss, and renal anomalies. *Am. J. Med. Genet.* **2:**241–252.

Halal, F., Gledhill, R. B., and Fraser, F. C. 1978. Dominant inheritance of Scheuermann's juvenile kyphosis. *Am. J. Dis. Child.* **132:**1105–1107.

Theodosis, D. T., and Fraser, F. C. 1978. Early changes in the mouse neuroepithelium preceding exencephaly induced by hypervitaminosis A. *Teratology* **18:**219–232.

Halal, F., and Fraser, F. C. 1979. Camptodactyly, cleft palate, and club foot (the Gordon syndrome): a report of a large pedigree. *J. Med. Genet.* **16:**149–150.

Biddle, F. G., and Fraser, F. C. 1979. Genetic independence of the embryonic reactivity difference to cortisone- and 6-aminonicotinamide-induced cleft palate in the mouse. *Teratology* **19:**207–212.

Lippman-Hand, A., and Fraser, F. C. 1979. Genetic counseling: provision and reception of information. *Am. J. Med. Genet.* **3:**113–127.

Fraser, F. C. 1979. The development of genetic counseling. *Birth Defects* **15**(2)**:**5–15.

Vekemans, M., and Fraser, F. C. 1979. Stage of palate closure as one indication of "liability" to cleft palate. *Am. J. Med. Genet.* **4:**95–102.

Lippman-Hand, A., and Fraser, F. C. 1979. Genetic counseling: the postcounseling period. I. Parents' perceptions of uncertainty. *Am. J. Med. Genet.* **4:**51–71.

Lippman-Hand, A., and Fraser, F. C. 1979. Genetic counseling: the postcounseling period. II. Making reproductive choices. *Am. J. Med. Genet.* **4:**73–87.

Lippman-Hand, A., and Fraser, F. C. 1979. Genetic counseling: parents' responses to uncertainty. *Birth Defects* **15**(5C):325–339.

Fraser, F. C. 1980. Diagnostic prénatal des désordres génétiques, in: *Le diagnostic prénatal, cahiers de bioéthique 2.* University of Laval, Quebec, pp. 3–19.

Juriloff, D. M., and Fraser, F. C. 1980. Genetic maternal effects on cleft lip frequency in A/J and CL/Fr mice. *Teratology* **21**:167–175.

Fraser, F. C. 1980. The William Allen Memorial Award Address: evolution of a palatable multifactorial threshold model. *Am. J. Hum. Genet.* **32**:796–813.

Fraser, F. C. 1980. Animal models for craniofacial disorders, in: *Etiology of Cleft Lip and Cleft Palate,* M. Melnick, D. Bixler, and E. D. Shields, eds. Liss, New York, pp. 1–23.

Hall, J. G., Pallister, P. D., Clarren, S. K., Beckwith, J. B., Wigglesworth, F. W., Fraser, F. C., Cho, S., Benke, P. J., and Reed, S. D. 1980. Congenital hypothalamic hamartoblastoma, hypopituitarism, imperforate anus, and postaxial polydactyly—a new syndrome? Part I: Clinical, causal, and pathogenic considerations. *Am. J. Med. Genet.* **7**: 47–74.

Fraser, F. C. 1980. The genetics of cleft lip and palate: yet another look, in: *Current Research Trends in Prenatal Craniofacial Development,* R. M. Pratt and R. L. Christiansen, eds. Elsevier/North-Holland, Amsterdam, pp. 357–366.

Fraser, F. C. 1980. The role of genetics in medicine. *Birth Defects* **16**(5):1–6.

Fraser, F. C., Sproule, J. R., and Halal, F. 1980. Frequency of the branchio-oto-renal (BOR) syndrome in children with profound hearing loss. *Am. J. Med. Genet.* **7**:341–349.

Fraser, F. C. 1981. The genetics of common familial disorders—major genes or multifactorial? Genetics Society of Canada, Award of Excellence Lecture. *Can. J. Genet. Cytol.* **23**:1–8.

Fraser, F. C., and Nussbaum, E. 1980. Neural tube defects in sibs of children with tracheo-oesophageal dysraphism. *Lancet* **2**:807.

Fraser, F. C., and Lytwyn, A. 1981. Spectrum of anomalies in the Meckel syndrome, or: "Maybe there is a malformation syndrome with at least one constant anomaly." *Am. J. Med. Genet.* **9**:67–73.

Fraser, F. C. 1981. The genetics of common birth defects and diseases, in: *Genetic Issues in Pediatric and Obstetric Practice,* M. M. Kaback, ed. Year Book Medical, Chicago, pp. 45–54.

Fraser, F. C., and Forse, R. A. 1981. On genetic screening of donors for artificial insemination. *Am. J. Med. Genet.* **10**:399–405.

Vekemans, M., Taylor, B. A., and Fraser, F. C. 1981. The susceptibility to cortisone-induced cleft palate of recombinant inbred strains of mice: lack of association with the *H-2* haplotype. *Genet. Res.* **38**:327–331.

Fraser, F. C. 1982. Letter to the editor: how psychotherapeutic should genetic counselling be? *Am. J. Med. Genet.* **11**:367–368.

Vekemans, M., and Fraser, F. C. 1982. Susceptibility to cleft palate and the major histocompatibility complex (H-2) in the mouse. *Teratology* 25:267–270.

Fraser, F. C., Czeizel, A., and Hanson, C. 1982. Increased frequency of neural tube defects in sibs of children with other malformations. *Lancet* **2**:144–145.

Aymé, S., and Fraser, F. C. 1982. Possible examples of the Goltz syndrome (focal dermal hypoplasia) without linear areas of skin hypoplasia. *Birth Defects* **18**(3B):59–65.

Glanz, A., and Fraser, F. C. 1982. Spectrum of anomalies in Fanconi anaemia. *J. Med. Genet.* **19**:412–416.

Cole, D. E., Fraser, F. C., Glorieux, F. H., Jequier, S., Marie, P. J., Reade, T. M., and Scriver, C. R. 1983. Panostotic fibrous dysplasia: A congenital disorder of bone with unusual facial appearance, bone fragility, hyperphosphatasemia and hypophosphatemia. *Am. J. Med. Genet.* **14**:725–735.

Fraser, F. C., Aymé, S., Halal, F., and Sproule, J. R. 1983. Autosomal dominant duplication of the renal collecting system, hearing loss, and external ear anomalies: a new syndrome? *Am. J. Med. Genet.* **14**:473–478.

Preus, M., Fraser, F. C., and Wigglesworth, F. W. 1983. An oculocerebral hypopigmentation syndrome. *J. Genet. Hum.* **31**:323–328.

Fraser, F. C., Maldoff, S., and Lippman-Hand, A. 1983. Evidence against a female specific class of neural tube defect. *J. Med. Genet.* **20**:78.

Glanz, A., and Fraser, F. C. 1984. Risk estimates for myotonic dystrophy. *J. Med. Genet.* **21**: 186–188.

Fraser, F. C. 1985. Genetic counselling, in: *Cecil Textbook of Medicine*, 17th ed., J. B. Wyngaarden and L. H. Smith, eds. Saunders, Philadelphia, pp. 147–149.

Forse, R. A., Ackman, C. F. D., and Fraser, F. C. 1985. Possible teratogenic effects of artificial insemination by donor. *Clin. Genet.* **28**:23–26.

Fraser, F. C., and Rex, A. 1985. Excess of parental non-righthandedness in children with right-sided cleft lip: a preliminary report. *J. Craniofac. Genet. Dev. Biol.* **1**:85–88.

Martin, J. R., Huang, S.-N., Lacson, A., Payne, R. H., Bridger, S., Fraser, F. C., Neary, A. J., McLaughlin, E. A., Hobeika, C., and Lawton, L. J. 1985. Congenital contractual deformities of the fingers and arthropathy. *Ann. Rheum. Dis.* **44**:826–830.

O'Leary, E., Slaney, J., Bryant, D. G., and Fraser, F. C. 1986. A simple technique for recording and counting sweat pores on the dermal ridges. *Clin. Genet.* **29**:122–128.

Fraser, F. C., Frecker, M., and Allderdice, P. 1986. Seasonal variation of neural tube defects in Newfoundland and elsewhere. *Teratology* **33**:299–303.

Biddle, F. G., and Fraser, F. C. 1986. Major gene determination of liability to spontaneous cleft lip in the mouse. *J. Craniofac. Gen. Dev. Biol. Suppl. 2:* 67–88.

Pena, S. D. J., Karpati, G., Carpenter, S., and Fraser, F. C. 1987. The clinical consequences of X-chromosome inactivation: Duchenne muscular dystrophy in one of monozygotic twins. *J. Neurol. Sci.* **79**:337–344.

Fraser, F. C., Anderson, R. A., Mulvihill, J. I., and Preus, M. 1987. An aminopterin-like syndrome without aminopterin (ASSAS). *Clin. Genet.* **32**:28–34.

Frecker, M., and Fraser, F. C. 1987. Epidemiological studies of neural tube defects in Newfoundland. *Teratology* **36**:355–361.

Fraser, F. C. 1987. Genetic counselling: the changing scene, in: *Frontiers in Genetic Medicine*, M. M. Kaback and L. J. Shapiro, eds. Ross Laboratories, Columbus, Ohio, pp. 190–193.

Frecker, M. F., Fraser, F. C., and Heneghan, W. D. 1988. Are "upper" and "lower" neural tube defects aetiologically different? *J. Med. Genet.* **25**:503–504.

Fraser, F. C. 1988. This week's Citation Classic. *Current Contents* **31**:18.

Fraser, F. C. 1988. Genetic counseling: using the information wisely. *Hosp. Pract.* **23**:245–266.

Fraser, F. C. 1988. Thalidomide retrospective: what did we learn? *Teratology* **38**:201–202.

Fraser, F. C. 1988. Genetic counseling, in: *Cecil Textbook of Medicine*, 18th ed., J. B. Wyngaarden and L. H. Smith, eds. Saunders, Philadelphia, pp. 171–174.

Fraser, F. C. 1989. Research revisited. *Cleft Palate J.* **26**:255–257.

Fraser, F. C. 1989. Rapping the cleft-lip genes: the first fix? *Am. J. Hum. Genet.* **45**: 345–347.

Fraser, F. C., Ronen, G. M., and O'Leary, E. 1989. Pectoralis major defect and Poland sequence in second cousins: extension of the Poland sequence spectrum. *Am. J. Hum. Genet.* **33**: 468–470.

Books

Nora, J. J., and Fraser, F. C. 1974. *Medical Genetics: Principles and Practice*. Lea & Febiger, Philadelphia.

Fraser, F. C., and Nora, J. J. 1975. *Genetics of Man.* Lea & Febiger, Philadelphia.
Wilson, J. G., and Fraser, F. C., eds. 1977. *Handbook of Teratology,* Volume 1. Plenum Press, New York.
Wilson, J. G., and Fraser, F. C., eds. 1977. *Handbook of Teratology,* Volume 2. Plenum Press, New York.
Wilson, J. G., and Fraser, F. C., eds. 1977. *Handbook of Teratology,* Volume 3. Plenum Press, New York.
Wilson, J. G., and Fraser, F. C., eds. 1978. *Handbook of Teratology,* Volume 4. Plenum Press, New York.
Nora, J. J., and Fraser, F. C. 1983. *Medical Genetics: Principles and Practice,* 2nd ed. Lea & Febiger, Philadelphia.
Fraser, F. C., and Nora, J. J. 1986. *Genetics of Man,* 2nd ed. Lea & Febiger, Philadelphia.
Nora, J. J., and Fraser, F. C. 1989. *Medical Genetics: Principles and Practice,* 3rd ed. Lea & Febiger, Philadelphia.

Issues and Reviews in Teratology **5**:77–113
Plenum Press, New York, 1990, 978-1-4612-7847-4

The Concept of Homology in Comparative Mammalian Teratology

2

JAMES R. MILLER

1. INTRODUCTION

Naturally occurring animal models provide a fashionable topic in current mammalian biology. The subject is not a fad; the ancient Egyptians recognized the unity of human and veterinary medicine (Sigerist, 1951), and a considerable degree of interdependence has marked the development of the two disciplines. More recently, extensive reviews (Patterson *et al.,* 1982; Leiter *et al.,* 1987), proceedings of symposia (Lindsey and Capen, 1976; Kawamata and Melby, 1987) and workshops (Hackel, 1980), monographs (Andrews *et al.,* 1979), and numerous research reports devoted to spontaneous animal models have appeared at regular intervals; the entry "Disease models, animal," which first appeared in *Index Medicus* in 1970, regularly includes a number of references related to naturally occurring diseases; since 1969 the Institute of Laboratory Animal Resources (ILAR), with the support of the Animal Resources Branch of the National Institutes of Health, has maintained a registry of animal models and genetic stocks (Anonymous, 1972); and some of the techniques of molecular biology are being used to create new models (Evans *et al.,* 1985; Hooper *et al.,* 1987; Kuehn *et al.,* 1987).

As might be expected from Held's (1983) definition of an animal model—"a living organism in which normative biology or behavior can be studied, or in which a spontaneous or induced pathological process can be investigated, and in which phenomenon in one or more respects resembles the same phenomenon in other species of animals"—interest

JAMES R. MILLER • Central Research Division, Takeda Chemical Industries, Juso Honmachi, Osaka, Japan; *present address:* 3744 West 12th Avenue, Vancouver, British Columbia V6R 2N6, Canada.

in the subject stems from diverse modern biological disciplines; anatomy, biochemistry, genetics, hematology, immunology, neurology, pharmacology, and physiology are only a few that contribute to the investigation of animal models. Although Held's definition refers to other species of animals, the aim of most animal model research is to better understand and alleviate human disease. As understandable and admirable as this goal may be, the one-way flow of practical information from model organism to man tends to obscure comparisons between various nonhuman species that might throw light on fundamental issues of mammalian evolution and development.

Teratology investigators, while slightly less guilty of these shortcomings, still possess a "how close to the human disorder is this model?" attitude, and tend to put aside other potentially relevant questions—How do apparently similar spontaneous defects or syndromes in different mammalian species resemble or differ from each other? What is the significance of such resemblances or differences? On the basis of this knowledge, what can be inferred about homologous relations between such defects or syndromes? In this essay I want to address these issues, and I will focus on spontaneous disorders because I believe they present the greatest challenge for understanding the evolutionary and developmental significance of studies in comparative teratology. I do not mean to suggest that such issues in comparative induced teratogenesis have been resolved, but the induced anomalies involved have a definite etiology, there is a body of information concerned with comparative data (Shepard, 1986), and criteria for models have been set forth and used practically (e.g., Chernoff, 1977).

2. HOMOLOGY

"Homology was introduced and differentiated from mere analogy by a famous superintendent of the Natural History Department of the British Museum, Richard Owen (1804–1892). Organs A, B, and C of genetically cognate structures are homologous if the development of one can be construed as a modification or variant of the development of another, of if the developments of all three are so many variants of the development of an evolutionary precursor of them all" (Medawar and Medawar, 1985). [It is interesting that the *Oxford English Dictionary* (1933) recognizes Owen as the first to use "homologous," in 1846, but cites R. B. Todd, a British physician, as using "homology" 10 years earlier, in *The Encyclopedia of Anatomy and Physiology*.] Periodic attempts to alter the definition of "homology" by qualification, expansion, or simple misuse have often precipitated vigorous debate; see Reek *et al.* (1987) for the most recent example. "Homology" is frequently used rather glibly by those who are unaware of how passionately its precise meaning is still debated among evolutionary biologists and comparative anatomists who, with reason, have strong feelings of proprietorship about the word. Van Valen (1982) believes that, strictly speaking, "[homology] now refers to a family of related concepts . . . because different people take different criteria as defining and the criteria do not always coincide." He defines homology, in a general way, as "correspondence caused by a continuity of information."

McKusick (1980), in his Wilhelmine E. Key lecture, "The anatomy of the human genome," points out that our chromosomes and the genes they contain are part of the human anatomy, and "Furthermore, there is a morbid anatomy, a comparative anatomy (with information relevant to evolution), a functional anatomy, a developmental anatomy, and even, if not a surgical anatomy, at least the beginnings of an applied anatomy." All of this would not have surprised J. B. S. Haldane who was one of the few geneticists to maintain an interest in comparative mammalian genetics long before there was much to compare aside from coat colors. In 1927 he published a remarkable paper setting forth criteria on genetic homology, and showed how these could be demonstrated by examining the coat color genetics of six species of rodents and nine species of carnivores (Haldane, 1927). His five criteria for homology are presented, together with those of Lalley and McKusick (1985), in Table I. The language and details of the two sets of criteria understandably differ, but the agreement on the fundamentals required to make a decision on homology is impressive.

Haldane, and those who, like Little (1958) and Searle (1968, 1969), maintained an interest in comparative studies, recognized that the closer the phenotypes being compared were to the primary action of the respective genes, the stronger the evidence for homology. As expected, the 1985 criteria reflect the rapid advances in the dissection of the mammalian genome made possible by new techniques in molecular biology. While these techniques are useful for analyzing homologous nucleotide sequences in DNA, and amino acid sequences in enzymes and other proteins, the analyses of phenotypes further removed from the site of primary gene action remain as difficult as ever, and revolve about the dilemma of determining what criteria must be met before concluding with some degree of comfortable certainty that similar phenotypes in two or more mammalian species are true homologies.

Haldane (1927) pointed out that there is no absolute criterion of genetic homology because we cannot study ancestral forms. Therefore, for practical purposes, it seems reasonable to use the term to describe genes with the same fundamental structure, action, or effect in two or

Table I. Comparison of the Criteria as Evidence for Gene Homology Proposed by Haldane (1927) and Lalley and McKusick (1985)

Haldane

Genes are homologous if they produce the same effects when brought in from either side in a species cross. (Experiments *in vitro* might yield equally definite results.)

Homology may be suspected when any or all of the following criteria are satisfied:

1. The genes produce similar but not necessarily identical effects.
2. If a certain effect in two species is produced by only one gene in each.
3. The genes have undergone several parallel mutations into more or less corresponding multiple alleles.
4. The genes exhibit similar linkages in different species.

Lalley and McKusick

The following criteria are recommended as evidence for identifying gene homologies:

1. Similar nucleotide or amino acid sequences
2. Similar immunological cross-reaction
3. Formation of functional heteropolymeric molecules in interspecific somatic cell hybrids in cases of multimeric proteins
4. Similar tissue distribution
5. Similar developmental time of appearance
6. Similar pleiotropic effects
7. Similar substrate specificity
8. Similar response to specific inhibitors
9. Cross hybridization to the same molecular probe

Mapping or linkage homologies[a] in two or more species are considered well established if:

1. Two or more pairs of homologous loci are linked or at least syntenic;
2. The order of three or more loci is the same; and
3. The map distances among the loci are roughly equivalent.

[a]If the order is not the same, if another locus or segment is inserted in one species, or if the map distances are very different, then the term "syntenic homology" is preferred. Homologous linkage implies a conserved, undisrupted chromosome segment in two or more species since the species separated during evolution; "syntenic homology" does not imply conservation for the observed association of the loci.

more different species without any evidence that a common ancestor of the species possessed the gene, providing the convention and its potential hazards are acknowledged; some of the hazards of relying solely on phenotypic similarities are discussed in the following section.

3. CONCEPT OF PHENOTYPE

Wilhelm Johannsen coined the term "phenotype" in 1909,* and its meaning has remained essentially unchanged. Because we tend to study

*Wilhelm Johannsen (1857–1927), a Danish plant physiologist, coined three vital genetic terms: gene, phenotype, and genotype. He never appeared satisfied with his original definitions, particularly that of genotype, which he constantly reexamined. (Churchill, 1974; Wanscher, 1975).

one stage of development at a time, we have a static view of phenotype, as exemplified by the frequent use of the definite article. However, for Johannsen, phenotype was a dynamic concept; the term was not intended to describe the final (adult) stage—or indeed any specific stage—of an organism, but the structural and functional attributes at all phases of development produced by the interaction of the genome and the environment to which it is exposed as development unfolds. The usual concept of phenotype is static for convenience.

The animal modelist and the traditional homologist, the comparative anatomist, are concerned primarily with the end products, the structures; the former may also be concerned to a degree with pathogenesis. This focus on endpoints is understandable, particularly in the case of the comparative anatomist who, in looking for homologies, often had little else to work with other than fossil specimens. However, this focus should not detract from a point of fundamental importance made by Danforth (1925, p. 66) over 60 years ago: the real basis of homology is to be sought in causal factors, not structures. Since the ultimate causal factors with which we are concerned here involve genetic elements and comparative norms of reaction, homology of complex phenotypes is invariably partial, not absolute.

Most analysis of gene action has understandably focused on the processes intervening between a mutation and its phenotypic effects during various developmental stages; in most instances this has been challenge enough without being concerned with the interaction with other genes, aside from perhaps noting that the major gene expressed itself (or was expressed) differently on different genetic backgrounds, i.e., that the gene was interacting in some way with other genes. In one area, however, some attempt has been made to understand the complexity of interaction of various loci on a final adult phenotype—coat color. The most detailed studies were made by Sewall Wright over a period of many years on the quality, intensity, and pattern of coat color in the guinea pig; the results were reviewed by Wright in 1963. Searle (1968, p. 57) commented on Wright's magnificent work: "His findings show clearly the dangers of trying to homologize genes on the basis of phenotypic effect, because of the extent to which the expressed phenotype depends on a large number of genetic and environmental variables, which will differ in the two species concerned." And Wright (1963), commenting not on coat color but on hair direction, noted "the extraordinary diversity in the relation of genes to characters, and in particular in kinds of interaction, that is to be found . . . in even a rather limited genetic system" pp. 175–176). Silvers (1979) reviewed the extensive information available on the genetics of coat color in the mouse to illustrate the complexities of gene action and interaction in another mammalian species.

Developments in molecular biology will yield much information about the homologous relations of many normal or wild-type DNA sequences and molecular variants of these sequences that may take the form of alleles or gene families of varying complexity in diverse mammalian species (Bodmer, 1981; Raff *et al.*, 1987). In addition, the application of an array of new techniques for studying biological structure and function has made it possible to study phenotypes that are close to the level of primary gene action. At this (molecular) level, determining homologous relations is easier although not entirely free of uncertainty; however, determining homologous relations for the vast majority of mammalian mutant forms—those of prime interest to the developmental geneticist and teratologist—depends on evaluating the congruence of phenotypes far removed from the level of primary gene action. The difficulty of making a decision is most critical in the case of complex phenotypes (syndromes) stemming from pleiotropic effects. Searle's (1968, p. 25) warning about such effects remains valid 20 years after it was made: "Pleiotropic effects tend to be variable and dependent on the genetic background, so no strict correspondence between phenotypes in different species can be expected even if the genes concerned are homologous." Despite the rapid advances in developmental and molecular biology, the field of comparative developmental pathogenesis remains a boggy challenge.

4. SOME ISSUES IN COMPARATIVE PATHOGENESIS

The current dogma of "reverse" or "bottom-up" genetics would lead us to believe that elucidating the fine structure of DNA will reveal the relation between specific genes and their phenotypes. A blueprint is a frequently used metaphor; the DNA is a blueprint that, if read correctly, will yield the final phenotype. Given what is known of mammalian development, this proposed relation between the one-dimensional DNA code and the three-dimensional morphological complexities of mammalian development makes no sense. But the search for a point-to-point agreement between genes and morphological features is eagerly pursued even by those who are not using molecular techniques. For example, Bailey (1986) stated, in connection with a study on genes affecting the morphogenesis of the mouse mandible, ". . . for surely the genetic basis of morphological detail would need to be as fine grained as the morphological detail itself." Surely it would not need to be. The "fine grained" detail of the human brain serves as an example. Young (1987) stated that there may be as many as 10^{15} synaptic points in the human cerebral cortex, an astronomic number that vastly exceeds the number of postulated genes, and makes any attempt to determine a point-to-point relation between a set of genes and the development of a set of synapses ridiculous. Rakik is quoted (Barnes, 1986) as saying, "Even if we knew all of the genes and what they coded for, we still could not predict in detail what kind of brain you would have." Certainly, the example of brain complexity is not unique.

Dawkins (1986, pp. 294–297) called the blueprint analogy "modern preformationism" and stated that the "recipe or 'cookery book' theory" (or analogy) is more apt. The recipe for a cake is a set of instructions that, if followed properly, will result in the desired product. There is no simple one-to-one relation between parts of the recipe and parts of the cake; just as there is no one-to-one mapping between specific genes and specific parts of the body. However, an error in the recipe itself, or in the interpretation or execution of it, can lead to predictable effects on the final appearance, texture, or taste of the cake; just as a gene error can result in certain pathological states.

The validity of the analogy is not altered by the fact that the "recipe" is becoming more complex as the details of the relation between the genic fine structure and the primary gene products are revealed. Although the principle of one gene (one cistron)–one polypeptide chain still seems valid in most instances, a number of intriguing variations are now known: in immunoglobulin synthesis, more than one gene is involved in the formation of one polypeptide (Tonegawa, 1983); two or more products can be transcribed from a single gene (Grima *et al.*, 1987); cellular polyproteins (composite, multihormone precursors, each of which is the product of a single gene), each giving rise to several peptide hormones—pro-opiomelanocortin, from which eight hormones are synthesized, is a striking example (Herbert, 1981); gene families occur extensively in the mammalian genome (Bodmer, 1981; Raff *et al.*, 1987); and two distinct genes can be transcribed from opposite strands of the same piece of DNA (Adelman *et al.*, 1987). These examples indicate that even at the level of primary gene action, hitherto unanticipated complexities exist. We are ignorant of the relevance, if any, of this molecular complexity and mutations in it to gross developmental defects.

Stent (1985) maintained that the linear conception of the molecular biologist will not solve the mysteries of normal embryogenesis. [I'm not sure why Stent restricted his accusations to molecular biologists—maybe he feels safer challenging those he knows best—because linear concepts have been used by developmental geneticists for decades; the challenge of complexity was met by resorting to multilinear concepts such as Grüneberg's (1943, p. 7) "pedigree of causes." Some embryologists have attempted to convey the multidimensional nature of the challenge (Weiss, 1955).] Stent contended that the complexity of interactions of primary, secondary, and higher-order gene products cannot be understood and resolved by linear concepts, and that indeed there is no apparent theory or set of hierarchical rules governing development. [See also Sydney Brenner's comments on the absence of neat sequential processes in development (Levin, 1984).] Although this view has been challenged by Laurence (1985), who cited several recent discoveries in developmental biology as examples of "lights" indicating that some developmental patterns spanning great phylogenetic distances do exist, most individuals would agree that, paraphrasing Trevelyan (1964), the pattern of embryogenesis is indeed a tangled web. No simple diagram will explain its infinite complication. The effects of individual mutations must be considered within the context of this web, which may vary considerably even in closely related species. Three examples will suffice to demonstrate this point.

4.1. Ectodermal Dysplasia in Man and Dog

Several forms of ectodermal dysplasia are known in man, other mammals, and birds. The species-specific expression of these disorders is determined by the characteristics of ectodermal derivatives in each species (Selmanowitz, 1979). Patterns in the coat or plumage that characterize the defects in nonhuman forms will not be present, except in very restricted areas, in man. In an animal whose skin is normally protected by fur or feathers, the full loss of such protection leads to reduced tolerance to cold, photosensitivity, irritability, scaliness, hyperkeratosis, and crusting, the extent and degree of which may vary from that observed in human skin, which is adapted for exposure. Species differences in sweat gland morphology also affect the expression of ectodermal dysplasia.

In man, exocrine glands play a crucial role in cooling the body. Males with anhydrotic ectodermal dysplasia [see entry 30510 in McKusick, 1986a; subsequently, reference to entries in this catalogue will be preceded by "MIM" (*Mendelian Inheritance in Man*), e.g., MIM 30510] lack these glands; consequently, they cannot tolerate heat and suffer febrile episodes. In dogs, sweat glands, of the apocrine type over most of the body, do not play a major role in heat dissipation at normal temperatures, and dogs with widespread absence of these glands function well even at reasonably high temperatures. Although the adult phenotypes of the human and canine forms of the disease may differ in certain important features, the two may be true homologues in the sense that the mutation giving rise to each occurs in a region of the X chromosome derived from a common ancestor of both.

4.2. Glycogen Storage Disease Type VII in Man and Dog

Mammalian phosphofructokinase (PFK; ATP:D-fructose-6-phosphate 1-phosphotransferase, EC 2.7.1.11) exists in multimolecular forms resulting from random tetramerization of three distinct subunits, M (muscle-type), L (liver-type), and P (platelet-type), each controlled by a separate locus. In man, muscle contains homotetramer M4, liver contains homotetramer L4, and erythrocytes contain a mixture of M4, M3L, M2L2, ML3, and L4 isozymes. A mutation at the M-type locus, when present in the homozygous state, causes a total absence of the M subunit and a consequent glycogen storage disease (GSD, type VII), characterized by exertional muscle weakness and compensated hemolysis (MIM 23280). Patients with this disorder totally lack muscle PFK and partially lack erythrocyte PFK; the residual erythrocyte PFK consists exclusively of the L4 isozyme.

Giger *et al.* (1985) described an autosomal recessive disorder among English springer spaniel dogs that is characterized by PFK deficiency associated with isolated anemia; the animals have no muscle weakness and do not develop myoglobinuria on exertion. Vora *et al.* (1985) demonstrated that these dogs apparently have the same genetic defect present in patients with GSD VII: they completely lack the M subunit. The differences in the phenotypic consequences in the two species reflect a species difference in the isozyme distribution in tissues and in muscle physiology. In the dog, as in man, muscle and liver PFK consists of M4 and L4 homotetramers, respectively; erythrocyte PFK consists of a three- or four-member set composed of M and P subunits. Affected dogs have the L4 isozyme in muscle; P4 and L4 isozymes and their hybrids in erythrocytes; and an increase of L-containing isozymes in platelets. The L subunit, confined to the liver in normal dogs, is present in other tissues in affected animals. Because the M subunit contributes about 50% to the human erythrocyte PFK, but 80–90% to the dog erythrocyte PFK, its total absence results in a large enzyme deficiency in the dog erythrocytes and presumably in a more severe hemolytic disease. The apparent absence of muscle disease in the dog can be explained by the high oxidative potential of canine muscle in general, and the presence of liver PFK in muscle lacking the M subunit.

4.3. HPRT Deficiency in Man and Mouse

In 1964 Lesch and Nyhan described a syndrome characterized by mental retardation, spastic cerebral palsy, choreoathetosis, uric acid urinary stones, and self-destructive biting of fingers and lips. The patients were males and the disorder was shown to be X-linked. Seegmiller *et al.* (1967) found that affected males were deficient in hypoxanthine-guanine phosphoribosyltransferase (HPRT), which is controlled by an X-linked locus (MIM 30800). The locus is known in over 24 mammalian species, but no examples of a spontaneously occurring Lesch–Nyhan-like syndrome have been reported.

Recently, however, two separate research groups in England (Hooper *et al.*, 1987; Kuehn *et al.*, 1987) reported success in producing $Hprt^-$ mice by injecting embryonal stem (ES) cells, which had been selected as $HPRT^-$ in tissue culture, into normal embryos. The injected cells mingled with the normal embryonic cells and contributed to all the differentiated tissues, including the germ cells, of the chimeric adults. Gametes derived from the ES cells passed the $Hprt^-$ allele to the next generation. Although the resulting $Hprt^-$ males were not very old when the reports were prepared, they were developing normally. The significance of this apparent differences between the human and mouse phenotypes is not clear; the disorder in humans is genetically heterogeneous, and the outcome is much less severe in patients who have a small amount of enzyme activity (Harris, 1980). However, the difference noted above may be real and reflect differences in purine metabolism between man and mouse (Carson *et al.*, 1980); this may be another example of the same mutation producing quite different phenotypes in two species.

These three examples involve phenotypes remote from those present in malformation syndromes, but they demonstrate how species differences in normal biochemistry, physiology, and anatomy—the complex milieu within which the effects of mutations must be studied—can alter the phenotypic consequences of the same mutation in different species. There is no reason to doubt that similar complexities exist in the development of malformation syndromes determined by homologous mutations in two or more mammalian species.

Despite the significant contributions of molecular biologists and the prodigious efforts of mammalian embryologists, the vast intricate network of developmental events between a specific gene(s) and consequent phenotype(s) remains *terra incognita*. Ironically, the application of some of the new techniques in molecular and developmental biology has resulted in findings that, while illuminating some general features, make certain details more obscure. Kollar and Fisher (1980) showed that the chicken still retains genes for enamel synthesis and that the genes can be made to express themselves under certain circumstances, and Mintz and Illmensee (1975) found that teratocarcinoma cells from the cores of embryoid bodies grown *in vivo* for many years are totipotent and can express hitherto "silent" genes when inoculated into blastocytes subsequently raised to be mature mice. These examples strikingly illustrate that the existence of homologous genes does not necessarily guarantee the existence of homologous phenotypes, and provide support for the prediction of Zuckerkandl and Pauling (1962, p. 207) who stated, "It is therefore quite likely that there exist in every organism numerous structural genes that do not find in any of the existing tissues conditions favorable to their expression and thus remain permanently dormant." These genes may play a significant role in major steps in evolution.

5. HOMOLOGOUS COMMON CONGENITAL MALFORMATIONS

Searching for homologues among the vast array of congenital defects known in mammals is not a popular pastime for teratologists; a cursory examination of back issues of *Teratology* indicates that few comparative studies of the required type have been published. The situation is not much better when the large veterinary literature is scanned.

The search for truly homologous congenital defects is frustrating and suffers from several serious handicaps. First, the phenotypes studied at the time of birth, when most defects are detected, are far removed from the site of the primary etiology. Second, although phenotypes strikingly similar to many defects known in man have been described in many mammalian species, the number actually available for comparative studies is meager because, in many instances, the descriptions do not go much beyond a case report; often no attempt is made to preserve a strain in which specific defects occur with a high frequency. Third, evidence suggests that the etiology of the common defects in man is heterogeneous and involves, in the majority of instances, genetic and environmental components; three examples are discussed below.

5.1. Neural Tube Defects (NTD)

These comprise a group of nervous system defects including craniorachischisis totalis, anencephaly, myelomeningocele, and, with less certainty, encephalocele. Campbell *et al.* (1986), after thoroughly reviewing the subject, concluded that "No animal model has been convincingly established as the equivalent of human neural tube defects." There are several mouse mutations that result in NTD similar to those in man, and a strain of hamsters is known to have a comparatively high frequency (17%) of spontaneous spina bifida. However, Campbell and her colleagues correctly pointed out that such models, although useful for investigating the pathogenesis of naturally occurring animal defects, cannot as yet be termed probable homologues because the necessary genetic analyses have not been undertaken in man.

5.2. Congenital Heart Defects

Various types of congenital heart defects have been reported in nonhuman mammals; the most extensive studies have been carried out in the dog by Patterson (1976, for review). As a group, congenital heart defects are found significantly more frequently in purebred dogs than in mongrels. In addition, there is strong evidence for breed-specific lesions; e.g., patent ductus arteriosus in the poodle, collie, Pomeranian, and Shetland sheepdog; tetralogy of Fallot in the keeshond; atrial septal defect in the Samoyed; and ventral septal defect in the English bulldog. As Patterson (1976) pointed out, "Most breeds of dogs, as they exist today, are essentially genetic isolates whose gene pools were derived from small samples of individuals from one or more preexisting lines or breeds" (p. 5). While today's breeds are not genetically pure in the sense that inbred strains of laboratory animals are, their members do have many genes in common. ". . . the nonrandom breed distributions of different anatomic forms of congenital heart disease in the dog suggest the hypothesis that each defect represents a specific developmental disturbance which is determined by lesion-specific genetic factors occurring predominantly in certain breeds" (p. 6). Patterson and his colleagues extended these initial epidemiological studies by formal genetic analyses and essentially confirmed this hypothesis.

Patterson drew attention to the close similarity between many features of canine and human congenital heart disease, and concluded that in both species, "At least a sizeable proportion of congenital heart defects are produced by polygenic sets that are lesion-specific" (p. 27). There is no evidence for major gene effects in either species. Hence, while the defects in the dog may serve admirably as models to provide a better understanding of mammalian cardiovascular pathophysiology, no firm evidence exists on which to base a definitive statement on the homologous relation of these defects and those in man.

5.3. Cleft Lip and Palate

Spontaneous cleft palate has been described in several species of domestic animals (Mulvihill *et al.*, 1980); the occurrence of considerable interbreed variation suggests that the defect has a major genetic component in these species, but detailed analyses are lacking. Juriloff (1980) summarized the genetics of clefting in the mouse. Biddle and Fraser (1986) and Juriloff (1986) analyzed the genetics of spontaneous cleft lip in the A/J strain and A/WySn strain, respectively, and concluded that a single major gene, which is close to being mapped (Juriloff, 1986), underlies the embryonic liability to the defect. However, the direct relation between gene and phenotype is obscured by maternal effects and other modifiers. In the mouse, the use of sublines of the A strain has made it possible to tease out the effects of the single gene from the interaction of a complex of factors. In man, the nature of the genetic component is controversial, but the results of studies on at least five quite different populations (Chung *et al.*, 1974; Demenais *et al.*, 1984; Marazita *et al.*, 1984a,b, 1986) suggest that a single major gene underlies the defects. Modifiers, such as maternal age (Khoury *et al.*, 1983), are probably also involved. Hence, the situation in the two species may not be as different as it would appear on cursory examination.

In summary, while the common congenital defects are not currently a rewarding source of information on mammalian homologues (aside perhaps from cleft lip and palate in mice and man), it is reasonable to expect that further careful, detailed genetic and phenotypic studies in man and experimental mammals, notably the mouse, will yield relevant data on which to make more positive conclusions.

6. SEARCHING FOR HOMOLOGOUS CONGENITAL DEFECTS

Given our limited knowledge of the homologous relations of complex phenotypes and the genetic fine structure influencing them, it would seem to make sense to focus attention on specific phenotypes determined with some certainty by single gene mutations whose location in a genome has been reasonably well determined. This latter qualification severely restricts the phenotypes available for study because, although much publicity surrounds efforts to map mammalian genomes—the uninitiated might believe that splendidly detailed maps of many mammalian species already exist—the number of species for which gene maps exist is small, and the information available on any of them, those of man and mouse excepted, is meager.

The mapping information that is available concerns the location of genes determining enzymes and other proteins. The explosion of information about the fine structure of mammalian genomes and details of gene mapping has provided a rich source of data for those interested in the evolutionary studies of proteins. However, the comparative teratologist is concerned with morbid anatomy (McKusick, 1980), and hard data on comparative aspects of the morbid anatomy of the mammalian genome are slight.

Genetic Maps 1987 (O'Brien, 1987) lists information on 24 placental mammalian species in addition to man and on several marsupial species; the placental species include 11 primates, eight domestic mammals, and five rodents. As might be expected, the map information on the mouse far exceeds that of any other species. This information includes not only loci for which the wild-type (normal) phenotype is known but also many loci at which mutations resulting in abnormal phenotypes occur; many of these loci were summarized by Kalter (1980). In the remaining 20 species the linkage maps consist overwhelmingly or, in most instances, exclusively of "biochemical" loci. Comparative studies of any depth, i.e., those based on sound genetic data and detailed phenotype descriptions, are restricted to mouse and man. This situation is far from ideal.

6.1. The Mammalian X Chromosome

The X chromosome might seem to be a fertile source of true homologies because of the evidence that the mammalian X chromosome has been essentially "frozen" for close to 100 million years (Ohno, 1973). However, a close examination does not yield much of value related to developmental defects.

As of October 1986, McKusick (1986b) recognized 130 proved and 164 possible but not proved X-linked loci in man. Of these 294 loci, 32 are known to occur in the X chromosome of at least one other mammalian species (four loci—*Gla, G6pd, Hprt,* and *Pgk*—are each known to occur in over 25 species), and another 15 can be considered as probably or possibly homologous loci. At first sight these 47 loci might seem to be rewarding material, but only seven involve phenotypes that are of teratological interest. These are summarized in Tables II and III.

The apparently true homologies (Table II) involve three disorders—hypophosphatemia, Menkes syndrome, and testicular feminization—that have been intensively studied in man. The pathogenesis in hypophosphatemia has not been elucidated, but severe skeletal defects in male mice can be prevented by phosphate supplementation of the diet from weaning age (Marie *et al.,* 1981). Menkes syndrome in man and the

Table II. Apparently True X-Linked Homologues Involving Developmental Defects in Man and Other Mammals

MIM[a] number	MIM name	Species	Name of locus or disorder (gene symbol)	References
30295	Chondrodysplasia punctata, X-linked	Mouse	Bare patches (*Bpa*)[b]	Happle *et al.* (1983)
30780	Hypophosphatemia, X-linked	Mouse	Hypophosphatemia (*Hyp*)	Eicher *et al.* (1976)
30940	Menkes syndrome	Hamster	Mottled (*Mo*)[c]	Magalhaes (1954)
		Mouse		Green (1981; summary)
31370	Testicular feminization syndrome	Cow	Testicular feminization (*Tfm*)	Nes, 1966; Kent *et al.* (1986)
		Horse		Lyon and Hawkes (1970)
		Mouse		Lojda (1975)
		Pig(?)		Allison *et al.* (1965)
		Rat		

[a]*Mendelian Inheritance in Man* (McKusick, 1986a).
[b]In addition to fur and skin disorders, many heterozygous females have hind limb abnormalities in which the toes are bent and sometimes shortened; in a few cases the forefeet are affected and other skeletal anomalies may occur. Hemizygous males die *in utero*.
[c]A series of alleles exist at the *Mo* locus in the mouse. In homozygous males, some cause death in mid- (11 days) or late gestation, or in the early postnatal period, while others produce viable phenotypes characterized by light-colored fur, curly whiskers, abnormal hairs, neurological disturbances, small body size, aortic aneurysms, and, occasionally, hind limb defects. (The males that die *in utero* often have blood vessel aneurysms or skeletal defects.) In heterozygous females, the alleles produce a variety of phenotypes characterized by mottled and slightly waved coats, curved vibrissae, calcareous deposits, aortic lesions, and skeletal defects of the limbs; the viability and fertility vary. In the Syrian hamster, the mutation is lethal in males; heterozygous females have mottled coats, and are smaller, less vigorous, and more nervous than their normal littermates.

Table III. Probable or Possible X-Linked Homologues Involving Developmental Defects in Man and Other Mammals

MIM[a] number	MIM name	Species	Name of locus or disorder (gene symbol)	References
30610	Gonadal dysgenesis, XY female type	Horse	XY sex reversal	Kent *et al.* (1986)
30970	Microphthalmia	Mouse	Eye–ear reduction (*Ie*)	Hunsicker (1974)
31350	Teeth, absence of	Mouse	Irregular teeth (*It*)	Phipps (1969)

[a]*Mendelian Inheritance in Man* (McKusick, 1986a).

series of disorders arising from mutations at the mottled (*Mo*) locus in the mouse appear to result from an error in copper metabolism, the details of which have not been clarified in either species. The diversity of phenotypes arising from the presence of multiple alleles at the mouse locus (Green, 1981, pp. 162–166) strongly suggests that at least some of the variation observed in the Menkes syndrome in man can be attributed to multiple alleles. The testicular feminization syndrome has been described in several species; however, only in man and mouse has the basic error been traced to a mutation on the locus for the dihydrotestosterone receptor. Based on clinical and pathological investigations the homologous relation between the mutations for X-linked chondrodysplasia punctata in man and bare patches in the mouse seems convincing, although the extent of the investigations is limited. Recently, Yang-Feng *et al.* (1986) pointed out that the relative map positions of the two responsible loci differ.

Three probable or possible X-linked homologous loci of teratological interest are listed in Table III. Kent *et al.* (1986) described an XY sex-reversal syndrome in which a phenotypic mare had the karyotype of a stallion. The syndrome is both genetically and clinically heterogeneous. One form appears to be X-linked, although autosomal sex-limited transmission cannot be ruled out. Phenotypic expression ranges from a feminine mare with a reproductive tract within normal limits to a greatly masculinized mare. The relation of this disorder(s) to similar ones in the human is not clear. XY gonadal dysgenesis (MIM 30610) is a human disorder in which phenotypic females with a 46,XY karyotype show bilateral streak gonads and sexual infantilism. Although it too exhibits genetic heterogeneity (Simpson *et al.*, 1981), its phenotypic expression does not encompass the extremes in the horse.

Females homozygous and males hemizygous for the *Ie* mutation in the mouse have anophthalmia and very small external ears with thickened, crinkled edges. The phenotype in heterozygous females may vary from normal to that of typically hemizygous or homozygous animals (Hunsicker, 1974). Lyon (1974) suggested that the mutation might be homologous with that for an X-linked microphthalmia in man (MIM 30970); however, McKusick stated that there is little evidence that such a "pure" X-linked microphthalmia exists. In females heterozygous for the *It* mutation in the mouse, both lower incisors are markedly reduced or absent; the upper incisors may be absent also. Expression is variable and may range to normal. Viability and fertility are low. Hemizygotes die *in utero* (Phipps, 1969). Lyon (1974) suggested homology with a human disorder characterized by oligodontia or hypodontia (MIM 31350); however, neither disorder has been investigated extensively and there is no basis for a decision.

Among the X-linked loci in man for which no proved or potential homologues are known in another mammal, there are at least 40 that are associated with phenotypes of teratological interest. Conversely, among 41 loci described in mammals for which no human homology is known, six yield phenotypes characterized by developmental defects; these are presented in Table IV. Therefore, the mammalian X chromosome may still provide a valuable source of material for those interested in comparative teratology.

6.2. Autosomally Linked Homologies

For reasons already stated, the search for autosomally linked homologies is formidable; there is a dearth of both sound genetic information and adequate phenotypic detail on which to make any definite conclusions. Therefore, we must lower our standards somewhat and concentrate on situations where the detailed examination strongly supports the

Table IV. X-Linked Loci Associated with Developmental Defects Described in Nonhuman Mammals but Not Known in Man

Species	Name of locus or disorder (gene symbol)	Reference
Mouse	Bent-tail (*Bn*)[a]	Garber (1952), Grüneberg (1955)
	Broad headed (*Bhd*)	Phillips and Fisher (1978)
Dog	Carpal subluxation[b]	Pick *et al.* (1967)
	Cleft palate, polydactyly, syndactyly, tibiofibular shortening, brachygnathia, and scoliosis[c]	Sponenberg and Bowling (1985)
Mouse	Gyro (*Gy*)[d]	Lyon *et al.* (1986)
	Polydactyly, preaxial, with hemimelia and urogenital defects, X-linked (*Xpl*)	Sweet and Lane (1980)

[a]Although the major defect is in the tail, there is evidence that the mutation may have a more generalized effect; hemizygous males and homozygous females have been described with an open neural tube in the sacral region and cranioschisis *in utero*.
[b]The defect is limited to the carporadial joints and always occurs bilaterally; the locus is closely linked to that for hemophilia A.
[c]MIM lists several X-linked syndromes of which cleft palate is a part; however, polydactyly is described in none of them, and syndactyly is part of the oral–facial–digital syndrome (MIM 31120) only.
[d]This is a second hypophosphatemic mutation; hemizygous males have abnormalities of the long bones and ribs.

view that strikingly similar phenotypes in two species are homologues. I will discuss one such example in detail because it illustrates how little interest there is in comparative studies even when the opportunity for them is ripe and potentially rewarding.

6.2.1. White (Mi^{wh}) in the Mouse, and Anophthalmic White (*Wh*) in the Syrian Hamster

Twenty years ago Searle (1968, Table II) suggested that these two mutations are probably homologous on the basis of the similarity of expression of two aspects, achromia and microphthalmia, of a complex phenotype that results from each mutation. Despite numerous subsequent studies, no definite conclusion about homology is possible at present because the primitive state of the hamster gene map (Robinson, 1984) makes it impossible to use comparative linkage data; any conclusions must be based on comparisons of phenotypes.

Mi^{wh} is one of several mutant alleles that are known to occur at the microphthalmia locus in the mouse; the primary function of the wild-type allele is unknown. *Wh* is the only mutation to have been described at the locus in the hamster. The locus is complex in the mouse (Grüneberg, 1953; Hollander, 1968; Konyukhov and Sazhina, 1980; West *et al.*, 1985), and the effects of several mutations and their interactions were reviewed by Silvers (1979, pp. 268–279). The literature on the locus is extensive, and I will cite only a few studies here to illustrate points relevant to homology.

Several principal features of the phenotypes in the mouse and hamster are presented in Table V. In only one of six characteristics examined, achromia, is there known to be close agreement in the two species. In the other five, either the information on which to make a comparison is lacking (deafness in the hamster, endocrine abnormalities in the mouse) or there are striking differences in the phenotypes (microphthalmia, fertility, and growth retardation). There is little to support a strong statement on homology; however, the differences recorded in Table V may reflect species modification of gene expression. It must also be remembered that several alleles are known at the mouse locus; *Wh* and Mi^{wh} may be nonhomologous mutations at homologous loci.

My contention that there is little interest in homological studies is supported by the fact that, despite Searle's suggestion of 20 years ago, investigators using one species make no reference to studies on the other. To my knowledge, Asher and James (1982) are the only authors who refer to a possible homologous relation between the two mutations, and unfortunately they don't know the meaning of homology. [Reek, *et*

Table V. A Comparison of the Principal Phenotypic Characteristics in the White (Mi^{wh}) Mouse and Anophthalmic White (*Wh*) Syrian Hamster

Phenotype characteristics	Mouse (Mi^{wh})	Hamster (*Wh*)	References
Achromia	Mi^{wh}/+: belly spot; coat diluted; eye pigment reduced; inner ear unpigmented or diluted or spotted	*Wh*/+: white; black eyes; reduced ear, eye, and skin pigmentation	Silvers (1979; summary), Robinson (1964)
	Mi^{wh}/Mi^{wh}: white; eye pigment much reduced; inner ear unpigmented	*Wh*/*Wh*: white; no ear or skin pigmentation	See above
Microphthalmia	Mi^{wh}/+: none Mi^{wh}/Mi^{wh}: moderate	*Wh*/+: slight *Wh*/*Wh*: extreme or anophthalmic	Konyukhov and Sazhina (1980), Robinson (1962), Jackson (1981)
Deafness	Mi^{wh}/+: severe inner ear abnormalities	*Wh*/+: ?	Deol (1970, 1973), Asher (1981)
	Mi^{wh}/Mi^{wh}: severe inner ear abnormalities	*Wh*/*Wh*: yes (degree and nature?)	
Fertility	Mi^{wh}/+: normal Mi^{wh}/Mi^{wh}: slightly reduced	*Wh*/+: normal *Wh*/*Wh*: usually sterile; hypoplastic and aspermic testes; ovulation inhibited	Grüneberg (1953), James *et al.* (1980), Hagen and Asher (1983)
Growth retardation	Mi^{wh}/+: none Mi^{wh}/Mi^{wh}: slight	*Wh*/+: none *Wh*/*Wh*: severe	James *et al.* (1980)
Endocrine abnormalities	No information	*Wh*/+: none *Wh*/*Wh*: thyroid, adrenal, and pituitary	Asher (1981), James and Asher (1981)

al. (1987) recently defended the position that the concept of homology, meaning having a common evolutionary origin, still plays a useful role in defining evolutionary relations, and contended that its meaning should be carefully guarded; see further discussion below.] Asher and James referred to Searle (1968) to support the statement that "*Wh* is a dominant spotting mutation and shares many of its characteristics with numerous other mutations in other mammalian species." A careful reading of Searle shows that this statement is false: the only mutation that Searle considered a possible homologue of *Wh* is Mi^{wh} (Table V). Asher and James further confused the issue by listing several mutations (including Mi^{wh}) that "appear to be homologous with *Wh*." The inclusion of Mi^{wh} is fortuitous because these authors really meant that the mutations cited produce phenotypes that share certain pathological features. Similarity is not synonymous with homology (Reek *et al.*, 1987) and, as I have already noted, gross phenotypic similarity is a precarious basis for making decisions on homology.

This brief overview of the possible homologous relations between Mi^{wh} and *Wh* illustrates the primitive state of homological studies of the type with which we are concerned in mammals. Even when seemingly ideal conditions for comparative studies exist, they are ignored rather than exploited; when homology is introduced into the discussion, it is usually misunderstood.

6.2.2. Other Examples Based on Clinical and Pathological Studies: The Chondrodysplasias

There are several other types of disorders that merit consideration as possible homologues on the basis of close phenotypic similarity in two or more species. Among these the dwarfing conditions provide many complex phenotypes for speculation; however, their numbers and varieties in many species tend to clutter the picture and make it difficult to extract potentially significant interspecific comparisons. Careless terminology also adds to the confusion. For example, in man, although there are several conditions that simulate achondroplasia, the "classic" disorder is a well-delineated autosomal trait (MIM 10080). Rimoin (1970) described well-organized endochondral ossification with apparent reduction in rate of cartilage growth, and pointed out the significance of these findings in the search for homologues. However, for many years a wide variety of genetically determined disorders in animals have been labeled "achondroplasia" on the basis of a phenotype of short limbs and the finding of disorganized endochondral ossification. Actually, the term is incorrectly used in man as it does not accurately describe or define the pathogenesis. But it is so firmly entrenched in the literature that it will persist, as will its inappropriate use in other animals. Despite this confusion and the multiplicity of complex phenotypes grouped as disproportionate short stature (dwarfism) in man and other mammals, this broad category of disorders merits close examination for possible homologues.

In man the chondrodysplasias form a subgroup of the osteochondrodysplasias, and are characterized by growth defects of the tubular bones, or the spine, or both (Rimoin *et al.*, 1979; Rimoin and Lachman, 1983, appendix). In this subgroup there are approximately 45 individual entities, divided into those identifiable at birth (25) and those identifiable later in life (20) (Rimoin *et al.*, 1979). These numbers are minimal because any one entry (e.g., thanatophoric dysplasia) may have several distinct subtypes (Horton *et al.*, 1979). In many of these disorders, anomalies other than those of the spine and limbs occur characteristically or frequently: a disproportionately large head with frontal bossing and flattening of the nose in achondroplasia and thanatophoric dysplasia; congenital cataracts in chondrodysplasia punctata; myopia in the Kniest dysplasia and spondyloepiphyseal dysplasia congenita; cleft palate, bifid uvula, or high-arched palate in the Kniest dysplasia, spondyloepiphyseal dysplasia congenita, and diastrophic dysplasia; postaxial polydactyly in chondroectodermal dysplasia and the lethal rib–polydactyly syndromes; preaxial polydactyly in chondroectodermal dysplasia and the short rib–polydactyly syndrome II (Majewski); short, broad hands and proximally inserted and abducted hypermobile thumbs in diastrophic dysplasia; trident hands in achondroplasia; hypoplastic nails in chondroectodermal dysplasia; short, broad nails in the McKusick type of metaphyseal dysplasia; clubfeet in diastrophic dysplasia; anomalies of the thorax in asphyxiating thoracic dysplasia, chondroectodermal dysplasia, metatropic dysplasia, thanatophoric dysplasia, and the short rib–polydactyly syndromes; and cardiac defects in chondroectodermal dysplasia and the short rib–polydactyly syndromes (Rimoin and Lachman, 1983).

The chondrodysplasias can be classified on the basis of their clinical and radiological features, chondro-osseous morphology, and mode of inheritance. Although the use of these criteria has resulted in a reasonably clear delineation of these disorders, details of chondro-osseous morphology have not been described in all of them and the underlying basic defects in collagen or proteoglycan chemistry have not been delineated in any; partial abnormalities have been described in several.

Chondrodysplastic anomalies occur in many nonhuman mammals, but, to my knowledge, they have never been systematically analyzed as a group to search for possible homologues. Attempts at analysis are confounded by the careless terminology mentioned above; e.g., "achondroplasia" has been described in the mouse (Lane and Dickie, 1968), dog (Gardner, 1959), rabbit (Brown and Pearce, 1945), cow (Gregory *et al.*, 1966), and sheep (Shelton, 1968). In some instances the term may be apt because the pathological features closely resemble those found in the human disorder, but in others the term derives from a perceived gross phenotypic similarity to the human condition without regard to the chondro-osseous pathogenesis.

Details of three pairs of potential homologues are presented in Table VI. In each case the information available on both disorders is

Table VI. Three Pairs of Possible Homologous Mutants among the Mammalian Chondrodysplasias

	Example 1	
	Man	Mouse
Dysplasia	Achondrogenesis II (Langer–Saldino) (MIM 20061)	Cartilage matrix deficiency
Clinical features		
Head/neck	Round, flat face; short neck	Short snout; cleft palate; protruding tongue
Chest/trunk	Short, barrel-shaped	Short trunk
Limbs	Very short	Short
Others	Distended abdomen; fetal hydrops	Short tail; distended abdomen
Skull	Normal	Short
Radiological features		
Ribs	Short	—
Vertebrae	Absence of centers for vertebral bodies and sacrum	—
Pelvis	Small ilia with concave inner and inferior margins; absent pubic and ischial bones	—
Limb bones	Short, straight; metaphyseal flaring and cupping	—
Chondro-osseous morphology	Large ballooned chondrocytes with deficient matrix; growth plate hypercellular and irregular; sclerosed vascular channels	Deficiency of cartilage matrix; tightly packed chondrocytes; many pycnotic cells, often with one or more conspicuous vacuoles; unusual amount or forms of collagen fibers
Biochemical defect	—	Defect in synthesis of cartilage proteoglycan core protein
Mode of inheritance	AR	AR
Comments	Infants and pups die shortly after birth; evidence of homology is strong although there is discordancy for cleft palate, and the biochemical defect underlying the human disorder has not been investigated	
References	Sillence *et al.* (1978), Rittenhouse *et al.* (1978), Kimata *et al.* (1981)	

	Example 2	
	Man	Rabbit
Dysplasia	Metatropic dysplasia (MIM 25060)	Chondrodystrophy

(*continued*)

Table VI. (*Continued*)

	Example 2	
	Man	Rabbit
Clinical features		
Head/neck	Normal	Large, rounded head; short, broad face; small features
Chest/trunk	Taillike sacral appendage; trunk seems long and narrow at birth; severe scoliosis develops	Trunk short
Limbs	Short with prominent joints	Short with large joints; forelimbs often clasped on the chest; hind limbs clubbed
Other	Seem short limbed in infancy, short trunked later; joint movement restricted	Tongue protudes slightly; cleft palate; cardiovascular anomalies common; joint movement restricted
Skull	Normal	Normal
Radiological features		
Ribs	Short, flared, cupped anteriorly	Short and bowed with flared ends
Vertebrae	Very flattened; wide intervertebral spaces (early); platyspondyly and scoliosis (late)	Irregular and immature
Pelvis	Hypoplastic crescent-shaped ilia; low-set anterior iliac spines	—
Limb bones	Short, broad, clublike	Short, broad, clublike
Chondro-osseous morphology	Chondrocytes vacuolated with inclusions; growth plate—irregular vascularization	—
Biochemical defect	—	—
Mode of inheritance	AR and AD	AR
Comments	Human disorder is probably a heterogeneous group of disorders; it is compatible with life though many affected infants die of respiratory failure; disorder in rabbit is lethal in neonatal period; although Fox and Crary (1975) stated that cleft palate is a feature of the human disorder, it does not occur commonly—Rimoin and Lachman (1983) did not refer to it and Cohen (1978) did not list it among syndromes with cleft lip and palate; no histological investigations of the rabbit have been reported; the responsible mutation has not been mapped in either species	
References	Maroteaux *et al.* (1966), Beck *et al.* (1983), Fox and Crary (1975)	

(*continued*)

Table VI. (*Continued*)

	Example 3	
	Man	Dog
Dysplasia	Pseudoachondroplastic dysplasia (MIM 26415)	Pseudoachondroplastic dysplasia
Clinical features		
Head/neck	Normal	Normal
Chest/trunk	Trunk seems disproportionately long	Flattened rib cage (secondary feature)
Limbs	Very short with genu varum or valgum; hypermotility of joints; small, broad hands	Short, bent with enlarged joints; abnormal movement; abducted hind limbs
Other	Not manifested until at least 2 y of age	Not manifested until 3 weeks of age; by 2 y limb bones are twisted, malshaped, and larger in diameter than usual; hips are dysplastic and arthritic
Skull	Normal	Normal
Radiological features		
Ribs	Normal	Wide and flat; curved as result of flattened thoracic cage
Vertebrae	Platyspondyly; anterior tonguelike protrusion; endplate irregularity	Spinous processes short and broad; bodies short and only a small portion ossified, giving beaked appearance; at 2 y completely ossified
Pelvis	Acetabula irregular; hypoplastic ischium and pubis	Acetabula dysplastic
Limb bones	Epihyseal and metaphyseal dysplasia; striking hand involvement with shortening of tubular bones with irregular metaphyses and small round epiphyses; marked capital femoral epiphysis	Areas of stippling and patchy densities in the epiphyses; long bones shortened, deformed; bones are twice normal width at 2 y, completely ossified
Chondro-osseous morphology	Prominent inclusions in chondrocytes showing lamellar or granular rough endoplasmic reticulum dilation on electron microscopy	Radiographic densities associated with retarded ossification; matrix diluted, poorly stained; most cartilage cells in large halos; no inclusions; at 2 y epiphyses fully ossified; beaking effect gone

(*continued*)

Table VI. (*Continued*)

	Example 3	
	Man	Dog
Biochemical defect	—	—
Mode of inheritance	AR	AR
Comments	Human disorder is genetically heterogeneous; disorder in dog described only in miniature poodles; basic biochemical error not known in either species	
References	Hall (1975), Hall and Dorst (1969), Gardner (1959), Riser *et al.* (1980)	

[a]The general format of this table and the information on the human disorders are derived from Rimoin and Lachman (1983, appendix).

considerable; however, the basic biochemical error is known in only one, cartilage matrix deficiency in the mouse; none of the six responsible loci has been mapped. In addition, two of the three human disorders are probably heterogeneous. Consequently, no definite statement about homology can be made.

Several other chondrodysplastic mutants merit comment. Brachymorphic (*bm,* on chromosome 19), a form of chondrodysplasia detectable in the newborn mouse, is characterized by disproportionate dwarfing, a short cone-shaped skull, and a short thick tail (Land and Dickie, 1968). The growth plates are regular, but the columns are shorter than normal (Orkin *et al.,* 1977); in this respect the mutant resembles classical achondroplasia in man (Rimoin *et al.,* 1970). However, the observations of Orkin *et al.* have been disputed by Miller and Flynn-Miller (1976) and Wikström *et al.* (1984) who believed that the columnar arrangement is disturbed. The chondroitin sulfate proteoglycan in the cartilage is undersulfated as a consequence of a defect in the synthesis of the sulfate donor 3′-phosphoadenosine 5′-phosphosulfate (PAPS) caused by diminished activity of both ATP sulfurylase and adenosine 5′-phosphosulfate kinase; the reduced activity of the latter seems to be the primary defect (Sugahara and Schwartz, 1982). This mutant is important because its basic underlying biochemical error has been determined and the mutation has been mapped; although no homologue is immediately apparent, these features establish a framework within which potential homologues can be assessed.

In the Alaskan malamute dog, a form of disproportionate dwarfism occurs that resembles the human metaphyseal dysplasias (Sande *et al.,* 1982). The growth plates and metaphyses of all tubular bones are predominantly involved. The skull, vertebrae, and pelvis are normal. Four common forms of metaphyseal dysplasia are recognized in man (Rimoin and Lachman, 1983); two are autosomal dominant (MIM 15640, 15650) and two are autosomal recessive traits (MIM 25025, 26040). The canine disorder is an autosomal dominant. The histopathological changes in all four human disorders are similar: large chondrocytes in a fibrillar matrix, clusters of hypertrophic cells at the growth plate, and an irregular line of ossification with tongues of cartilage in the metaphyses (Rimoin and Lachman, 1983). In the dwarfed malamutes, chondrocytes in the upper half of the zone of chondrocyte proliferation have bizarre shapes. These chondrocytes contain profiles of markedly dilated rough endoplasmic reticulum, which is not oriented parallel to the long axis of the cells but consists of irregularly dilated cisternae (Bingel *et al.*, 1983). The proteoglycans in these growth plates differ significantly from those in growth plates of normal malamute dogs (Bingel *et al.*, 1985). Affected dogs also have a hemolytic anemia with stomatocytosis (Fletch and Pinkerton, 1973); the developmental relation between the skeletal and hematological disorders has not been analyzed. To my knowledge, no similar hematological disturbance has been described in any form of human metaphyseal dysplasia. However, the two recessive forms have phenotypic features—cartilage hair hypoplasia (MIM 25025) and pancreatic insufficiency (MIM 26040)—that are not present in the canine disorder. Neither the canine nor the human disorder has been genetically mapped.

This brief examination of clinical, pathological, and genetic heterogeneity among the mammalian chondrodysplasias vividly illustrates some of the difficulties in searching for and defining homologies. Two final points merit comment: the potential importance of normal interspecies variation and the issue of dominance. In Section 4 several examples were presented demonstrating how the action of the same mutation in two species can produce different phenotypes as a consequence of the modifying effects of the two different "normal" backgrounds. To my knowledge, the significance of interspecies variation in normal growth plate morphogenesis on the expression of the mammalian chondrodysplasias has never been thoroughly examined. However, Rimoin (1975, p. 97) noted in passing, apropros of the achondroplastic (*ac*/*ac*) rabbit, "The type of histopathologic abnormality does not appear to resemble that seen in human thanatophoric dwarfism; however, the cellular appearance of the growth plate in the normal rabbit differs from that in the normal human, making a comparison difficult."

At first glance it may seem reasonable to conclude that an autosomal trait in one species and a recessive trait in another cannot be homologues. However, as the observations of Juriloff *et al.* (1987) on the murine *far* mutation (see Section 8) aptly demonstrate, the dominance of a trait, which depends on modifiers, can vary dramatically among strains of the same species. Dominance is not a fixed characteristic. Among the achondrodysplasias there is evidence, of varying degrees of substance, for changes in dominant/recessive relations that may alter perceptions in assigning homologous status. Two examples will be cited here. Achondroplasia in man (MIM 10080) is an autosomal dominant trait featuring rhizomelic dysplasia, characteristic facies with midface hypoplasia, exaggerated lumbar lordosis, limitation of hip and elbow extension, genu varum or valgum, and trident hand (Rimoin and Lachman, 1983). The chondro-osseous morphology is striking: the chondrocytes are normal and the growth plate is regular. A few homozygotes for the mutant allele have been described; they are much more severely affected than heterozygotes, and the histological changes are dramatically different (Aterman *et al.*, 1983). So it is more correct to say that the disorder, as characteristically seen in heterozygotes, is partially dominant. The second example is found in cattle in which a variety of "achondroplastic" mutants have been described; the literature on the subject is full of contradictions and confusion. In an attempt to resolve some of the confusion, Gregory *et al.* (1966) analyzed the interrelations of several distinct mutants in the Hereford, shorthorn, and Angus breeds, and concluded that ". . . all achondroplastic mutants, both 'dominant' and recessive, are interrelated and components of the same genetic complex."

Finally, it should be pointed out that a mechanism resulting in dominant inheritance of a trait in one species and recessive inheritance of the homologue in another may reside in the nature of the responsible gene and the errors occurring in it. Investigations (Baas *et al.*, 1984; van Ommen, 1987) of errors in thyroglobulin (TG) synthesis in man and goats suggest that differences in the subunits of oligomeric proteins resulting from amorphic or hypomorphic alleles can give rise to recessive or dominant phenotypes. TG, the major protein of the thyroid, is a dimeric glycoprotein and is the precursor protein for the thyroid hormones T_4 and T_3. Errors of TG synthesis giving rise to congenital goiter have been described in merino sheep (Mayo and Mulhearn, 1969), man (Lever *et al.*, 1983, pp. 226–228), Afrikander cattle (Ricketts *et al.*, 1985), goat (Kok *et al.*, 1987), and mouse (Beamer *et al.*, 1987; Taylor and Rowe, 1987). Most of the resultant phenotypes in man and all those in the nonhuman mammals are autosomal recessives. However, Baas *et al.* (1984) reported a family in which there is strong evidence for autosomal dominant inheritance of an error in TG synthesis, and Couch *et al.*

(1986) described an autosomal dominant form of euthyroid adolescent multinodular goiter that most likely results from an abnormality in TG structure and function.

van Ommen (1987) explained the apparent contradiction posed by dominant and recessive forms of what appears to be the same trait as follows. When the mutation results in an amorph (null allele), the gene product will be absent; heterozygotes will have only normal subunits in their cells and will usually function normally. However, when the mutation results in a hypomorph (minus allele), the gene product will be present but functionally defective. In the case of a dimeric protein, such as TG, 75% of the dimers in heterozygotes will contain at least one abnormal subunit and metabolism will be seriously disturbed. Thus, two apparently very similar disorders, one exhibiting dominance in one species, the other recessiveness in a second species, could be homologues in that they arise from mutations within a gene that each species has inherited from a common ancestor.

7. DO NAMES AND DEFINITIONS MATTER?

Does it matter if the terms "homologue" and "model" are indiscriminately interchanged, or if the generally accepted use of "homologue" is ignored and abused, as mentioned in a previous section? Scientists pride themselves on being precise when the occasion demands it, and most would probably agree that proper science demands proper and accurate definition of terms. However, the Medawars (1985, p. 66) contend that "it is simply not true that no discourse is possible unless terms are precisely defined." In genetics it is not necessary to search far for an example of this view: the word "gene" serves the purpose admirably. A precise definition of "gene" has eluded the efforts of several generations of geneticists since Johannsen coined the term in 1909. Falk (1984), after reviewing the history of efforts to define "gene," conceded that the concomitant flexibility and usefulness of the gene concept supports Kramer's contention that "the most fruitful concepts are those to which it is impossible to attach a well defined meaning."

Is "homology" another example of such a word? Hardly. Although, as noted at the outset, the term has periodically been the focus of dispute (see Reek *et al.*, 1987, as an example of the most recent), the original intent of the definition—having a common evolutionary origin—is quite clear and is still useful. And there is no need for confusion with the concept of model; Held's definition of the latter (Section 1) does not refer explicitly or implicitly to homologue. The terms "congruence of phenotypes" or "congruence of pathogenesis" might be useful to describe a phenomenon where the two concepts might be considered to overlap. This suggestion is in keeping with the ideas of a group of evolutionary-oriented biologists who recently deplored the confusion that is "clogging the literature on protein and nucleic acid sequence comparisons" because "homology" is used carelessly as a synonym for "similarity" (Reek *et al.*, 1987). These authors contend that the confusion interferes with clear thinking about evolutionary relations.

8. SOME CONCLUDING THOUGHTS

Anyone who has followed the discussion this far may have a feeling of futility about the search for truly homologous developmental defects. Why is there so little substantial evidence for such homologies when there is such a prodigious amount of information available on mammalian genomes and relevant phenotypes? This deficiency seems to arise from a lack of interest in comparative studies of complex phenotypes or an unwillingness to extend such comparative studies to the extent necessary. The Mi^{wh}–Wh story presented above is an outstanding example of this point. No one would dispute that a remarkably keen interest in comparative mammalian gene mapping exists (Lalley and McKusick, 1985), and Bulfield (1980) documented the splendid advances in the comparative studies of genetically determined metabolic disease in mammals. However, Bulfield cautioned about the confusion introduced by the uncritical use of animal models and the lack of understanding of the meaning of homology. This is a particularly glaring feature of comparative studies involving complex morphological phenotypes. Problems arise from uncritically equating comparative pathogenesis with homology, a broader concept—a problem of confusing or equating a part with the whole. Comparative pathogenesis is an essential part of the understanding of the homologous relations of similar teratological phenotypes in two or more mammalian species, but it is only a part, and additional genetic information is required to complete the picture. The overview of Campbell *et al.* (1986) of CNS defects in mammals demonstrates this superbly. As noted in the discussion on Mi^{wh}, the terms "congruent phenotypes" or "congruent pathogenesis" might be used to describe phenotypes in which similar multiple system defects occur in two or more mammalian species.

Some may consider the issue of determining homology of complex syndromes a fruitless exercise given that the proliferation of syndromes in one species, man, has elicited cries of despair from those who see only chaos emerging from the unstructured, uncritical publication of vast numbers of syndromes (Warkany, 1971). Even commendable and extremely valuable attempts to create order from this mess are unable to stem the flood (Pinsky, 1974, 1975). The problems created by variable expressivity are also formidable. The intrafamilial variation in the expression of a single gene in human studies can be demonstrated even in the "controlled" conditions employed in experimental animal studies. In inbred strains of mice maintained under well-described environmental conditions, some genes seem to behave in mysterious ways, and their expression can be altered by varying the genetic milieu in which they function or by interaction with other genes. Recent investigations by Juriloff *et al.* (1987) on the *far* (first arch) mutation in the mouse serve as a striking example of how gene expression can be modified by altering the genetic background. On the BALB/cGaBc background, *far* homozygotes have bilaterally deficient maxillae, cleft palate, deficient lower eyelids, and facial skin tags; most heterozygotes are normal—5% have deficient lower eyelids and 1% have cleft palate (Juriloff and Harris, 1983). When the gene was transferred to the ICR/Bc background, about 38% of +/*far* animals developed hemifacial deficiency, i.e., *far* became an incomplete dominant on this background. The homozygous phenotype was no more severe than that in strain BALB/cGaBc mice. Juriloff *et al.* aptly commented that, "in human pedigrees, where the equivalents of the dominance modifiers in BALB/cGaBc and ICR/Bc would segregate within families, it would be difficult to recognize that sporadic hemifacial deficiency and severe bilateral maxillary deficiency were due to the same gene."

Despite the considerable challenges, the search for and ultimate understanding of homologies of developmental defects are important aspects of evolutionary and developmental biology. The task will not be easy, and answers will come from the combined efforts of many disciplines, among which teratology is uniquely placed, because of its multifaceted nature, to make a major contribution.

NOTE. After this chapter was submitted for publication. Winter (1988) published a report in which he reviewed known and possible homologies among multiple congenital anomaly syndromes in mouse and man.

In addition, an eighth edition of *Mendelian Genetics in Man: Catalogs of Autosomal Dominant, Autosomal Recessive, and X-Linked Phenotypes* was published in the autumn of 1988.

ACKNOWLEDGMENTS. This chapter is dedicated with respect to Dr. Ei Matsunaga on the occasion of his retirement as the director of the National Institute of Genetics of Japan.

I appreciate the constructive criticisms I received from Drs. Diana Juriloff and Muriel Harris on an early draft of this essay. I also thank Dr. Kazuo Fujikawa, Takeda Chemical Industries, Osaka, Japan, who critically reviewed an early draft, and Dr. Meredith Runner who generously gave me an important reference.

REFERENCES

Adelman, J. P., Bond, C. T., Douglass, J., and Herbert, E. 1987. Two mammalian genes transcribed from opposite strands of the same DNA locus. *Science* **235:**1514–1517.

Allison, J. E., Stanley, A. J., and Gumbreck, L. G. 1965. Sex chromatin and idiograms from rats exhibiting anomalies of the reproductive organs. *Anat. Rec.* **153:**85–92.

Andrews, E. J., Ward, B. C., and Altman, N. H., eds. 1979. *Spontaneous Animal Models of Human Disease.* Academic Press, New York.

Anonymous. 1972. Animal models and genetic stocks program of the National Academy of Sciences. *J. Hered.* **63:**25.

Asher, J. H., Jr. 1981. Concerning the primary defect leading to the pleiotropic effects caused by anophthalmic white (*Wh*) in the Syrian hamster *Mesocricetus auratus. J. Exp. Zool.* **217:**159–169.

Asher, J. H., Jr., and James, S. C. 1982. The primary ultrastructural defect caused by anophthalmic white (*Wh*) in the Syrian hamster. *Proc. Natl. Acad. Sci. USA* **79:**4371–4375.

Aterman, K., Welch, J. P., and Taylor, P. G. 1983. Presumed homozygous achondroplasia: a review and report of a further case. *Pathol. Res. Pract.* **178:**27–39.

Baas, F., Bikker, H., van Ommen, G. J. B., and de Vijlder, J. J. M. 1984. Unusual scarcity of restriction site polymorphism in the human thyroglobulin gene: a linkage study suggesting autosomal dominance of a defective thyroglobulin allele. *Hum. Genet.* **67:** 301–305.

Bailey, D. W. 1986. Genes that affect morphogenesis of the murine mandible: recombinant-inbred strain analysis. *J. Hered.* **77:**17–25.

Barnes, D. M. 1986. Brain architecture: beyond genes. *Science* **233:**155–156.

Beamer, W. G., Maltais, L. J., DeBaets, M. H., and Eicher, E. M. 1987. Inherited congenital goiter in mice. *Endocrinology* **120:**838–840.

Beck, M., Roubicek, M., Rogers, J. G., Naumoff, P., and Spranger, J. 1983. Heterogeneity of metatropic dysplasia. *Eur. J. Pediatr.* **140:**231–237.

Biddle, F. G., and Fraser, F. C. 1986. Major gene determination of liability to spontaneous cleft lip in the mouse. *J. Craniofac. Genet. Dev. Biol.* Suppl. **2:**67–88.

Bingel, S. A., Sande, R. D., and Newbrey, J. 1983. Dwarfism in the Alaskan malamute: ultrastructural feature of dwarf growth plate chondrocytes. *Calcif. Tissue Int.* **35:**216–224.

Bingel, S. A., Sande, R. D., and Wight, T. N. 1985. Chondrodysplasia in the Alaskan malamute: characterization of proteoglycans dissociatively extracted from dwarf growth plates. *Lab. Invest.* **53:**479–485.

Bodmer, W. F. 1981. Gene clusters, genome organization, and complex phenotypes: when the sequence is known, what will it mean? *Am. J. Hum. Genet.* **33:**664–682.

Brown, W. H., and Pearce, L. 1945. Hereditary achondroplasia in the rabbit. I. Physical appearance and general features. *J. Exp. Med.* **82:**241–260.

Bulfield, G. 1980. Inherited metabolic disease in laboratory animals: a review. *J. Inher. Metab. Dis.* **3:**133–143

Campbell, L. R., Dayton, D. H., and Sohal, G. S. 1986. Neural tube defects: a review of human and animal studies on the etiology of neural tube defects. *Teratology* **34:**171–187.

Carson, D. A., Kaye, J., and Wasson, D. B. 1980. Differences in desoxyadenosine metabolism in human and mouse lymphocytes. *J. Immunol.* **124:**8–12.

Chernoff, G. F. 1977. The fetal alcohol syndrome in mice: an animal model. *Teratology* **15:** 223–230.

Chung, C. S., Ching, G. H. S., and Morton, N. E. 1974. A genetic study of cleft lip and palate in Hawaii. II. Complex segregation analysis and genetic risks. *Am. J. Hum. Genet.* **26:**177–188.

Churchill, F. B. 1974. William Johannsen and the genotype concept. *J. Hist. Biol.* **7:**5–30.

Cohen, M. M., Jr. 1978. Syndromes with cleft lip and cleft palate. *Cleft Palate J.* **15:**306–328.

Couch, R. M., Hughes, I. A., DeSa, D. J., Schiffrin, A. C., Guyda, H., and Winter, J. S. D. 1986. An autosomal dominant form of adolescent multinodular goiter. *Am. J. Hum. Genet.* **39:**811–816.

Danforth, C. H. 1925. Hair in its relation to questions of homology and phylogeny. *Am. J. Anat.* **36:**47–68.

Dawkins, R. 1986. *The Blind Watchmaker.* Longman, London, pp. 294–297.

Demenais, F., Bonaïti-Pellié, C., Briard, M. L., and Feingold, J. 1984. An epidemiological and genetic study of facial clefting in France. II. Segregation analysis. *J. Med. Genet.* **21:**436–440.

Deol, M. S. 1970. The relationship between abnormalities of pigmentation and of the inner ear. *Proc. R. Soc. London Ser. A* **175:**201–217.

Deol, M. S. 1973. The role of tissue environment in the expression of spotting genes in the mouse. *J. Embryol. Exp. Morphol.* **30:**483–489.

Eicher, E. M., Southard, J. L., Scriver, C. R., and Glorieux, F. H. 1976. Hypophosphatemia: mouse model for human familial hypophosphatemic (vitamin D-resistant) rickets. *Proc. Natl. Acad. Sci. USA* **73:**4667–4671.

Evans, R. M., Swanson, L., and Rosenfeld, M. G. 1985. Creation of transgenic animals to study development and as models for human disease. *Recent Prog. Horm. Res.* **41:**317–337.

Falk, R. 1984. The gene in search of an identity. *Hum. Genet.* **68:**195–204.

Fletch, S. M., and Pinkerton, P. H. 1973. Inherited hemolytic anemia with stomatocytosis in the Alaskan malamute dog. *Am. J. Pathol.* **71:**447–480.

Fox, R. R., and Crary, D. D. 1975. Hereditary chondrodystrophy in the rabbit: genetics and pathology of a new mutant, a model for metatropic dwarfism. *J. Hered.* **66:**271–276.

Garber, E. D. 1952. "Bent-tail", a dominant, sex-linked mutation in the mouse. *Proc. Natl. Acad. Sci. USA* **38:**876–879.

Gardner, D. L. 1959. Familial canine chondrodystrophia foetalis (achondroplasia). *J. Pathol. Bacteriol.* **77:**243–247.

Giger, U., Harvey, J. W., Yamaguchi, R. A., McNulty, P. K., Chiapella, A., and Beutler, E. 1985. Inherited phosphofructokinase deficiency in dogs with hyperventilation-induced hemolysis: increased in vitro and in vivo alkaline fragility of erythrocytes. *Blood* **65:**345–351.

Green, M. C. 1981. Catalog of mutant genes and polymorphic loci, in: *Genetic Variants and Strains of Laboratory Mice,* M. C. Green, ed. Fischer, Stuttgart, pp. 162–166.

Gregory, P. W., Tyler, W. S., and Julian, L. M. 1966. Bovine achondroplasia: the reconstitution of the Dexter components from non-Dexter stock. *Growth* **30:**393–418.

Grima, B., Lamouroux, A., Boni, C., Julien, J.-F., Javoy-Agid, F., and Mallet, J. 1987. A single human gene encoding multiple tyrosine hydroxylases with different predicted functional characteristics. *Nature* **326**:707–711.

Grüneberg, H. 1943. Congenital hydrocephalus in the mouse, a case of spurious pleiotropism. *J. Genet.* **45**:1–21.

Grüneberg, H. 1953. The relations of microphthalmia and white in the mouse. *J. Genet.* **51**: 359–362.

Grüneberg, H. 1955. Genetical studies on the skeleton of the mouse. XVII. Bent-tail. *J. Genet.* **53**:551–562.

Hackel, D. B., ed. 1980. A workshop on needs for new animal models of human disease. *Am. J. Pathol.* **101**:s1–s271.

Hagen, S. C., and Asher, J. H., Jr. 1983. Effects of pinealectomy on reproduction in the Syrian hamster mutant anophthalmic white (*Wh*). *Am. J. Anat.* **167**:523–538.

Haldane, J. B. S. 1927. The comparative genetics of colour in rodents and Carnivora. *Biol. Rev.* **2**:199–212.

Hall, J. G. 1975. Pseudoachondroplasia. *Birth Defects* **11**(6):187–202.

Hall, J. G., and Dorst, J. P. 1969. Four types of pseudoachondroplastic spondylo-epiphyseal dysplasia (SED). *Birth Defects* **4**(5):242–259.

Happle, R., Phillips, R. J. S., Roessner, A., and Jünemann, G. 1983. Homologous genes for X-linked chondrodysplasia punctata in man and mouse. *Hum. Genet.* **63**:24–27.

Harris, H. 1980. *The Principles of Human Biochemical Genetics*, 3rd ed. Elsevier/North-Holland, Amsterdam, pp. 418–422.

Held, J. R. 1983. Appropriate animal models. *Ann. N.Y. Acad. Sci.* **406**:13–19.

Herbert, E. 1981. Discovery of pro-opiomelanocortin—a cellular polyprotein. *Trends Biochem. Sci.* **6**:184–188.

Hollander, W. F. 1968. Complementary alleles at the *mi*-locus in the mouse. *Genetics* **60**:189 (abstr.).

Hooper, M., Hardy, K., Handyside, A., Hunter, S., and Monk, M. 1987. HPRT-deficient (Lesch–Nyhan) mouse embryos derived from germline colonization by cultured cells. *Nature* **326**:292–295.

Horton, W. A., Rimoin, D. L., Hollister, D. W., and Lachman, R. S. 1979. Further heterogeneity within lethal neonatal short-limbed dwarfism: the platyspondylic types. *J. Pediatr.* **94**:736–742.

Hunsicker, P. 1974. Private communication. *Mouse News Lett.* **50**:51–52.

Jackson, C. G. 1981. Prenatal development of the microphthalmic eye in the golden hamster. *J. Morphol.* **167**:65–90.

James, S. C., and Asher, J. H., Jr. 1981. A histological examination of the pars distalis from the Syrian hamster mutant anophthalmic white (*Wh*). *J. Exp. Zool.* **218**:335–350.

James, S. C., Hooper, G., and Asher, J. H., Jr. 1980. Effects of the gene *Wh* on reproduction in the Syrian hamster, *Mesocricetus auratus*. *J. Exp. Zool.* **214**:261–275.

Johannsen, W. 1909. *Elemente der Exakten Erblichkeitslehre,* 2nd ed. Fischer, Jena.

Juriloff, D. M. 1980. The genetics of clefting in the mouse, in: *Etiology of Cleft Lip and Cleft Palate,* M. Melnick, D. Bixler, and E. D. Shields, eds. Liss, New York, pp. 39–71.

Juriloff, D. M. 1986. Major genes that cause cleft lip in mice: progress in the construction of a congenic strain and in linkage mapping. *J. Craniofac. Genet. Dev. Biol.* Suppl. **2**:55–66.

Juriloff, D. M., and Harris, M. J. 1983. Abnormal facial development in the mouse mutant first arch. *J. Craniofac. Genet. Dev. Biol.* **3**:317–337.

Juriloff, D. M., Harris, M. J., and Froster-Iskenius, U. 1987. Hemifacial deficiency induced by a shift in dominance of the mouse mutation *far:* a possible genetic model for hemifacial microsomia. *J. Craniofac. Genet. Dev. Biol.* **7**:27–44.

Kalter, H. 1980. A compendium of the genetically induced congenital malformations of the house mouse. *Teratology* **21**:397–429.

Kawamata, J., and Melby, E. C., eds. 1987. *Animal Models: Assessing the Scope of Their Use in Biomedical Research.* Liss, New York.

Kent, M. G., Shoffner, R. N., Buoen, L., and Weber, A. F. 1986. XY sex-reversal syndrome in the domestic horse. *Cytogenet. Cell Genet.* **42**:8–18.

Khoury, M. J., Erickson, J. D., and James, L. M. 1983. Maternal factors in cleft lip with or without palate: evidence from interracial crosses in the United States. *Teratology* **27**: 351–357.

Kimata, K., Barrach, H.-J., Brown, K. S., and Pennypacker, J. P. 1981. Absence of proteoglycan core protein in cartilage from *cmd/cmd* (cartilage matrix deficiency) mouse. *J. Biol. Chem.* **256**:6961–6968.

Kok, K., van Dijk, J. E., Sterk, A., Baas, F., van Ommen, G. J. B., and de Vijlder, J. J. M. 1987. Autosomal recessive inheritance of goiter in Dutch goats. *J. Hered.* **78**:298–300.

Kollar, E. J., and Fisher, C. 1980. Tooth induction in chick epithelium: expression of quiescent genes for enamel synthesis. *Science* **207**:993–995.

Konyukhov, B. V., and Sazhina, M. V. 1980. Allelic complementation at the *mi*-locus of the mouse: partial normalization of the retinal pigment of epithelial development. *Biol. Zentralbl.* **99**:571–584.

Kuehn, M. R., Bradley, A., Robertson, E. J., and Evans, M. J. 1987. A potential animal model for Lesch–Nyhan syndrome through introduction of HPRT mutations into mice. *Nature* **326**:295–298.

Lalley, P. A., and McKusick, V. A. 1985. Report of the committee on comparative mapping. *Cytogenet. Cell Genet.* **40**:536–566.

Lane, P. W., and Dickie, M. M. 1968. Three recessive mutations producing disproportionate dwarfing in mice: achondroplasia, brachymorphic, and stubby. *J. Hered.* **59**:300–308.

Laurence, P. A. 1985. Molecular development: is there a light burning in the hall? *Cell* **40**: 221.

Leiter, E. H., Beamer, W. G., Schultz, L. D., Barker, J. E., and Lane, P. W. 1987. Mouse models of genetic diseases. *Birth Defects* **23**:221–257.

Lesch, M., and Nyhan, W. L. 1964. A familial disorder of uric acid metabolism and central nervous system function. *Am. J. Med.* **36**:561–570.

Lever, E. G., Medeiros-Neto, G. A., and DeGroot, L. J. 1983. Inherited disorders of thyroid metabolism. *Endocr. Rev.* **4**:213–239.

Levin, R. 1984. Why is development so illogical? *Science* **224**:1327–1329.

Lindsey, J. R., and Capen, C. C. (Chairmen) 1976. Symposium: animal models for biochemical research VI—metabolic disease. *Fed. Proc.* **35**:1192–1236.

Little, C. C. 1958. Coat color genes in rodents and carnivores. *Q. Rev. Biol.* **33**:103–137.

Lojda, L. 1975. The cytogenetic pattern in pigs with hereditary intersexuality similar to the syndrome of testicular feminization in man. *Doc. Vet.* **8**:71–82.

Lyon, M. F. 1974. Mechanisms and evolutionary origins of variable X-chromosome activity in mammals. *Proc. R. Soc. London Ser. B* **187**:243–268.

Lyon, M. F., and Hawkes, S. G. 1970. *X*-linked gene for testicular feminization in the mouse. *Nature* **227**:1217–1219.

Lyon, M. F., Scriver, C. R., Baker, L. R. I., Tenenhouse, H. S., Kronick, J., and Mandla, S. 1986. The *Gy* mutation: another cause of X-linked hypophosphatemia in mouse. *Proc. Natl. Acad. Sci. USA* **83**:4899–4903.

Magalhaes, H. 1954. Mottled-white, a sex-linked lethal mutation in the golden hamster, *Mesocricetus auratus. Anat. Rec.* **120**:752 (abstr.).

Marazita, M. L., Spence, M. A., and Melnick, M. 1984a. Genetic analysis of cleft lip with or without cleft palate in Danish kindreds. *Am. J. Med. Genet.* **19**:9–18.

Marazita, M. L., Smalley, S. L., and Spence, M. A. 1984b. Cleft lip with or without cleft palate: reanalysis of a three-generation family study in England. *Am. J. Hum. Genet.* **36**: 174S.

Marazita, M. L., Spence, M. A., and Melnick, M. 1986. Major gene determination of liability to cleft lip with or without cleft palate: a multiracial view. *J. Craniofacial Genet. Dev. Biol.* **2**(Suppl.):89–97.

Marie, P. J., Travers, R., and Glorieux, F. H. 1981. Healing of rickets with phosphate supplementation in the hypophosphatemic male mouse. *J. Clin. Invest.* **67**:911–914.

Maroteaux, P., Spranger, J., and Weidemann, H.-R. 1966. Der metatropische Zwergwuchs. *Arch. Kinderheilk.* **173**:211–226.

Mayo, G. M. E., and Mulhearn, C. J. 1969. Inheritance of congenital goitre due to a thyroid defect in Merino sheep. *Aust. J. Agric. Res.* **20**:533–547.

McKusick, V. A. 1980. The anatomy of the human genome. *J. Hered.* **71**:370–391.

McKusick, V. A. 1986*a*. *Mendelian Inheritance in Man: Catalogs of Autosomal Dominant, Autosomal Recessive, and X-Linked Phenotypes*, 7th ed. Johns Hopkins University Press, Baltimore.

McKusick, V. A. 1986*b*. The human gene map (footnote, Table III). *Cold Spring Harbor Symp. Quant. Biol.* **51**:1123–1205.

Medawar, P., and Medawar, J. 1985. *Aristotle to Zoos: A Philosophical Dictionary of Biology.* Oxford, Oxford, p. 145.

Miller, W. A., and Flynn-Miller, K. L. 1976. Achondroplastic, brachymorphic and stubby chondrodystrophies in mice. *J. Comp. Pathol.* **86**:349–363.

Mintz, B., and Illmensee, K. 1975. Normal genetically mosaic mice produced from malignant teratocarcinoma cells. *Proc. Natl. Acad. Sci. USA* **72**:3585–3589.

Mulvihill, J. J., Mulvihill, C. G., and Priester, W. A. 1980. Cleft palate in domestic animals: epidemiologic features. *Teratology* **21**:109–112.

Nes, N. 1966. Testikulaer feminisering hos storfe. (Norwegian) *Nord. Vet.-Med.* **18**:19–29.

O'Brien, S. J., ed. 1984. *Genetic Maps 1984: A Compilation of Linkage and Restriction Maps of Genetically Studied Organisms*, Volume 3. Cold Spring Harbor Laboratory, Cold Spring Harbor, New York, pp. 343–415.

Ohno, S. 1973. Ancient linkage groups and frozen accidents. *Nature* **244**:259–262.

Orkin, R. W., Williams, B. R., Cranley, R. E., Poppke, D. C., and Brown, K. S. 1977. Defects in the cartilaginous growth plates of brachymorphic mice. *J. Cell Biol.* **73**:287–299.

Patterson, D. F. 1976. Congenital defects of the cardiovascular system of dogs: studies in comparative cardiology. *Adv. Vet. Sci. Comp. Med.* **20**:1–37.

Patterson, D. F., Haskins, M. E., and Jezyk, P. F. 1982. Models of human genetic disease in domestic animals. *Adv. Hum. Genet.* **12**:263–339.

Phillips, R. J. S., and Fisher, G. 1978. Private communication. *Mouse News Lett.* **58**:43–44.

Phipps, E. L. 1969. Private communication. *Mouse News Lett.* **40**:41–42.

Pick, J. R., Goyer, R. A., Graham, J. B., and Renwick, J. H. 1967. Subluxation of the carpus in dogs: an X chromosomal defect closely linked with the locus for hemophilia A. *Lab. Invest.* **17**:243–248.

Pinsky, L. 1974. A community of human malformation syndromes involving the Müllerian ducts, distal extremities, urinary tract, and ears. *Teratology* **9**:65–80.

Pinsky, L. 1975. The community of human malformation syndromes that shares ectodermal dysplasia and deformities of the hands and feet. *Teratology* **11**:227–242.

Raff, E. C., Diaz, H. B., Hoyle, H. D., Hutchens, J. A., Kimble, M., Raff, R. A., Rudolph, J. A., and Subler, M. A. 1987. Origins of multiple gene families: are there both functional and regulatory constraints? in: *Development as an Evolutionary Process,* R. A. Raff and E. C. Raff, eds. Liss, New York, pp. 203–238.

Reek, G. R., de Haën, C., Teller, D. C., Doolittle, R. F., Fitch, W. M., Dickerson, R. E., Chambon, P., McLachlan, A. D., Margoliash, E., Jukes, T. H., and Zuckerkandl, E. 1987. "Homology" in proteins and nucleic acids: a terminology muddle and a way out of it. *Cell* **50:**667.

Ricketts, M. H., Schulz, K., van Zyl, A., Bester, A. J., Boyd, C. D., Meinhold, H., and van Jaarsveld, P. P. 1985. Autosomal recessive inheritance of congenital goiter in Afrikander cattle. *J. Hered.* **76:**12–16.

Rimoin, D. L. 1975. The chondrodystrophies. *Adv. Hum. Genet.* **5:**1–118.

Rimoin, D. L., and Lachman, R. S. 1983. The chondrodysplasias, in: *Principles and Practice of Medical Genetics,* A. E. H. Emery and D. L. Rimoin, eds. Churchill-Livingstone, New York, Volume 2, pp. 703–735.

Rimoin, D. L., Hughes, G. N., Kaufman, R. L., Rosenthal, R. E., McAlister, W. H., and Silverberg, R. 1970. Endochondral ossification in achondroplastic dwarfism. *N. Engl. J. Med.* **283:**728–735.

Rimoin, D. L., Hall, J., and Maroteaux, P. 1979. International nomenclature of constitutional diseases of bone with bibliography. *Birth Defects* **15**(10).

Riser, W. H., Haskins, M. E., Jezyk, P. F., and Patterson, D. F. 1980. Pseudoachondroplastic dysplasia in miniature poodles: clinical, radiologic, and pathologic features. *J. Am. Vet. Med. Assoc.* **176:**335–341.

Rittenhouse, E., Dunn, L. C., Cookingham, J., Calo, C., Spiegelman, M., Dooher, G. B., and Bennett, D. 1978. Cartilage matrix deficiency (*cmd*): a new autosomal recessive lethal mutation in the mouse. *J. Embryol. Exp. Morphol.* **43:**71–84.

Robinson, R. 1962. Anophthalmic white: a mutant with unusual morphological and pigmentary properties in the Syrian hamster. *Am. Nat.* **96:**183–185.

Robinson, R. 1964. Genetic studies of the Syrian hamster. VI. Anophthalmic white. *Genetica* **35:**241–250.

Robinson, R. 1984. Linkage in the Syrian hamster (*Mesocricetus auratus*), in: *Genetic Maps 1984: A Compilation of Linkage and Restriction Maps of Genetically Studied Organisms,* Volume 3, S. J. O'Brien, ed. Cold Spring Harbor Laboratory, Cold Spring Harbor, New York, p. 384.

Sande, R. D., Alexander, J. E., Spencer, G. R., Padgett, G. A., and Davis, W. C. 1982. Dwarfism in Alaskan malamutes: a disease resembling metaphyseal dysplasia in human beings. *Am. J. Pathol.* **106:**224–236.

Searle, A. G. 1968. *Comparative Genetics of Coat Colour in Mammals.* Logos Press, London.

Searle, A. G. 1969. Coat color genetics and problems of homology, in: *Haldane and Modern Biology,* K. R. Dronamraju, ed. Johns Hopkins University Press, Baltimore.

Seegmiller, J. E., Rosenbloom, F. M., and Kelley, W. N. 1967. Enzyme defect associated with a sex-linked human neurological disorder and excessive purine synthesis. *Science* **155:** 1682–1684.

Selmanowitz, V. J. 1979. Ectodermal dysplasias including epitheliogenesis imperfecta, icthyoses, and follicular/glandular anomalies, in: *Spontaneous Animal Models of Human Disease,* Volume II, E. J. Andrews, B. C. Ward, and N. H. Altman, eds. Academic, New York, p. 4.

Shelton, M. 1968. A recurrence of the Ancon dwarf in Merino sheep. *J. Hered.* **59:**267–268.

Shepard, T. H. 1986. *Catalog of Teratogenic Agents,* 5th ed. Johns Hopkins University Press, Baltimore.

Sigerist, H. E. 1951. *A History of Medicine,* Volume I. Oxford, New York, pp. 300–302.

Sillence, D. O., Rimoin, D. L., and Lachman, R. 1978. Neonatal dwarfism. *Pediatr. Clin. North Am.* **25**:453–483.

Silvers, W. K. 1979. *The Coat Colors of Mice: A Model for Mammalian Gene Action and Interaction.* Springer-Verlag, Berlin.

Simpson, J. L., Blagowidow, N., and Martin, A. O. 1981. XY gonadal dysgenesis: genetic heterogeneity based upon clinical observations, X-Y antien status and segregation analysis. *Hum. Genet.* **58**:91–97.

Sponenberg, D. P., and Bowling, A. T. 1985. Heritable syndrome of skeletal defects in a family of Australian shepherd dogs. *J. Hered.* **76**:393–394.

Stent, G. S. 1985. Thinking in one dimension: the impact of molecular biology on development. *Cell* **40**:1–2.

Sugahara, K., and Schwartz, N. B. 1982. Defect in 3′-phospho-adenosine 5′-phosphosulfate synthesis in brachymorphic mice. I. Characterization of the defect. *Arch. Biochem. Biophys.* **214**:589–601.

Sweet, H. O., and Lane, P. 1980. X-linked polydactyly (*Xpl*), a new mutation in the mouse. *J. Hered.* **71**:207–209.

Taylor, B. A., and Rowe, L. 1987. The congenital goiter mutation is linked to the thyroglobulin gene in the mouse. *Proc. Natl. Acad. Sci. USA* **84**:1986–1990.

Tonegawa, S. 1983. Somatic generation of antibody diversity. *Nature* **302**:575–581.

Trevelyan, G. M. 1964. *Illustrated English Social History: 1.* Penguin, Baltimore, p. 192.

van Ommen, G. J. B. 1987. Merging autosomal dominance and recessivity. *Am. J. Hum. Genet.* **41**:689–690.

Van Valen, L. M. 1982. Homology and causes. *J. Morphol.* **173**:305–312.

Vora, S., Giger, U., Turchen, S., and Harvey, J. W. 1985. Characterization of the enzymatic lesion in inherited phosphofructokinase deficiency in the dog: an animal analogue of human glycogen storage disease type VII. *Proc. Natl. Acad. Sci. USA* **82**:8109–8113.

Wanscher, J. H. 1975. The history of Wilhelm Johannsen's genetical terms and concepts from the period 1903–1926. *Centaurus* **19**:125–147.

Warkany, J. 1971. Syndromes. *Am. J. Dis. Child.* **121**:365–370.

Weiss, P. 1955. Nervous system (neurogenesis), in: *Analysis of Development,* B. H. Willier, P. Weiss, and V. Hamburger, eds. Saunders, Philadelphia, pp. 346–401 (Fig. 144).

West, J. D., Fisher, G., Loutit, J. F., Marshall, M. J., Nisbet, N. W., and Perry, V. H. 1985. A new allele of microphthalmia induced in the mouse: microphthalmia-defective iris (*mi^{di}*). *Genet. Res.* **46**:309–324.

Wikström, B., Gay, R., Gay, S., Hjerpe, A., Mengarelli, S., Reinholt, F. P., and Engfeldt, B. 1984. Morphological studies of the epiphyseal growth zone in the brachymorphic (*bm/bm*) mouse. *Virchows Arch. [Cell Pathol.]* **47**:167–176.

Winter, R. M. 1988. Malformation syndromes: a reveiw of mouse/human homology. *J. Med. Genet.* **25**:480–487.

Wright, S. 1963. Genetic interaction, in: *Methodology in Mammalian Genetics,* W. J. Burdette, ed. Holden-Day, San Francisco, pp. 159–192.

Yang-Feng, T. L., DeGennaro, L. J., and Francke, U. 1986. Genes for synapsin I, a neuronal phosphoprotein, map to conserved regions of human and murine X chromosomes. *Proc. Natl. Acad. Sci. USA* **83**:8679–8683.

Young, J. Z. 1987. *Philosophy and the Brain.* Oxford, Oxford, p. 157.

Zucherkandl, E., and Pauling, L. 1962. Molecular disease, evolution and genic heterogeneity, in: *Horizons in Biochemistry,* M. Kasha and B. Pullman, eds. Academic, New York, p. 207.

Issues and Reviews in Teratology **5**:115–153
Plenum Press, New York, 1990, 978-1-4612-7847-4

Short-Term Methods of Assessing Developmental Toxicity Hazard

3

Status and Critical Evaluation

FRANK WELSCH

1. INTRODUCTION

"*In vitro*" methods will refer here to means of assessing developmental toxicological hazards of chemicals in material other than pregnant mammals. Materials to be considered range from invertebrates (Hydra, Drosophila) to mammalian cells to isolated intact rodent embryos to chick embryos in ovo (though, strictly, the last is not an *in vitro* object). Such work has been used to examine the consequences of perturbing developing systems (Kochhar, 1975) and to investigate the mechanism of action of specific agents (Bass *et al.*, 1970; Ebert and Marois, 1976). Some of the chemicals so used were known to induce gross structural malformations in laboratory animals under *in vivo* conditions and therefore were designated as "teratogens" based on the conventional definition of the term. Indiscriminate use of the terms "teratogen" and "teratogenic," however, has caused considerable confusion in prenatal toxicology and at times has led to ill-founded arguments (IRLG, 1986).

In this chapter I will adhere to the recommendation of the IRLG (1986) workshop that endpoints of toxicity applicable to the conceptus be referred to by the term "developmental toxicity," which includes teratogenicity in the strictest sense of its definition. In this broad context, perturbed prenatal development encompasses death, altered growth, gross structural malformations, and functional impairment. However,

FRANK WELSCH • Department of Experimental Pathology and Toxicology, Chemical Industry Institute of Toxicology, Research Triangle Park, North Carolina 27709.

the terminology used by the original authors will not be altered when reviewing the literature cited.

Early on it was hoped that *in vitro* studies would yield an understanding of the "mechanism" of action of chemical teratogens and lead to shortcuts that would simplify the task of anticipating the teratogenic risk associated with hitherto untested chemicals (Wilson, 1973, 1977, 1978a). In this context, "mechanism" commonly refers to any or all of a series of events between the application of a chemical and its developmental consequences. But progress in understanding the mechanism of action of chemical teratogens has been slow, and for even the most intensely studied agents this knowledge remains elusive. Because of the urgent need to assess the developmental toxicity hazard of large numbers of chemicals, it was recognized that action needed to be taken independent of understanding mechanism (Johnson, 1981). This perception has led to the devising of numerous *in vitro* and other short-term teratology testing methods and to keen competition among them. Individual investigators are committed to the merits of their favorite test, but only a few of the tests have undergone comparative studies that might establish the relative reliability of the various test systems to be considered in this chapter.

Numerous reviews have dealt with these many assay methods. Some have been descriptive (Perraud, 1982; Johnson and Kochhar, 1983; Beaudoin, 1985; Gross and Sabourin, 1985; Daugherty *et al.*, 1985; Brown *et al.*, 1986a; Schmid, 1987), some claim to provide critical assessment without actually doing so (Collins, 1987; Flynn, 1987; Friedman, 1987; Welsh, 1987; Whitby, 1987; Whitby and Flynn, 1987; Johnson, 1987), and others outline shortcomings (Kochhar, 1980; Flint, 1981; Neubert, 1981, 1982; Brown and Fabro, 1982; Shepard *et al.*, 1983; Neubert and Barrach, 1983; Bleyl, 1984; Sadler and Warner, 1984; WHO, 1984; Brown and Freeman, 1984; Johnson, 1985; Neubert, 1985a,b, 1989; Brown, 1985, 1987; Daston and D'Amato, 1989). The present review will only consider test methods that have been described in more than one publication or that, though first proposed some time ago, recent publications indicate are still being investigated.

2. THE DRIVE TO DEVELOP NEW METHODS OF DETECTING DEVELOPMENTAL HAZARD POTENTIAL

2.1. Basis of the Drive

Since about 1960, heightened perceptions regarding danger to humans caused by exposure to toxic chemicals have led in industrialized societies to legislation regulating the chemical industry. For prenatal toxicological concerns in the United States, the important legislation was the Toxic Substances Control Act (TSCA) of 1976, which drew attention to the vast number of chemicals for which there were little or no data regarding their potential hazard.

The number of chemicals that might be of toxicological concern has been said to be in the tens if not hundreds of thousands, not counting the ones that are added each year. The implication is that they require some hazard potential assessment, including that of developmental toxicity hazard. So far there have been no critical recommendations as to how to handle these seemingly alarming numbers. But it has been suggested that the figure needing early developmental toxicity testing may only be several hundred and that the task may not be as enormous as it is commonly presented (Neubert, 1982). A step in the needed direction was the publication of a priority list of hazardous substances of 100 compounds divided into four priority groups based on lists provided by U.S. federal regulatory agencies (Federal Register, 1987).

TSCA made it clear that conventional methods used to screen pharmaceutical chemicals for "teratogenic potential" would not be suitable for industrial chemicals. Those tests are too labor intensive and expensive. Furthermore, determination of potential developmental toxicity across the board is most likely unnecessary, because based on past experience only some would be expected to pose a developmental hazard. The search for and development of other test methods was intensified, guided in part by the perceived requirements detailed by Wilson (1977, 1978a,b).

A number of proposed assay systems were soon in various stages of "validation," but chemicals to be tested were chosen indiscriminately and based on available data that designated them as "teratogens" or "nonteratogens," despite the admonition that "validation is the most critical process in selecting any *in vitro* system for the purposes of preliminary screening for adverse effects on development" (Wilson, 1978a). Preliminary data were published and a workshop convened in 1981 whose aim was "to evaluate the applicability of *in vitro* test systems as screens of potential teratogens" (Kimmel *et al.*, 1982).

2.2. Alternative Methods as a Driving Force

"Alternative methods" of assessing developmental toxicity means different things to various constituencies (Goldberg, 1985, 1987; Rowan and Goldberg, 1985), but essentially they would reduce or eliminate altogether the use of mammalian test animals. Two groups have advocated the pursuit of alternative methods. One is the chemical manufacturing industry, which is interested in faster approaches for determining the health hazard potential of their products. The second is the interest group pleading for more humane treatment and use of fewer laboratory animals.

It appears as if more moderate views that advocate the use of fewer animals could be accommodated in testing for developmental toxicity. This aim may be reached by modifications in the design of the Segment II studies, which is the phase of reproductive toxicity evaluations that tests the effect of chemicals on embryonic development *in vivo* (Neubert, 1982). But replacement of pregnant laboratory animals by *in vitro* tests is not contemplated even by outspoken advocates of alternative methods. With respect to developmental toxicity, pregnant women may react entirely different than do animal models. The uneasy reliance on *in vivo* animal studies rests on the fact that all chemical agents that have caused congenital malformations in humans affect at least one of the three endpoints recorded in Segment II tests (Schardein, 1985), yet the predictive value of pregnant animals with respect to human risk has been poor (Schardein, 1983, 1987) and supports the arguments that efforts toward screening chemicals by other systems may be worth serious consideration. The questions that must first be asked of the existing *in vitro* approaches are: have they already provided trustworthy answers; and how close are we to putting such test systems to practical use in assessing chemicals of unknown biological activity?

3. GOALS OF *IN VITRO* AND ALTERNATIVE METHODS IN DEVELOPMENTAL TOXICOLOGY

3.1. Elucidation of Normal and Abnormal Development

In vitro methods offer advantages for examining specific biochemical and morphological events during embryonic development and for studying their perturbation by chemicals. When *in vivo* investigations have revealed embryotoxic effects, it may be advantageous to isolate the embryo or parts thereof for more detailed experiments regarding the initial adverse effects. Some of the advantages and disadvantages are summarized in Table I. For instance, if metabolism of the chemical as a prerequisite for embryotoxic activity appears likely, then it may be possible to test the hypothesis *in vitro* with small quantities of authentic metabolites.

3.2. Detection of Developmental Toxicity Potential

Although designing of tests for developmental toxicity would seem to require understanding of how chemicals induce adverse effects, the

Table I. Advantages and Disadvantages of *in Vitro* Techniques in Teratology[a]

Advantages	Disadvantages
Confounding factors excluded—maternal, nutritional, hormonal (stress), uterine position	Complex *in vivo* interactions are ignored
Direct embryotoxicity may be studied	Whole-animal pharmacokinetics are ignored
Variables of placental transfer are eliminated	Maternal xenobiotic metabolism is absent
Exact concentrations can be defined	Some *in vitro* systems are quite sensitive to culture conditions
Concentration–response relationships can be examined	Only limited periods of mammalian embryogenesis may be studied
Exact time of exposure can be defined	Responses may result that the same chemical never causes *in vivo*
Biochemical and radioisotopic studies are facilitated	Limited water solubility of test chemical
Small quantities of authentic metabolites can be studied	Physiochemical stability of test chemical
Exact developmental stage can be defined	Potential for artifacts if reconstituted systems for metabolism are added
Microsurgical manipulation is facilitated	
Continuous observation of embryos is possible	
Human tissues may be used	
Embryonic/fetal metabolism can be studied	

[a]Modified from Brown and Fabro (1982).

search for definitive mechanisms, as already noted, has been disappointing and a unifying mechanism is most unlikely (Wilson, 1977); this has suggested two possible routes in pursuit of new testing methods. First is a single complex test system which would include many primary target sites, and the second is a battery of tests with each contributing one or several cellular events that are believed to be among the initial sites of embryotoxic insult. The latter approach has been proposed (Shepard *et al.*, 1983), but could possibly jeopardize the objectives of the *in vitro* tests since neither time nor cost would be saved, depending on the complexity of the test battery.

3.3. Replacement of *in Vivo* by *in Vitro* Testing

The point has been made (Neubert, 1981, 1982) that reliance upon the outcome of unproven *in vitro* tests for potential prenatal toxicity which are much less complex than conventional *in vivo* studies would, in a society striving for more safety, require fundamental attitudinal change and willingness to accept greater risk. The unavoidable conclusion is that at present and for the foreseeable future, mammalian species will remain an essential part of the hazard assessment approach.

3.4. Reduction of Expenses Associated with *in Vivo* Testing

Conventional developmental toxicity testing requires much manpower and large numbers of animals, thus is costly and limits the number of chemicals that can be evaluated. It has been estimated that worldwide only about 200 chemicals could be screened per year (Johnson, 1983) at considerable expense (Johnson, 1987). But based on the experience gained over the past 20 years, it appears that not all chemicals require the same intensity of testing since it seems realistic to expect that only relatively few of the vast number alluded to earlier would pose a developmental hazard. This view formed the basis for probing *in vitro* systems to uncover such potential (see Section 4.1).

4. TESTING CHEMICALS OF UNKNOWN EMBRYOTOXIC ACTIVITY

4.1. Terminology Applied

In vitro tests are meant to screen a large number of chemicals and identify those displaying the activity the test was designed to uncover. This step is followed by examination of the suspect chemicals *in vivo*, since *in vitro* tests are not designed to replace such conventional testing. When the latter itself is used as a screening test for developmental toxicity, then the designation "prescreening" seems most appropriate for the *in vitro* phase. A WHO (1984) panel designated tests that serve "for the selection of chemicals for possible future developmental toxicity testing" as "short-term selection tests (STST)."

4.2. Selective Embryotoxicity and Developmental Hazard

It has been repeatedly recognized that the conventional developmental toxicity screen is not a good test for teratogenicity, because the customary repeated treatment with the test chemicals often kills the embryos or injures the pregnant females (e.g., Johnson and Christian, 1984). Nevertheless, the test is still often considered to be a "Segment II" or "test for teratogenicity," which is misleading since it was designed to uncover various forms of prenatal toxicity. Palmer (1976, 1981) thus proposed that it should more appropriately be called a test for selective embryopathy (or selective embryotoxicity) as was seconded by the WHO (1984).

The importance of selective embryotoxicity and the question of how to detect it were recently reemphasized and the concept of "developmental hazard" introduced (Johnson, 1980, 1981). Other investigators have proposed the term "teratogenic hazard" and used it synonymously with "developmental hazard" (Fabro *et al.*, 1982), although strictly speaking one could argue about this. Calculation of a ratio called "relative teratogenic index," from the minimum teratogenic dose (tD_{05}) and the LD_{01}, was proposed to be indicative of the teratogenic hazard (Fabro *et al.*, 1982). The ability of an *in vitro* prescreen to reveal developmental hazard is its critical feature (Brown and Freeman, 1984; Brown, 1987).

The concept of developmental hazard has been widely discussed in reviews of maternal and developmental toxicity observed in conventional animal testing (Khera, 1984, 1985), of human studies (Khera, 1987a), and in a workshop (Kimmel *et al.*, 1987). Published records regarding maternal toxicity are unreliable; such phenomena are vital and should be collected in future studies (Schwetz and Moorman, 1987). Participants in the workshop just cited (Chernoff *et al.*, 1987; Khera, 1987b; DeSesso, 1987; Rogers, 1987; Schardein, 1987) considered chemical-induced maternal toxicity and associated developmental toxicity.

5. *IN VITRO* SYSTEMS EXPLORED FOR PRESCREENING PURPOSES

5.1. Scientific Basis for Test Design

The shortcoming of all *in vitro* systems for developmental toxicity prescreening is that they cannot mimic the intricacies of mammalian *in vivo* developmental processes. This produces uncertainty as to the validity of the results, whether positive or negative, for the endpoint or endpoints about which the assay system is supposed to be informative. But if the outcome is untrustworthy, the test system cannot be considered a prescreen (Neubert, 1982), since as defined above, *in vitro* prescreening reduces the number of chemicals requiring further evaluation by *in vivo* testing. Evaluation of the criteria that must enter into design of *in vitro* prenatal toxicology prescreening tests (Neubert *et al.*, 1980, 1985; Neubert, 1981, 1982, 1985a,b; Neubert and Barrach, 1983) has led to the conclusion that even if they are successful in reducing the number of chemicals for subsequent *in vivo* screening, prescreening tests cannot provide information relevant to human risk assessment (Neubert, 1985a; Kochhar and Hickey, 1985; Neubert *et al.*, 1985).

In the real world of prescreening, a decision about the hazard potential of a chemical has to be based on a single test system or a battery of simple tests that are highly reproducible. The battery approach is only justified if (1) the overall results are trustworthy, and (2) definitive conclusions can be reached either in a shorter time or at less expense than in conventional *in vivo* screening.

Can one trust a negative outcome of prescreening and proceed without *in vivo* testing to marketing a product? Trust in the reliability of a test derives from its "validation." This all-important consideration will be assessed when individual test methods are described.

5.2. Actively Explored Test Systems

The current status of validation studies in various test systems was recently examined by Brown (1987). In most instances, particular methods have been explored only by their original proponents. In this chapter, prescreening tests are discussed only if they have been the subject of at least one publication in the past 3 years aside from the one originally describing the test. The reason for this selection is that more recent publications provide clues as to which systems are receiving continuing attention, thus indicating that they are still felt to be potentially useful.

6. MOST FREQUENTLY PUBLISHED *IN VITRO* PRESCREENING SYSTEMS

6.1. Primary Cell Cultures of Phylogenetically Lower Cells

6.1.1. *Drosophila melanogaster* Embryo Cells

The use of cultured *D. melanogaster* early gastrula cells in an *in vitro* teratogen assay was proposed by Bournias-Vardiabasis and Teplitz (1982). In the 24 h after plating these cells, some of them proliferate and differentiate as myoblasts and neurons, producing myotubes and ganglia, respectively. The number of these formations in control cultures is compared with that in chemical-exposed dishes by automated image analysis. Critical evaluation (Brown, 1987) of preliminary studies made with "teratogens" and "nonteratogens" (Bournias-Vardiabasis and Teplitz, 1982) selected from literature compilations (Schardein, 1985; Shepard, 1986) indicated that they gave about 35% false-negatives and false-positives. Further study with 100 compounds found that the Drosophila data compared favorably to those from animal and human developmental toxicity reports (Bournias-Vardiabasis *et al.*, 1983). The assay was said to have several desirable features defined during a consensus workshop (Kimmel *et al.*, 1982) since Drosophila embryo cells proliferate, differentiate, and undergo morphogenesis, all of which occur during mammalian embryogenesis.

For this test a fraction of the LD_{50} in adult female flies was used in embryo cell cultures. Therefore, it seems that an index of the developmental hazard of the chemical could be obtained, but the use of the test for this purpose has apparently not been considered (Bournias-Vardiabasis and Buzin, 1987). Another advantage claimed for the assay is potential for endogenous drug metabolism (Bournias-Vardiabasis and Flores, 1983). It seems that the emphasis has now shifted to mechanistic studies, and the study of so-called heat shock proteins whose synthesis is induced by chemical teratogens is being pursued (Buzin and Bournias-Vardiabasis, 1984; Bournias-Vardiabasis and Flores, 1986; Bournias-Vardiabasis and Buzin, 1987). Assessment of this endpoint by two-dimensional gel electrophoresis was advocated to assess teratogenic potential (Bournias-Vardiabasis and Flores, 1986), but is too elaborate for inclusion in a prescreen.

6.1.2. *Hydra attenuata* Cells

The coelenterate *H. attenuata* has been proposed as a developmental hazard prescreen (Johnson, 1980, 1981, 1983, 1985, 1987; Johnson *et al.*, 1982, 1984, 1986, 1987; Johnson and Gabel, 1982, 1983; Johnson and Christian, 1984, 1985). This organism can regenerate its entire body from mechanically dissociated cells of adult Hydra when the cells are randomly reaggregated into pellets (Johnson *et al.*, 1982). Apparently, close cell contacts are required, a feature that resembles the need of limb bud mesenchymal cells for high cell density as a prerequisite for biochemical and morphogenetic differentiation (see Section 6.2).

Adult Hydra and reaggregated cells are continuously exposed to the test chemical for 90 h and the morphology of the silhouette of the adult and the regenerate are scored according to a standard system. The ratio of the minimal concentration of the chemical toxic to adults (= *A*) and that adversely affecting the regenerates (i.e., development of new Hydra = *D*) determines the developmental hazard, a high value supposedly revealing selective action on regenerating Hydra, and one near unity indicating lack of selectivity, since regeneration was only disrupted at concentrations also affecting adults (Johnson, 1980, 1981).

There are only three other independent short reports about Hydra as a prescreen (Sabourin *et al.*, 1985; Wiger and Stottum, 1985, 1986). One deals with ethylene glycol ethers and their alkoxy acid metabolites which were tested in Hydra and chick limb bud mesenchyme cell cultures. The outcome in Hydra was compared with that in other cells and *in vivo*, and there was poor correlation (Wiger and Stottum, 1986). Several glycols and their ethers and acetates showed no "marked predilection for the embryo" in Hydra when the parent compounds were tested (Johnson *et al.*, 1984). However, enzymatic oxidation to the corresponding alkoxy acids is a prerequisite for embryotoxic activity *in vivo* (Yonemoto *et al.*, 1984; Ritter *et al.*, 1985; Scott *et al.*, 1987; Sleet *et al.*, 1987, 1988). Use of the assay with compounds requiring metabolism is undocumented (Johnson, 1985) and therefore uncertainty arises as to whether developmental hazard potential or unspecific cytotoxicity is being measured. In the conventional screen, some glycol ethers elicited selective adverse effects on the conceptus (George *et al.*, 1987; Price *et al.*, 1987); and metabolism to a common alkoxy acid derivative appears to be the unifying link that caused embryotoxicity (Hardin and Eisenmann, 1987). Another shortcoming of the assay may be the inability to test substances with limited water solubility (Johnson, 1985; Johnson *et al.*, 1986). However, one concern about the human habitat is that the health hazard posed by the environmental chemical contaminants with low water solubility lacks assessment.

In a recent modification of the assay (Wilby *et al.*, 1986), the concentration of a test chemical affecting intact adult Hydra (= toxicity, *T*) is related to that inhibiting regeneration (= inhibition, *I*) of an isolated region of the Hydra body, the data being expressed as a *T*/*I* ratio. This variation prompted an exchange of opinions (Johnson and Dansky, 1987; Wilby and Tesh, 1987).

As of this writing, no independent validation studies of the Hydra assay have been published, as was noted by Kochhar and Hickey (1985). Supposedly the Hydra assay is the long-sought single test system ready to serve as a prescreen (Johnson, 1987). Several aspects of the test, such as *A*/*D* ratios, their consistency across phyla, and the comparability of the ratio solely to mammalian conventional test data, have undergone refinements, also noticed by Schardein (1987). The *A*/*D* ratio indicative of selective action of a chemical on regeneration has been decreased (Johnson *et al.*, 1984 versus Johnson *et al.*, 1987), but remains the critical datum yielded by this prescreen. The Hydra *A*/*D* ratios were initially compared with calculated mammalian ones gleaned from the NIOSH registry of listings of the lowest teratogenic and lowest available adult lethal dose determined in nonpregnant animals (Johnson, 1980). However, mammalian adult toxicity data must be derived from pregnant animals since pregnancy can markedly affect the toxicity of a chemical (Rogers, 1987). A comment is indicated regarding chemicals designed to treat human diseases, because considerations different from those applicable to environmental chemicals enter into their evaluation for developmental toxicity. Drugs (e.g., antimetabolites, some anticonvulsants) claimed to cause human congenital malformations have a small therapeutic ratio, i.e., the dose eliciting adverse effects divided by the one required for therapeutic efficacy; experimentally this ratio is commonly determined as LD_{50}/ED_{50}. Doses needed for therapy are often similar to those causing toxic side effects in the pregnant woman and in her conceptus.

The predictive value of the Hydra *A/D* ratio for prescreening chemicals of unknown embryotoxic potential can presumably be based on the successful recapitulation through Hydra testing of *A/D* ratios calculated from published data of "state-of-the-art Segment II" protocols (Johnson, 1987). This conclusion appears open to subjective judgment, as illustrated by isoniazid and dinocap. The *A/D* ratio for the former was initially 500 (Johnson, 1980), then 20 (Johnson and Gabel, 1983), and more recently 100 (Wiger and Stottum, 1985). Regardless of the differences, developmental hazard is indicated. However, when one examines the recapitulation by Hydra of the mammalian data base concerning isoniazid, a striking discrepancy becomes apparent. This chemical is one of 47 on a list suggested for *in vitro* teratogenesis test validation and was included because an adequate mammalian data base existed (see criteria for selection in Smith *et al.*, 1983) and shows isoniazid as negative in *in vivo* teratology studies conducted in mice, rats, and rabbits. Other calculations, arrived at retrospectively from the published mammalian data base on isoniazid, give an *A/D* ratio of 46 and thus confirm that the Hydra assay had correctly recapitulated the drug as an agent hazardous to embryonic development (Johnson and Gabel, 1983). Concern about the reliability of Hydra *A/D* ratios compared with those calculated from an existing mammalian data base was also expressed by Brown (1987). Therefore, the critical questions are: Are published mammalian *A/D* ratios reliable? Are they uniform across species for comparable test conditions? Can they be extrapolated across phyla?

In reviewing reports of maternal and fetal toxic dose–response curves, Rogers (1987) detected a potential for considerable error when using such data and proposed systematic studies designed to provide more precise determinations of mammalian *A/D* ratios, as did Schwetz and Moorman (1987). A comparison of *A/D* ratios with the fungicide dinocap in mice, rats, and hamsters revealed striking discrepancies (Rogers *et al.*, 1986; Rogers, 1987). A survey of 28 two-species comparisons revealed *A/D* ratio differences in two studies (Schardein, 1987). Such discrepancies between mammalian species undermine confidence that extrapolation across phyla is valid, and cast doubt on the claims that in most instances the Hydra *A/D* ratio has recapitulated the mammalian one.

Last, overt maternal toxicity may be too crude to provide a useful criterion, as studies with diflunisal (Clark *et al.*, 1984) and acetazolamide (Ellison and Maren, 1972; Hirsch and Scott, 1983; Weaver and Scott, 1984) suggest. Adverse effects on maternal physiological homeostasis which would remain undetected in conventional studies may be related to developmental toxicity.

6.2. Culture of Mammalian Cells

6.2.1. Primary Cell Cultures

6.2.1a. Avian Embryo Limb Mesenchyme Cells. Avian limbs are a favorite system for studying the normal development of the extremities, including cell migration, differentiation, interaction, and pattern formation. These events are very similar in avian and mammalian limb development (Kelley *et al.*, 1984) and, further, it is well established that dissociated avian and mammalian limb bud cells behave similarly *in vitro* and *in vivo*.

Limb bud mesenchyme cells differentiate into chondrocytes which produce extracellular cartilaginous matrix (sulfated proteoglycans) when grown at cell densities greater than confluency (for reviews see Solursh, 1983, 1984, 1986). This transformation was used to test chemicals for "teratogenic activity." Among 14 compounds tested, some were positive for both teratogenicity *in vivo* and alterations in growth and differentiation of the cultured cells (Wilk *et al.*, 1980). The assay was presumably able to differentiate between teratogens that specifically alter cell differentiation and those inhibiting growth processes nonspecifically. The test seemed compatible with a reconstituted metabolic conversion system because cyclophosphamide became activated (Wilk *et al.*, 1980). A short report presented at the 1981 consensus workshop (Hassel and Horigan, 1982) indicated that limb bud cells from chick and mouse embryos were equally suitable for the test.

6.2.1b. Mouse Embryo Limb Mesenchyme Cells. In a modified mouse limb bud cell culture test system (Guntakatta *et al.*, 1983), cells are grown at high density (micromass) in the presence of two isotopes: $^{35}SO_4$ is incorporated into sulfated proteoglycans and [^{3}H]thymidine labels DNA. The former relates to reduction of cartilage proteoglycan matrix induced by specific disruption of matrix protein and glycoprotein synthesis, and the latter assesses chemical-induced cytostasis. Thus, this provides the ratio of $^{3}H/^{35}S$; a ratio >1.0 indicates differential inhibition of ^{35}S incorporation and one <1.0, reduction of [^{3}H]thymidine incorporation into DNA.

Preliminary data appeared to indicate that the test can detect selective effects on sulfated proteoglycan synthesis (Guntakatta *et al.*, 1984). Although "developmental hazard" was not mentioned, the assay may be able to provide this information when one compares the concentrations of a chemical that affect the $^{3}H/^{35}S$ ratio. The test is noted here in spite of lack of further publications, because limb cells are being explored in the pharmaceutical industry as a prescreen for teratogenic hazard potential (see Sections 6.2.1c and 6.2.1e).

6.2.1c. Rat Embryo Limb Mesenchyme Cells in Conjunction with Rat Embryo Midbrain Cells. Rat embryo mesencephalic cells differentiate in high-density culture into neurons, which develop specific transport capacity for the neurotransmitter γ-aminobutyric acid (GABA; Flint, 1983). This phenomenon in conjunction with differentiation of rat limb mesenchyme cells into chondrocytes was used as an assay system combining *in vivo* treatment of pregnant rats with teratogens and nonteratogens (classified according to the secondary literature) and explantation of embryonic brain and limb cells from these animals into cell culture (Flint *et al.*, 1984). Preliminary data regarding foci of differentiated cells and incorporation of radioactive GABA into the CNS or $^{35}SO_4$ into sulfated proteoglycans synthesized by differentiating chondrocytes suggested that the test was suitable as a short-term *in vitro* assay.

In later studies, limb and CNS cells responded just as well when only exposed to chemicals *in vitro,* with maternal drug metabolism (Flint *et al.*, 1984) substituted for by a liver subfraction with cofactors added to cell cultures (S9 mix; Flint and Orton, 1984). Morphological endpoints were monitored by counting the number of stained foci of differentiated cells in both types of cultures with an automated image analyzer. It was said that additional studies with monoclonal antibodies against products synthesized by differentiated cells together with simple measures of cell death allowed conclusions about the unique hazard potential of a compound (Girling and Flint, 1984). Our own unpublished observations warrant a note of caution: simple, commonly used methods of detecting cytotoxicity such as dye exclusion of trypan blue or neutral red uptake by surviving cells should be viewed with suspicion. We found that [^{3}H]leucine incorporation by V79 Chinese hamster cells was a much more sensitive indicator of chemical-induced cytotoxicity than trypan blue. Furthermore, cytofluorometry revealed that DNA synthesis was rapidly blocked by 6-thioguanine, yet the cells remained impermeable to trypan blue. Thus, it is conceivable that the cells that Flint *et al.* (1984) called "surviving cells" might reveal cytotoxicity with a more sensitive method. The original assay and the rationale for using [^{3}H]GABA and $^{35}SO_4$ which was similar to the double isotope protocol of Guntakatta *et al.* (1984) were changed (Flint *et al.*, 1984). It seems that the test, when used with two isotopes, has the potential to assess hazard associated with a chemical. However, no systematic work has been published and there is no convincing evidence that hazard can be detected by the assay.

The practical application of the simplified limb cell assay without isotopes was demonstrated in structure–activity relationship tests with triazole antifungals (Flint and Boyle, 1985). The antifungal activity and developmental toxicity of some of the triazoles were determined in several species before the cell differentiation assay was performed. The *in vitro* test ranked 16 analogues according to their teratogenic hazard; for the three compounds also examined in pregnant animals, there was good correlation between the *in vivo* and *in vitro* outcomes. It was concluded that it should "be possible to use the *in vitro* test with some confidence in predicting the relative teratogenic hazard of new triazole antifungals."

6.2.1d. Assessment of Rat Embryo Limb Cells Alone and in Combination with CNS Cells. The limb cell test has been used to make product development decisions (Flint, 1986). Metabolism as a prerequisite for activity in these cells has been studied with cyclophosphamide and diphenylhydantoin (Brown *et al.*, 1986b), as well as with synthetic retinoids and ethylene glycol ethers (Flint and Brown, 1987). Although useful immunocytochemical data concerning inducible isozymes of cytochrome P-448 in cultured midbrain and limb cells were obtained, metabolic activation requires further investigation. One of the puzzling *in vitro* observations is that limb cells from embryos of female rats administered β-naphthoflavone (an inducer of cytochrome P-448) converted cyclophosphamide to a product toxic to the limb cells (Flint and Brown, 1987), whereas the metabolic activation of that drug is commonly attributed to cytochrome P-450 and its isozymes (Mirkes, 1987). It seems unresolved what the drug metabolizing potential of cultured rat embryo midbrain and limb cells is. Regarding the effects of retinoids in the limb cell test, there were striking discrepancies when the developmental (teratogenic) hazard of vitamin A (Flint and Brown, 1987) was compared with that determined in the Hydra assay (Johnson and Gabel, 1983). Limb cells indicated high hazard (Flint and Brown, 1987), while in

Hydra vitamin A had an *A/D* ratio of 2.0, not indicative of developmental hazard (Johnson and Gabel, 1983). Experience from human medicine demonstrates that vitamin A analogues cause major congenital malformations (Rosa *et al.*, 1986).

6.2.1e. Rat Limb Bud Cells Alone. High-density cultures of rat limb mesenchyme cells have been used to study effects of retinoic acid on chondrogenesis (Gallandre *et al.*, 1980). Several retinoids were tested *in vitro* and their effects compared with the developmental toxicity caused *in vivo,* to examine the usefulness of the test as a measure of teratogenic potential. Quantitative spectrophotometric data of alcian blue extracted from stained cartilage proteoglycans were well correlated *in vivo* and *in vitro,* provided that the tested retinoid had a free carboxyl end-group. But lack of chemical metabolic capacity was a serious drawback, and in evaluating the assay Kistler (1985) concluded that "the usefulness of limb bud cell cultures for the prediction of teratogenic potential of compounds with no experience of their *in vivo* teratogenicity is questionable."

The test has been further explored focusing exclusively on a series of retinoids (Kistler, 1987). The starting premise followed a suggestion of Neubert's (1985b) for rational test validation: evaluate a series of chemicals of unknown biological activity simultaneously *in vivo* and *in vitro.* This was done in pilot experiments with retinoids that had not been previously screened *in vivo.* For compounds with a free carboxylic end-group there were good quantitative correlations between *in vivo* and *in vitro* results, but the serious drawback of lack of metabolic capacity of the limb cells was again revealed. Addition of drug-metabolizing enzymes present in S9 fractions was considered but was thought to raise the possibility that unphysiological metabolites might be generated (Kistler, 1987), a concern previously voiced (Neubert *et al.*, 1985; Neubert, 1987). Results obtained with the retinoids illustrate that the problem of "metabolic activation" in limb cell cultures (as well as in all other *in vitro* systems for that matter) is unresolved and will require much more systematic investigation.

It is apparent that two scientists associated with pharmaceutical companies—Flint then at ICI in the United Kingdom and Kistler at Hoffmann–LaRoche in Switzerland—have different outlooks on the applicability of the limb cell assay method. It may be that this reflects observations on national character made by Homburger (1985), who stated about his native Switzerland: ". . . the Swiss are slow and cautious in accepting change, preferring evolution to revolution not only in politics but in science as well. Their first reaction when faced with innovation tends to be 'it cannot be done.'"

6.2.2. Established Cell Lines

6.2.2a. Mouse Ovarian Tumor Cells. The belief that cell–cell interactions are important in development and that chemicals interfering with such processes, regardless of the interference mechanism, are potential teratogens, were the rationale for using a simple *in vitro* cell-to-surface recognition test as a screen (Braun *et al.*, 1979). The ability of "55 drugs with known teratogenic properties," selected almost exclusively from one secondary literature source, to alter the attachment of [^{3}H]thymidine-labeled mouse ovarian tumor (MOT) cells to a polyethylene surface coated with concanavalin A was the basis of the test. The assay looked promising since known teratogens inhibited cell attachment, while chemicals with no teratogenic properties failed to do so. The number of chemicals to be validated was then increased to 102 (selected from the same literature source), 74 of which were designated by the investigators as "teratogenic" and 28 "nonteratogenic" (Braun *et al.*, 1982a). The system was said to be 79% accurate, with 19% false negatives and 25% false positives. Nearly linear relations between cell attachment-inhibiting concentrations and the lowest reported teratogenic dose were found, with a correlation between *in vivo* and *in vitro* of 0.69. It was thus proposed that "attachment inhibition may be a useful member of a battery of simple and rapid *in vitro* assay systems" for detecting environmental teratogens (Braun *et al.*, 1982b).

Preliminary data of another assay (Pratt *et al.*, 1982) involving proliferation of an established line of human embryonic cells (see Section 6.2.2b for details) were also presented at the 1981 consensus workshop. This test in conjunction with the above-cited tumor cell attachment assay was concurrently validated in two laboratories independent of those in which the methods were developed. The project (DHHS, 1983), sponsored by the National Toxicology Program (NTP), began in October 1983 and was completed 3 years later (NTP, 1986; see also Section 6.2.2b).

6.2.2b. Human Embryonic Palatal Mesenchyme (HEPM) Cells. The HEPM cell line was established from a single human embryo's palatal shelves just prior to palatal closure, and presumably consists of undifferentiated fibroblastlike diploid cells with a stable chromosomal complement, high plating efficiency (>90%), and a cell cycle transit time of about 22 h (Pratt *et al.*, 1982; Welsch *et al.*, 1986). Preliminary testing focused on 13 of 102 "known teratogens" (Pratt *et al.*, 1982) that were false negatives in the MOT cell assay (Braun *et al.*, 1982a,b). The effect of these chemicals on proliferation of HEPM cells over 72 h (i.e., for about three complete cell cycles) was determined by simply counting the number of cells, and 12 of the 13 were positive. When each of the two systems was used by itself and the results were combined, there was 90% reliability with few false negatives (Pratt *et al.*, 1982). Further compounds (Pratt and Willis, 1985) were selected from the list of chemicals that had been tested in the MOT assay (Braun *et al.*, 1982b) until a total of 55 chemicals were assayed, of which 35 were designated "teratogenic" and 20 "non-teratogenic" (Pratt and Willis, 1985). Earlier conclusions (Pratt *et al.*, 1982) were corroborated, and what was called "predictability" of the two assays was 90% with a false negative rate of 3%.

Emphasis in the HEPM and MOT cell assays was placed on the hypnotic/sedative thalidomide. Investigators, in evaluating other *in vitro* teratogenesis prescreens, seemed to view this drug as the ultimate test for validation of their proposed system and made great efforts to have thalidomide come out positive (e.g., Flint and Orton, 1984; Flint and Brown, 1987), using unrealistic concentrations to obtain the desired result (Newall and Tesh, 1986). Thalidomide did not alter HEPM cell proliferation (Pratt and Willis, 1985), but did inhibit MOT cell attachment after modification with liver microsomes (Braun and Dailey, 1981; Braun and Weinreb, 1984, 1985). However, there is no unequivocal evidence, despite 25 years of effort, that enzyme-catalyzed metabolism of thalidomide is the source of its embryotoxic/teratogenic derivatives (see Chapter 16 in Welsch, 1987).

Another comment regarding thalidomide seems indicated, but is not meant to detract from the fact that thalidomide revolutionized our thinking about vulnerability of the human conceptus to chemical insult. Much of the teratology literature creates the impression that, until the teratogenic effects in human babies were established beyond a doubt in 1961, there was no indication whatever that thalidomide had any side effects in man. When first marketed, the drug, with its apparently high therapeutic ratio compared with barbiturates, was hailed as a breakthrough among the sedatives/hypnotics. However, the historical facts indicate that not long after thalidomide became commonly available in 1959 and was soon taken by many people of both sexes for its much appreciated soporific properties without morning hangovers, side effects became apparent. Some patients, using thalidomide habitually for 6 months or more, developed a peripheral neuropathy (Garfield, 1986).
The first extensive validation of developmental toxicity prescreens sponsored by the NTP involved the HEPM and MOT cell assays and was conducted in parallel in two laboratories with 44 of the 47 compounds suggested for validation (Smith *et al.*, 1983). The concordance with *in vivo* data turned out to be much lower than the >90% level that emerged when the HEPM assay focused on the false-negative chemicals of the MOT cell

MOT cell test (NTP, 1986). This outcome may be attributable in part to the fact that a good number of the chemicals selected by the NTP for assay validations were different from those chosen by Braun *et al.* (1979, 1982a,b) for validation of the MOT test and by Pratt and Willis (1985) when testing the MOT false-negative compounds in the HEPM cell proliferation assay.

6.3. Intact Lower Animals

6.3.1. Planaria

The planarian flatworm *Dugesia dorotocephala* has been suggested as a screening test of putative mammalian teratogens, including that of "behavioral teratogens," by exposing surgical fragments undergoing regeneration and intact flatworms to the test chemical and examining the morphological effects (Best and Morita, 1982). In the only other known study using this assay, four chemicals were tested in Hydra, frog embryos (see Section 6.3.3), and flatworms and a teratogenic index (TI_{50}) calculated. The values ranged from 0.8 to 1.2 in flatworms, while the other two species had considerable differences in the magnitude of the TI_{50}, suggesting that Dugesia was not a suitable discriminator worthy of further prescreening consideration (Sabourin *et al.*, 1985).

6.3.2. Frog Embryo Teratogenesis Assay: Xenopus (FETAX)

The South African clawed toad *Xenopus laevis* has been studied by developmental biologists for decades. The suggested use of Xenopus embryos as test objects for teratogenic potential came from water pollution and complex mixture testing needs (Dumont *et al.*, 1983; Dawson *et al.*, 1985). The only publications have been an abstract on the "validation" with 40 compounds (34 known mammalian teratogens and 6 compounds designated as nonteratogenic) (Dumont and Epler, 1984) and other tests with a few chemicals (Courchesne and Bantle, 1985; Sabourin *et al.*, 1985; Sabourin and Faulk, 1987; Dawson and Bantle, 1987). FETAX follows for 96-h frog development from the blastula stage through hatching to swimming tadpoles under control or chemical exposure conditions, using endpoints of function (swimming behavior) and morphology (pigmentation, length, general appearance) (Sabourin *et al.*, 1985). Positive or negative responses are classified on the basis of a teratogenic index (TI; Dumont *et al.*, 1983; Dumont and Eppler, 1984) derived by dividing the concentration of a chemical that causes 50%

abnormal surviving tadpoles (EC_{50}) by the concentration that kills 50% of the embryos (LC_{50}). The data are interpreted to indicate that "compounds whose TI's are 1.0 or less are considered embryotoxic or coeffective teratogens. Teratogenic compounds have TI's greater than 1.0" (Dumont and Eppler, 1984).

Since the teratogenic "index used for Xenopus was derived by comparing embryolethality with embryo malformation" (Sabourin *et al.*, 1985), and thus toxicity for adult Xenopus was not assessed, the index obviously differs from that used for Dugesia and Hydra, which compared toxicities of the chemical to regenerates and adults. But a teratogenic index derived solely from embryos has no relevance to the concepts of the *A/D* ratio and developmental hazard.

Other deficiencies of the Xenopus test come to mind, e.g., the assay is terminated after 96 h and does not monitor developmental events that occur later such as limb morphogenesis; and it is not clear how the system accommodates chemicals with poor water solubility or those whose developmental toxicity may depend on metabolism by cytochromes P-450 and P-448. Ongoing testing of compounds from the list suggested for validation (Smith *et al.*, 1983) was mentioned (Sabourin *et al.*, 1985), but no further data have been published.

6.3.3. Intact *D. melanogaster* Embryos

The approach using intact Drosophila embryos (Schuler *et al.*, 1982) is quite different from the Drosophila embryo cell test described above. Females deposit eggs on a nutrient medium containing a defined concentration of test chemical, and throughout metamorphosis from egg to pupa the developing fly feeds on this medium. Adult flies start hatching after 9 to 10 d, and flies emerging during the next 10 d are scored for morphological anomalies. The preliminary (Schuler *et al.*, 1982) and later reports (Schuler *et al.*, 1985) noted numerous abnormalities in flies hatched from chemical-containing nutrient medium, but the system still has no established criteria for labeling a result positive or negative. The subject of developmental hazard was mentioned, and two approaches were offered. One is a ratio of the concentration of chemical in the nutrient medium that causes malformations divided by that eliciting embryolethality, and the second relates the adult LD_{50} to embryotoxic levels, and hence appears more promising in view of the comments made above regarding Xenopus.

The concentration of chemical to which eggs are exposed appears poorly defined. The amount added to the nutrient medium ("applied concentration" or "administered dose") is difficult to relate to the actual embryo exposure concentration. Fly eggs, larvae, and pupae ingest ingredients from the culture medium ("delivered dose"; Starr and Gibson, 1984) over 10 and more d after eggs are deposited on the medium, but that is no measure of how much reaches the developing tissues.

Evaluation of this test with compounds on the list suggested for validation (Smith *et al.*, 1983) in a single laboratory independent of that of the original proponents has been completed under NTP sponsorship (Morrissey, personal communication). The results should become available in 1989, and judgment of the potential merits of the assay has to be reserved until then.

6.3.4. Chick Embryotoxicity Screening Text (CHEST)

Strictly speaking the chicken egg is not an *in vitro* system, but assessment of this test is included because chick embryos have been persistently advocated for developmental toxicity screening purposes. When concern over environmental pollution rose in the 1970s, Wilson (1978b) suggested reconsideration of the rejection of the use of chick embryos by a World Health Organization Study Group (Technical Report 364, 1967; cited in Wilson, 1978a). In the decade since, only one group has explored this possibility and refined a two-phase "chick embryotoxicity screening test" (CHEST) (Jelinek, 1982). The CHEST is a truncated method compared with classical morphological evaluation made to facilitate scoring large numbers. A data base on 130 very diverse chemicals was presented as the basis for considerations regarding the usefulness of CHEST as a developmental toxicity prescreening method (Jelinek *et al.*, 1985). From the assay description it appears that the evaluation for testing purposes requires considerable experience in chick embryo morphology and large numbers of embryos with no prospects for automated scoring.

In spite of Wilson's (1978a,b) encouragement and numerous publications from Jelinek's group, use of the chick embryo for prescreening has aroused no interest, as judged by citation analysis. It was said that the CHEST can disclose the embryotoxic potential of any soluble substance. However, the CHEST data review covering 130 chemicals reveals that 117 induced dose-dependent embryotoxic effects (Jelinek *et al.*, 1985). This high percentage seems to reflect the concerns of the WHO 1967 panel (Wilson, 1978a) that the chick embryo is oversensitive to developmental perturbation by chemicals. The CHEST does not provide an estimate of developmental hazard as defined by Johnson (1980, 1981) and apparently would yield too many positive responses which would lead to unnecessary *in vivo* testing.

6.4. Whole Embryo Culture (WEC) of Rodent Embryos

WEC is the most frequently used *in vitro* prenatal toxicology system. Mouse or rat postimplantation embryos are cultured for 24 or 48 h during early organogenesis when yolk sac placenta function is most important. The method has some attractive features (Table I). Those listed in the Advantages column explain why WEC has become so popular for studies on the mechanism of action of chemical teratogens. Preliminary data suggested that WEC might have utility in teratogen screening (Fantel, 1982; Sadler *et al.*, 1982), but the wide consensus at present is that it is best suited for basic studies of teratogenesis (Fantel, 1982; Shepard *et al.*, 1983; Warner *et al.*, 1984; Lewandowski *et al.*, 1986; Steele and Marlow, 1986; Juchau *et al.*, 1987; Mirkes, 1987). Investigators who were once optimistic about the prospects of WEC as a screening assay (Sadler *et al.*, 1982; Sadler and Warner, 1984; Sadler, 1985) have shifted the emphasis of their studies to embryotoxic mechanisms (Horton and Sadler, 1983, 1985; Sadler and Horton, 1983; Freeman and Steele, 1986; Hunter *et al.*, 1987). There are others, however, who maintain that WEC is applicable to screening potential teratogens (Kitchin *et al.*, 1986; Schmid and Cicurel, 1986, 1987). Some reports from these proponents emphasized the mode of action of specific compounds or classes of therapeutic agents (e.g., Schmid, 1984; Schmid *et al.*, 1985), while others supposedly demonstrated the screening ability of WEC for teratogenic activity (Schmid, 1985a,b; Bechter *et al.*, 1986; Kitchin *et al.*, 1986; Schmid and Cicurel, 1986, 1987; Schmid *et al.*, 1987; Bechter and Schmid, 1987).

Problems of various sorts have been encountered in the use of WEC as a prescreen. In WEC cyclophosphamide was embryotoxic only when exogenous enzymes were added; but procarbazine, an *in vivo* rodent teratogen, was not embryotoxic even when the medium was supplemented with drug-metabolizing cell fractions, yet dysmorphogenesis was induced when embryos were cultured in undiluted serum obtained from procarbazine-treated male rats (Schmid *et al.*, 1982). Apparently stable metabolites of procarbazine are formed only within body compartments (Schmid, 1984, 1985b). This failure again raises the question about subcellular fractions as providers of potential for chemical metabolism in any *in vitro* test. The example suggests that one cannot trust a negative WEC test outcome even with metabolic activation. How is one to know that a chemical of unknown embryotoxic activity does not belong in the procarbazine category? Concurrent *in vivo* dosing would have to be conducted to obtain serum from other rats (preferably pregnant!) to account for stable embryotoxic metabolites whose biosynthesis occurs only within body compartments. Labile, reactive products would most likely escape detection.

Recent studies with valproic acid have shown that pharmacokinetic factors cannot be neglected when WEC embryotoxicity data are interpreted. Thus, one cannot assume that the concentration of a chemical in the embryonic tissues is equal to that added to the culture medium (Lewandowski *et al.*, 1986). The comments regarding administered versus delivered dose (see Section 6.3.3) seem equally applicable. Such confounding factors lessen WEC as an ideal prescreen. Furthermore, substantial numbers of animals as serum donors are needed in addition to pregnant animals as well as special equipment and skills, making WEC fairly expensive. Finally, however, it is unlikely that WEC can prescreen for the developmental hazard of chemicals of totally unknown embryotoxic potential. In a critical status report, Brown (1985), who has many years of personal experience with WEC, judged the method unsuitable for screening large numbers of chemicals.

Nevertheless, WEC may still have some uses. It may be possible to screen for embryotoxic potential of close structural analogues when data about *in vivo* developmental toxicity of closely related compounds exist. Four antimycotic drugs with considerable differences in chemical structure were tested in WEC, and *in vitro* effects reflected the known *in vivo* activity (Bechter and Schmid, 1987). Under suitable conditions, poorly water-soluble chemicals may be tested, and drug metabolism as a prerequisite for embryotoxicity can be routinely addressed by supplementation with cell fractions (Kitchin *et al.*, 1986). Concentration–response data obtained with some compounds may distinguish between embryolethal and teratogenic effects (Schmid, 1984), i.e., between specific teratogenic insults and general toxic effects (Schmid, 1985b).

6.5. Short-Term Tests with Pregnant Laboratory Animals

An entirely different approach has been proposed for which, though it is not an *in vitro* system as defined above, many of the questions regarding validation of *in vitro* assays are applicable. The purpose of this abbreviated *in vivo* assay was to circumvent the disadvantages of all *in vitro* prescreens, particularly maternal–conceptus interactions and delivery of water-insoluble compounds. The rationale for the test was that most prenatal chemical insults might express themselves postnatally as reduced viability and/or impaired growth of the offspring (Chernoff and Kavlock, 1982).

The minimal toxic dose (determined by body weight loss, mortality, or other signs of toxicity) of 28 compounds with known developmental toxicity was first established in nonpregnant female mice and then given to pregnant ones during specific gestation times. The females were allowed to give birth, and litter size, body weights, and morphology of all offspring on postnatal day 1 as well as neonatal survival to and body weight on day 3 were recorded when the test was terminated. The 15 chemicals labeled teratogenic by conventional screens revealed some form of toxicity on the endpoints named, and for some of the other 13 the outcome was positive: four chemicals known to produce only fetal toxicity reduced offspring weight, while among nine compounds showing no effects *in vivo,* six were negative and three reduced either offspring weight or viability. Further testing of the same and additional compounds (total of 41) with evaluations at up to 250 days of age revealed that the extended test gave additional indications of prenatally induced toxicity not apparent on postnatal day 3 (Gray and Kavlock, 1984).

This procedure, commonly known as the "Chernoff/Kavlock Test for Developmental Toxicity" and more recently also called "a preliminary *in vivo* study for developmental toxicity," has attracted much attention and its usefulness has been variously explored (Seidenberg *et al.*, 1986; Kavlock *et al.*, 1987; Kimmel, 1987). Results with 15 glycol ethers suggested that the test could be helpful in establishing priorities for further conventional testing (Schuler *et al.*, 1984). Numerous other chemicals were tested and the results reported at a workshop (Hardin, 1987a,b; Hardin *et al.*, 1987; Seidenberg and Becker, 1987). An indirect assessment of the assay by retrospective analysis of dose-range findings which preceded *in vivo* developmental toxicity screening of about 600 conventional teratology studies (Palmer, 1987) strongly supported the rationale of the assay (Chernoff and Kavlock, 1982).

This *in vivo* prescreen has received much interest and further validations of the test are being sponsored (*Commerce Business Daily*, 1987). Recent modifications (Johnson, 1985) have made the test more useful and certain inadequacies for industrial application (Christian *et al.*, 1987) have been overcome. The Environmental Protection Agency has adapted the test to rats (Smith *et al.*, 1987), and the largest industrial toxicology laboratory in Europe has modified the test for use with rats and is quite satisfied with the information obtainable (Wickramaratne, 1986, 1987).

7. ASSESSMENT OF PROPOSED *IN VITRO* TESTS

7.1. Developmental Hazard Potential and *A*/*D* Ratio

The workshop on maternal toxicity (Kimmel *et al.*, 1987) and the various prescreening tests reviewed here have emphasized the increasing awareness of selective embryotoxicity and developmental hazard and the possible significance to them of the *A*/*D* ratio (see Section 4.2). As also noted by Brown (1987), the hazard aspect is relevant to the selection by prescreening of environmental chemicals for further *in vivo* testing. Although some of the proposed *in vitro* test systems may have the potential of providing a hazard index, others clearly do not (e.g., MOT and HEPM cells). *A/D* ratios calculated from published *in vivo* studies are apparently unreliable and uncertainty surrounds the extrapolation of these ratios to untested species and phyla. Therefore, it is not possible to decide which of the prescreening tests claiming successful recapitulation of established mammalian *A/D* ratios or ability to detect selective effects on regenerative or developmental processes in a given test can provide truly trustworthy data.

7.2. Physicochemical Properties of Test Chemicals

Most of the tests discussed above have had to grapple with the problem of chemicals having low water solubility, which potentially limits exposure concentrations. The solubility of compounds in a tissue or cell-compatible vehicle is usually addressed by cytotoxicity studies with the vehicle alone. But this does not provide assurance that the chemical remains in solution once the vehicle solvent has been delivered into the aqueous culture medium and does not account for the chemical stability of a compound at about neutral pH. A specified concentration of the chemical is added to the culture medium and is then presumed to remain at that level throughout the exposure period, an assumption that has seldom been confirmed. *In vitro* chemicals may undergo spontaneous or enzyme-catalyzed transformation to more or less active derivatives with entirely unknown half-lives. If the substance remained in its original chemical form, then the static nature of *in vitro* systems would raise the question of relevance of exposure length to adverse effects that may result. Knowledge of relevant *in vivo* exposure conditions has been the basis for the experimental design in some WEC studies which correlated *in vivo* and *in vitro* drug concentrations and exposure duration (Warner *et al.*, 1983; Schmid *et al.*, 1987).

7.3. Chemical Exposure and the Role of Metabolism

In vitro assays deliver a chemical to the tissues and cells in a way that is often irrelevant to the expected route of human exposure. However, this shortcoming does not preclude their usefulness for an assessment of developmental hazard potential. A more serious problem seems to be that of parent compound versus metabolite activity. Lack of metabolic capacity may prevent the generation of embryotoxic derivatives or the limited metabolic potential of added reconstituted subcellular fractions may give rise to unphysiologically persistent metabolites. Such artifacts could arise *in vitro* because it is possible that the product(s) generated cannot be converted further and detoxified *in vitro*. Often, reconstituted microsomal fractions are toxic to the cells of *in vitro* developmental toxicology systems, and the chemical-metabolizing enzymes remain active for a very limited time only (Wilk *et al.*, 1980; Kitchin *et al.*, 1986). Attempts to overcome this drawback were made in WEC by coculturing embryos with intact hepatocytes (Oglesby *et al.*, 1986; Kitchin *et al.*, 1986) or by adding purified cytochrome P-450 to organ cultures (Kastner *et al.*, 1987). The issue of toxification and detoxification *in vitro* in developmental toxicology is controversial and requires further investigation.

8. COMMENTS ON VALIDATION

8.1. Selection of Chemicals

A recent review of validation studies (Brown, 1987) found that the selection of chemicals seemed biased and included many substances lacking the adequate mammalian data base that is the most important consideration in the choice of chemicals for validation. Too few chemicals that unequivocally lack developmental toxicity at realistic exposure levels when given to pregnant laboratory animals were tested in all *in vitro* systems surveyed.

Only one list of chemicals is prefaced by specific comments about the selection process (Smith *et al.*, 1983). This compilation of 47 compounds is heavily weighted toward pharmaceutical chemicals (drugs) and, although based on the outcome of *in vivo* teratology studies, the selection criteria actually describe chemicals that induce developmental toxicity presumably in the absence of maternal toxicity. However, Brown (1987) cautioned that many of the studies analyzed in composing the list had flaws that seemingly violate the selection criteria. Concern about the compounds on the list was also voiced by Johnson (1985) who had labeled the effort counterproductive and inaccurate because of discrepancies between the inclusion criteria and the data in the original publications as regards selective embryotoxicity and developmental hazard (Section 4.2).

It seems that few published studies are able to pass the stringent standards applied now, with our heightened awareness about manifestations of maternal toxicity and its inadequate documentation. One must wonder whether it is possible at all to compile retrospectively a list of compounds with an adequate mammalian data base that is balanced between chemicals exerting selective embryotoxicity and those unequivocally lacking it. The compounds included must be acceptable to a panel of experienced investigators who agree that detection of developmental hazard is the aim of all prescreening methods. However, the shortcomings of the existing list seem compatible with the summary that the selection was to be considered preliminary and open to revision (Smith *et al.*, 1983). In fact, developmental toxicologists were invited to submit documented suggestions for additions or deletions, and the time to reconsider this offer seems ripe.

8.2. Endpoint Selection and Designation of Test Outcome

There has been no progress in the debate surrounding endpoints of *in vitro* assays (Kimmel *et al.*, 1982; Kimmel, 1985). Endpoints do not have to predict the types of *in vivo* responses, just as sites of manifestation of developmental toxicity in *in vivo* test systems do not predict the human response. Some experience has now been gained with fairly simple assays (MOT cells, Section 6.2.2a; HEPM cells, Section 6.2.2b) with disappointing prospects for future general usefulness. On the other hand, the much more complex *in vivo* prescreening test seems to be gaining favor (Section 6.5). This development is not entirely surprising in light of the many ways by which chemicals can disrupt mammalian development (Fig. 1) and the shortcomings of test methods with reduced complexity (Neubert, 1981, 1985a,b).

To validate their test systems, most investigators use chemicals chosen by their own criteria and label the assay outcome "positive" or "negative." The designations "teratogen" and "nonteratogen" for the selected chemicals are commonly used without critical definition. Since the relation between chemicals that may cause congenital malformations in humans and those that induce developmental toxicity in laboratory animals and the predictive value of the latter are still far from clear, we do not know what the validity of animal studies is for the human situation. Therefore, it was recommended that a more appropriate term to denote the outcome of *in vitro* developmental toxicity prescreening tests would be "concordance" or "nonconcordance" with animal studies (NTP, 1986).

8.3. Proposed Changes in the Approach to Validation

The poorly coordinated efforts to validate many assays have not markedly advanced the field of "alternative" or "*in vitro*" tests for developmental hazard prescreening. To some extent the worst-case predic-

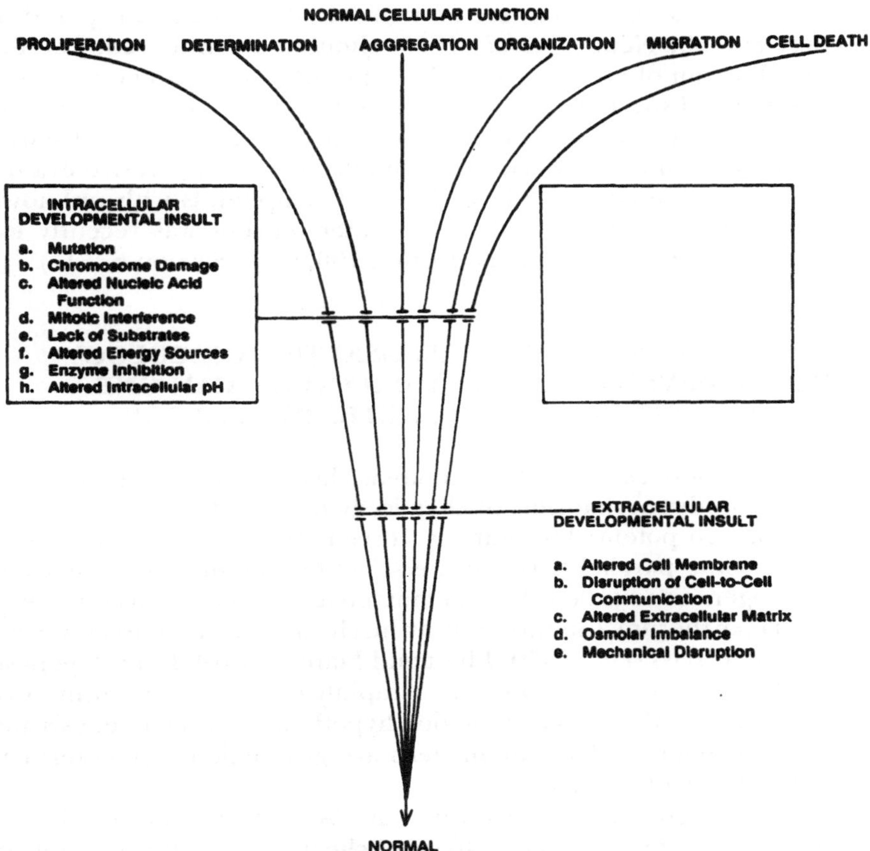

Figure 1. Mechanisms of teratogenesis. This scheme provides an overview of cellular processes that are recognized as essential for normal embryogenesis. Chemical insults capable of disrupting developmental events are differentiated by intra- and extracellular target sites and the resulting impact on cellular homeostasis. Modified from Brown and Fabro (1982).

tions of creating data that cannot be interpreted have come true (Neubert, 1982; Neubert *et al.*, 1985). We are nowhere near the point where any one of the *in vitro* tests has earned the trust of investigators (other than the test proponent) and where a specific test or a battery of tests could serve the desired prescreening goal. The "battery" approach was tried on a modest scale with disappointing results (NTP, 1986).

Since it may not be possible to compose a list of chemicals suitable for future validation efforts (see Section 8.1), the solution proposed by Neubert (1985b; Neubert *et al.*, 1985) should be considered. For unbiased validation of any new test, two approaches were suggested: (1) a series of pairs of substances should be used for which it is known from *in vivo* studies that one is clearly positive and the other clearly negative with respect to a given toxic effect; or (2) simultaneous comparative evaluation *in vivo* and *in vitro* of a large series of compounds with unknown developmental toxicity potential. The latter concept was recently explored with a series of retinoids (Kistler, 1987) (see Section 6.2.1e).

9. OUTLOOK FOR THE FUTURE: GENETIC TOXICOLOGY VERSUS DEVELOPMENTAL TOXICOLOGY. CAN ONE GLEAN ANYTHING FROM GENETIC TOXICOLOGY?

Short-term *in vitro* tests for genotoxic chemicals were originally developed to study mechanisms of chemically induced DNA damage and to assess hazard potential to humans. It seems that in their early history there was some parallel between the scientific rationale underlying *in vitro* teratogenesis tests and those in mutagenesis/carcinogenesis because in the former the aim also was to study mechanisms leading to abnormal development (Bass *et al.*, 1970; Ebert and Marois, 1976). In mutagenesis the emphasis on these tests increased rapidly in view of mounting evidence supporting the somatic mutation hypothesis of carcinogenesis and the fact that many rodent carcinogens are genotoxic in short-term *in vitro* tests (Tennant *et al.*, 1987).

There was great enthusiasm when the "Ames test," a bacterial assay system for detecting mutagenic effects of chemicals, became available in the early 1970s. Then the prevailing view became that carcinogens are mutagens (see review by Pool and Schmahl, 1987), and hope was therefore raised that this rapid and economic assay might be useful for determining carcinogenic potential. Intense validation efforts began in laboratories around the world, and a vast number of other short-term tests, many involving mammalian cells, were evaluated. Today more than 100 different tests are on record, all focusing on a single end: unmasking the potential of chemical carcinogens to induce a large variety of biological effects which result from the interaction of the compound or its reactive intermediates with the DNA of an indicator organism or biological material (Pool and Schmahl, 1987).

But the sobering realization has now set in that *in vitro* genotoxicity tests are of limited value for predicting carcinogenic potential, since it appears that the desire for simple solutions to complex problems has not been fulfilled (Flamm and Lorentzen, 1986/87). The recommendation is now that in the future these short-term tests should be increasingly used to understand the mechanism of action of chemical mutagens and carcinogens and to aid in the development of better scientific understanding of the human risk from chemical exposure (Flamm and Lorentzen, 1986/87; Pool and Schmahl, 1987). Opinions are divided over how to use the information accumulated since 1970. Some argue that the development and validation of short-term tests for application in screening new and existing chemicals for carcinogenic potential should be accelerated (e.g., Lave and Omenn, 1986), while others question that strategy (Tennant *et al.,* 1987). It is also proposed that a systematic test approach should combine *in vitro* with *in vivo* exposures of short duration and the use of human cells to elucidate specific mechanistic aspects of a compound's activity (e.g., Pool and Schmahl, 1987).

Compared with short-term tests in "genetic toxicology," the manpower involved in developmental toxicology prescreens is miniscule. The problems facing teratology testing are similar to those in genetic toxicology. Short-term tests in genetic toxicology have shown that the concordance with long-term rodent bioassays is inconsistent and that the tests cannot replace conventional bioassays (Tennant *et al.*, 1987). *In vivo* developmental toxicology testing shares with carcinogen bioassays interspecies divergences of response to the same chemical. Just as the complexity of mechanistic considerations in genetic toxicology has increased (e.g., genetic versus epigenetic) since the 1970s from a single mechanism of DNA alteration, the challenges to *in vitro* studies have also grown owing to the recognition of the multitude of mechanisms (Fig. 1). Results in mutagenesis/carcinogenesis have made it obvious that short-term *in vivo* tests with specific target tissues of tumorigenesis are necessary to take into account the metabolic complexity of the intact animal (Pool and Schmahl, 1987). Maybe the favorable reception of the short-term *in vivo* developmental toxicity test also reflects that appreciation.

Some *in vitro* developmental toxicity prescreens may turn out to be useful. Standardization of those methods that look promising in one laboratory is the first priority. The requirements are identification of developmental hazard and objective endpoints. Limited experience with the MOT and HEPM cell assays and Hydra already suggest that the next step, interlaboratory standardization, may not be a trivial task. If this phase can be mastered, then the next stage—validation—would be set. The crucial prerequisite for the latter is a critical reconsideration of the selection of chemicals. Much work remains to be done and we must be prepared to experience many failures before reaching a decision whether the objectives of *in vitro* prescreening can be reached.

REFERENCES

Bass, R., Beck, F., Merker, H.-J., Neubert, D., and Randhahn, B. 1970. *Metabolic Pathways in Mammalian Embryos During Organogenesis and Its Modification by Drugs.* Freie Universität, Berlin.

Beaudoin, A. R. 1985. Validity of in vitro methods in teratogenicity testing, in: *New Approaches in Toxicity Testing and Application in Human Risk Assessment,* A. P. Li, ed. Raven Press, New York, pp. 203–212.

Bechter, R., and Schmid, B. P. 1987. Teratogenicity in vitro: a comparative study of four antimycotic drugs using the whole-embryo culture system. *Toxicol. In Vitro* **1:**11–15.

Bechter, R., Schmid, B., and Mayer, F. K. 1986. Teratogenic potential of antimycotic drugs evaluated in the whole-embryo culture system. *Food Chem. Toxicol.* **24:**641–642.

Best, J. B., and Morita, M. 1982. Planarians as a model system for in vitro teratogenesis studies. *Teratog. Carcinog. Mutag.* **2:**277–291.

Bleyl, D. W. R. 1984. Grenzen und Möglichkeiten der in-vitro-Technik für pränatal toxikologische Untersuchungen. *Nahrung* **28:**1053–1063.

Bournias-Vardiabasis, N., and Buzin, C. H. 1987. Altered differentiation and induction of heat shock proteins in Drosophila embryonic cells associated with teratogen treatment, in: *Developmental Toxicology: Mechanisms and Risk, Banbury Report* 26, J. M. McLachlan, R. M. Pratt, and C. L. Markert, eds. Cold Spring Harbor Laboratory, Cold Spring Harbor, New York, pp. 3–16.

Bournias-Vardiabasis, N., and Flores, J. 1983. Drug metabolizing enzymes in *Drosophila melanogaster:* teratogenicity of cyclophosphamide in vitro. *Teratog. Carcinog. Mutag.* **3:** 255–262.

Bournias-Vardiabasis, N., and Flores, J. C. 1986. Response of Drosophila embryonic cells to tumor promoters. *Toxicol. Appl. Pharmacol.* **85:**196–206.

Bournias-Vardiabasis, N., and Teplitz, R. L. 1982. Use of Drosophila embryo cell cultures as an in vitro teratogen assay. *Teratog. Carcinog. Mutag.* **2:**333–341.

Bournias-Vardiabasis, N., Teplitz, R. L., Chernoff, G. F., and Seecof, R. L. 1983. Detection of teratogens in the Drosophila embryonic cell culture test: assay of 100 chemicals. *Teratology* **28:**109–122.

Braun, A. G., and Dailey, J. P. 1981. Thalidomide metabolite inhibits tumor cell attachment to concanavalin A-coated surfaces. *Biochem. Biophys. Res. Commun.* **98:**1029–1034.

Braun, A. G., and Weinreb, S. L. 1984. Teratogen metabolism: activation of thalidomide and thalidomide analogues to products that inhibit the attachment of cells to concanavalin A-coated plastic surfaces. *Biochem. Pharmacol.* **33:**1471–1477.

Braun, A. G., and Weinreb, S. L. 1985. Teratogen metabolism: spontaneous decay products of thalidomide analogues are not bioactivated by liver microsomes. *Teratog. Carcinog. Mutag.* **5:**149–158.

Braun, A. G., Emerson, D. J., and Nichinson, B. B. 1979. Teratogenic drugs inhibit tumour cell attachment to lectin-coated surfaces. *Nature* **282:**507–509.

Braun, A. G., Buckner, C. A., Emerson, D. J., and Nichinson, B. B. 1982a. Quantitative correspondence between the in vivo and in vitro activity of teratogenic agents. *Proc. Natl. Acad. Sci. USA* **79:**2056–2060.

Braun, A. G., Nichinson, B. B., and Horowicz, P. B. 1982b. Inhibition of tumor cell attachment to concanavalin A-coated surfaces as an assay for teratogenic agents: approaches to validation. *Teratog. Carcinog. Mutag.* **2:**343–354.

Brown, L. P., Flint, O. P., Orton, T. C., and Gibson, G. G. 1986a. Chemical teratogenesis: testing methods and the role of metabolism. *Drug Metab. Rev.* **17:**221–260.

Brown, L. P., Flint, O. P., Orton, T. C., and Gibson, G. G. 1986b. In vitro metabolism of teratogens by differentiating rat embryo cells. *Food Chem. Toxicol.* **24:**737–742.

Brown, N. A. 1985. *Alternative Tests for Teratogenicity of Petroleum Products.* American Petroleum Institute, Washington, D.C.

Brown, N. A. 1987. Teratogenicity testing in vitro: status of validation studies. *Arch. Toxicol.* Suppl. **11:**105–114.

Brown, N. A., and Fabro, S. E. 1982. The in vitro approach to teratogenicity testing, in: *Developmental Toxicology,* K. Snell, ed. Praeger, New York, pp. 31–57.

Brown, N. A., and Freeman, S. J. 1984. Alternative tests for teratogenicity. *Alternatives Lab. Anim.* **12:**7–23.

Buzin, C. H., and Bournias-Vardiabasis, N. 1984. Teratogens induce a subset of small heat shock proteins in Drosophila primary embryonic cell cultures. *Proc. Natl. Acad. Sci. USA* **81:**4075–4079.

Chernoff, N., and Kavlock, R. J. 1982. An in vivo teratology screen utilizing pregnant mice. *J. Toxicol. Environ. Health* **10:**541–550.

Chernoff, N., Kavlock, R. J., Beyer, P. E., and Miller, D. 1987. The potential relationship of maternal toxicity, general stress, and fetal outcome. *Teratog. Carcinog. Mutag.* **7:**241–253.

Christian, M. S., Hoberman, A. M., and Lochry, E. A. 1987. Currently used alternatives to the Chernoff–Kavlock short-term in vivo reproductive toxicity assay. *Teratog. Carcinog. Mutag.* **7:**65–71.

Clark, R. L., Robertson, T. R., Minsker, D. H., Cohen, S. M., Tocco, D. J., Allen, H. L., James, M. L., and Bokelman, D. L. 1984. Diflunisal-induced maternal anemia as a cause of teratogenicity in rabbits. *Teratology* **30:**319–332.

Collins, T. F. X. 1987. Teratological research using in vitro systems. V. Nonmammalian model systems. *Environ. Health Perspect.* **72:**237–249.

Commerce Business Daily. 30 November 1987. PSA-9475. Developmental Toxicity Testing and Research, NIH-ES-88-01.

Courchesne, C. L., and Bantle, J. A. 1985. Analysis of the activity of DNA, RNA, and protein synthesis inhibitors on Xenopus embryo development. *Teratog. Carcinog. Mutag.* **5:**177–193.

Daston, G. P., and D'Amato, R. A. 1989. In vitro techniques in teratology. *Toxicol. Indust. Health* **5:**555–585.

Daugherty, M. L., Ross, R. H., and Ryon, M. G. 1985. Scientific rationale for the selection of toxicity testing methods II. Teratology, immunotoxicology, and inhalation toxicology, in: M. J. Ryon and D. S. Sawhney, eds. U.S. Department of Energy, ORNL-6094, DE86-003507, pp. 1–67.

Dawson, D. A., and Bantle, J. A. 1987. Development of a reconstituted water medium and preliminary validation of the frog embryo teratogenesis assay—Xenopus (FETAX). *J. Appl. Toxicol.* **7:**237–244.

Dawson, D. A., McCormick, C. A., and Bantle, J. A. 1985. Detection of teratogenic substances in acidic mine water samples using the frog embryo teratogenesis assay—Xenopus (FETAX). *J. Appl. Toxicol.* **5:**234–244.

DHHS. 1983. *RFP No. NIH-ES-83-50018: Evaluation of two in vitro teratogenesis testing systems.*

DeSesso, J. M. 1987. Maternal factors in developmental toxicity. *Teratog. Carcinog. Mutag.* **7:** 225–240.

Dumont, J. N., and Epler, R. G. 1984. Validation studies on the FETAX teratogenesis assay (frog embryos). *Teratology* **29:**27A.

Dumont, J. N., Schultz, T. W., Buchanan, M. V., and Kao, G. L. 1983. Frog embryo teratogenesis assay Xenopus (FETAX)—a short-term assay application to complex environmental mixtures, in: *Short-Term Bioassays in the Analysis of Complex Environmental Mixtures III*, M. D. Waters, S. S. Sandhu, J. Lewtas, and L. Claxton, eds. Plenum Press, New York, pp. 393–405.

Ebert, J. D., and Marois, M. 1976. *Tests of Teratogenicity in vitro*. North-Holland, Amsterdam.

Ellison, A. C., and Maren, T. H. 1972. The effect of potassium metabolism on acetazolamide-induced teratogenesis. *Johns Hopkins Med. J.* **130:**105–114.

Fabro, S., Shull, G., and Brown, N. A. 1982. The relative teratogenic index and teratogenic potency: proposed components of developmental toxicity risk assessment. *Teratog. Carcinog. Mutag.* **2:**61–76.

Fantel, A. G. 1982. Culture of whole rodent embryos in teratogen screening. *Teratog. Carcinog. Mutag.* **2:**231–242.

Federal Register. 1987. Notice of first priority list of hazardous substances that will be the subject of toxicological profiles. *Department of Health and Human Services, Environmental Protection Agency* **52**(74)**:**12866–12874.

Flamm, W. G., and Lorentzen, R. J. 1986/87. The use of an in vitro method in safety evaluation. *In Vitro Toxicol.* **1:**1–4.

Flint, O. P. 1981. An assessment of the available in vitro techniques for detecting teratogens, in: *Culture Techniques: Applicability for Studies on Prenatal Differentiation and Toxicity*, D. Neubert and H.-J. Merker, eds. de Gruyter, Berlin, pp. 561–566.

Flint, O. P. 1983. A micromass culture method for rat embryonic neural cells. *J. Cell Sci.* **61:** 247–262.

Flint, O. P. 1986. An in vitro test for teratogens: its practical application. *Food Chem. Toxicol.* **24:**627–631.

Flint, O. P., and Boyle, F. T. 1985. An in vitro test for teratogens: its application in the selection of non-teratogenic triazole antifungals, in: *Concepts in Toxicology*, volume 3, F. Homburger, ed. Karger, Basel, pp. 29–35.

Flint, O. P., and Brown, L. P. 1987. Rat embryo limb cells in culture: ability to metabolize xenobiotics, in: *Approaches to Elucidate Mechanisms in Teratogenesis*, F. Welsch, ed. Hemisphere, Washington, D.C., pp. 185–196.

Flint, O. P., and Orton, T. C. 1984. An in vitro assay for teratogens with cultures of rat embryo midbrain and limb bud cells. *Toxicol. Appl. Pharmacol.* **76:**383–395.

Flint, O. P., Orton, T. C., and Ferguson, R. A. 1984. Differentiation of rat embryo cells in culture: response following acute maternal exposure to teratogens and non-teratogens. *J. Appl. Toxicol.* **4:**109–116.

Flynn, T. J. 1987. Teratological research using in vitro systems. I. Whole embryo culture. *Environ. Health Perspect.* **72:**203–210.

Freeman, S. J., and Steele, C. E. 1986. Post-implantation whole embryo culture and the study of teratogenesis. *Food Chem. Toxicol.* **24:**619–622.

Friedman, L. 1987. Teratological research using in vitro systems. II. Rodent limb bud culture system. *Environ. Health Perspect.* **72:**211–219.

Gallandre, F., Kistler, A., and Galli, B. 1980. Inhibition and reversion of chondrogenesis by retinoic acid in rat limb bud cell cultures. *Roux Arch. Dev. Biol.* **189:**25–33.

Garfield, E. 1986. Teratology literature and the thalidomide controversy. *Current Contents* **50:**3–11.

George, J. D., Price, C. J., Kimmel, C. A., and Marr, M. C. 1987. the developmental toxicity of triethylene glycol dimethyl ether in mice. *Fundam. Appl. Toxicol.* **9:**173–181.

Girling, L., and Flint, O. P. 1984. Inhibition of embryonic cell differentiation by teratogens in vitro: quantification using ELISA (enzyme linked immunosorbent assay). *Hum. Toxicol.* **3:**155–156.

Goldberg, A. M. 1985. Integration of fundamental knowledge and in vitro testing strategies, in: *Concepts in Toxicology,* Volume 3, F. Homburger, ed. Karger, Basel, pp. 1–5.

Goldberg, A. M. 1987. Guest editorial. *Toxicol. Lett.* **35:**VII–XI.

Gray, L. E., and Kavlock, R. J. 1984. An extended evaluation of an in vivo teratology screen utilizing postnatal growth and viability in the mouse. *Teratog. Carcinog. Mutag.* **4:**403–426.

Gross, L. B., and Sabourin, T. D. 1985. Utilization of alternative species for toxicity testing: an overview. *J. Appl. Toxicol.* **5:**193–219.

Guntakatta, M., Matthews, E. J., and Rundell, J. O. 1983. Development of a mouse embryo limb bud cell culture system for the estimation of chemical teratogenic potential. Technical Information Bulletin No. 5, Litton Bionetics, Kensington, Md.

Guntakatta, M., Matthews, E. J., and Rundell, J. O. 1984. Development of a mouse embryo limb bud cell culture for the estimation of chemical teratogenic potential. *Teratog. Carcinog. Mutag.* **4:**349–364.

Hardin, B. D. 1987a. Evaluation of the Chernoff/Kavlock test for developmental toxicity. *Teratog. Carcinog. Mutag.* **7:**7–127.

Hardin, B. D. 1987b. A recommended protocol for the Chernoff/Kavlock preliminary developmental toxicity test and a proposed method for assigning priority scores based on results of that test. *Teratog. Carcinog. Mutag.* **7:**85–94.

Hardin, B. D., and Eisenmann, C. J. 1987. Relative potency for four ethylene glycol ethers for induction of paw malformations in the CD-1 mouse. *Teratology* **35:**321–328.

Hardin, B. D., Schuler, R. L., Burg, J. R., Booth, G. M., Hazelden, K. P., Mackenzie, K. M., Piccirillo, V. J., and Smith, K. N. 1987. Evaluation of 60 chemicals in a preliminary developmental toxicity test. *Teratog. Carcinog. Mutag.* **7:**29–48.

Hassell, J. R., and Horigan, E. A. 1982. Chondrogenesis: a model developmental system for measuring teratogenic potential of compounds. *Teratog. Carcinog. Mutag.* **2:**325–331.

Hirsch, K. S., and Scott, W. J., Jr. 1983. Searching for the mechanism of acetazolamide teratogenesis, in: *Issues and Reviews in Teratology,* Volume 1, H. Kalter, ed. Plenum Press, New York, pp. 309–347.

Homburger, F. 1985. Perspective of an old-time toxicologist on new in vitro testing, in: *Concepts in Toxicology,* Volume 3, F. Homburger, ed. Karger, Basel, pp. 16–21.

Horton, W. E., and Sadler, T. W. 1983. Effects of maternal diabetes on early embryogenesis: alterations in morphogenesis produced by the ketone body, β-hydroxybutyrate. *Diabetes* **32:**610–616.

Horton, W. E., and Sadler, T. W. 1985. Mitochondrial alterations in embryos exposed to β-hydroxybutyrate in whole embryo culture. *Anat. Rec.* **213:**94–101.

Hunter, E. S., Sadler, T. W., and Wynn, R. E. 1987. A potential mechanism of DL-β-hydroxybutyrate-induced malformations in mouse embryos. *Am. J. Physiol.* **253:**E72–E80.

IRLG. 1986. Interagency regulatory liaison group workshop on reproductive toxicity risk assessment. *Environ. Health Perspect.* **66:**193–221.

Jelinek, R. 1982. Use of chick embryos in screening for embryotoxicity. *Teratog. Carcinog. Mutag.* **2:**255–261.

Jelinek, R., Peterka, M., and Rychter, Z. 1985. Chick embryotoxicity screening test—130 substances tested. *Ind. J. Exp. Biol.* **23:**588–595.

Johnson, E. M. 1980. A subvertebrate system for rapid determination of potential teratogenic hazards. *J. Environ. Pathol. Toxicol.* **4:**153–156.

Johnson, E. M. 1981. Screening for teratogenic hazards: nature of the problem. *Annu. Rev. Pharmacol. Toxicol.* **21**:417–429.

Johnson, E. M. 1983. Practical application of systems for rapid detection of potential teratogenic hazards, in: *Advances in Modern Environmental Toxicology: Assessment of Reproductive and Teratogenic Hazards,* Volume III, M. S. Christian, W. M. Galbraith, P. Voytek, and M. A. Mehlman, eds. Princeton Scientific, Princeton, N.J., pp. 77–91.

Johnson, E. M. 1984. A prioritization and biological decision tree for developmental toxicity safety evaluations. *J. Am. Coll. Toxicol.* **3**:141–147.

Johnson, E. M. 1985. A review of advances in prescreening for teratogenic hazards, in: *Progress in Drug Research,* Volume 29. Birkhäuser, Basel, pp. 121–154.

Johnson, E. M. 1987. A tier system for developmental toxicity evaluations based on considerations of exposure and effect relationships. *Teratology* **35**:405–427.

Johnson, E. M., and Christian, M. S. 1984. When is a teratology study not an evaluation of teratogenicity. *J. Am. Coll. Toxicol.* **3**:431–434.

Johnson, E. M., and Christian, M. S. 1985. The Hydra assay for detecting and ranking developmental hazards, in: *Concepts in Toxicology,* Volume 3, F. Homburger, ed. Karger, Basel, pp. 107–113.

Johnson, E. M., and Dansky, L. A. 1987. A Hydra assay as a pre-screen for teratogenic potential. *Food Chem. Toxicol.* **25**:637–638.

Johnson, E. M., and Gabel, B. E. G. 1982. Application of the Hydra assay for rapid detection of developmental hazards. *J. Am. Coll. Toxicol.* **1**:57–71.

Johnson, E. M., and Gabel, B. E. G. 1983. An artificial "embryo" for detection of abnormal developmental biology. *Fundam. Appl. Toxicol.* **3**:243–249.

Johnson, E. M., and Kochhar, D. M. 1983. *Teratogenesis and Reproductive Toxicology, Handbook Exp. Pharmacol.,* Volume 65, Springer, Berlin.

Johnson, E. M., Gorman, R. M., Gabel, B. E. G., and George, M. E. 1982. The *Hydra Attenuata* system for detection of teratogenic hazards. *Teratog. Carcinog. Mutag.* **2**:263–276.

Johnson, E. M., Gabel, B. E. G., and Larson, J. 1984. Developmental toxicity and structure/activity correlates of glycols and glycol ethers. *Environ. Health Perspect.* **57**:135–139.

Johnson, E. M., Gabel, B. E. G., Christian, M. S., and Scia, E. 1986. The developmental toxicity of xylene and xylene isomers in the Hydra assay. *Toxicol. Appl. Pharmacol.* **82**: 323–328.

Johnson, E. M., Christian, M. S., Dansky, L., and Gabel, B. E. G. 1987. Use of the adult developmental relationship in prescreening for developmental hazards. *Teratog. Carcinog. Mutag.* **7**:273–285.

Juchau, M. R., Harris, C., Beyer, B. K., and Fantel, A. G. 1987. Reactive intermediates in chemical teratogenesis, in: *Approaches to Elucidate Mechanisms in Teratogenesis,* F. Welsch, ed. Hemisphere, Washington, D.C., pp. 167–183.

Kastner, M., Blankenburg, G., and Schultz, T. 1987. Incorporation of an isolated and reconstituted cytochrome P-450 complex in an organ culture system, in: *Pharmacokinetics in Teratogenesis,* Volume II, H. Nau and W. J. Scott, eds. CRC, Boca Raton, Fla., pp. 145–152.

Kavlock, R. J., Short, R. D., and Chernoff, N. 1987. Further evaluation of an in vivo teratology screen. *Teratog. Carcinog. Mutag.* **7**:7–16.

Kelley, R. O., Fallon, J. F., and Kelly, R. E., 1984. Vertebrate limb morphogenesis: a review of normal development in a model experimental system with applications toward understanding abnormal limb formation, in: *Issues and Reviews in Teratology,* Volume 2, H. Kalter, ed. Plenum Press, New York, pp. 219–265.

Khera, K. S. 1984. Maternal toxicity: a possible factor in fetal malformations in mice. *Teratology* **29**:411–416.

Khera, K. S. 1985. Maternal toxicity: a possible etiological factor in embryo–fetal deaths and fetal malformations of rodent–rabbit species. *Teratology* **31**:129–153.

Khera, K. S. 1987a. Maternal toxicity of drugs and metabolic disorders—a possible etiologic factor in the intrauterine death and congenital malformation: a critique on human data. *CRC Crit. Rev. Toxicol.* **17**:345–375.

Khera, K. S. 1987b. Maternal toxicity in humans and animals: effects on fetal development and criteria for detection. *Teratog. Carcinog. Mutag.* **7**:287–295.

Kimmel, G. L. 1985. In vitro tests in screening teratogens: considerations to aid the validation process, in: *Prevention of Physical and Mental Congenital Defects, Part C: Basic and Medical Science, Education, and Future Strategies,* Volume 163C, M. Marois, H. S. Bennett, R. L. Brent, and M. A. Klingberg, eds. Liss, New York, pp. 259–263.

Kimmel, G. L. 1987. Short-term developmental toxicity testing: considerations in the validation process. *Teratog. Carcinog. Mutag.* **7**:1–6.

Kimmel, G. L., Smith, K., Kochhar, D. M., and Pratt, R. M. 1982. Overview of in vitro teratogenicity testing: aspects of validation and application to screening. *Teratog. Carcinog. Mutag.* **2**:221–229.

Kimmel, G. L., Kimmel, C. A., and Francis, E. Z. 1987. Evaluation of maternal and developmental toxicity. *Teratog. Carcinog. Mutag.* **7**:203–338.

Kistler, A. 1985. Inhibition of chondrogenesis by retinoids: limb bud cell cultures as a test system to measure the teratogenic potential of compounds, in: *Concepts in Toxicology,* Volume 3, F. Homburger, ed. Karger, Basel, pp. 86–100.

Kistler, A. 1987. Limb bud cell cultures for estimating the teratogenic potential of compounds. *Arch. Toxicol.* **60**:403–414.

Kitchin, K. T., Schmid, B. P., and Sanyal, M. K. 1986. Rodent whole-embryo culture as a teratogen screening method. *Methods and Findings Exp. Clin. Pharmacol.* **8**:291–301.

Kochhar, D. M. 1975. The use of in vitro procedures in teratology. *Teratology* **11**:273–288.

Kochhar, D. M. 1980. In vitro testing of teratogenic agents using mammalian embryos. *Teratog. Carcinog. Mutag.* **1**:63–74.

Kochhar, D. M., and Hickey, T. 1985. Goals and potential value of alternative teratogenicity tests, in: *Concepts in Toxicology,* Volume 3, F. Homburger, ed. Karger, Basel, pp. 6–15.

Lave, L. B., and Omenn, G. S. 1986. Cost-effectiveness of short-term tests for carcinogenicity. *Nature* **324**:29–34.

Lewandowski, C., Klug, S., Nau, H., and Neubert, D. 1986. Pharmacokinetic aspects of drug effects in vitro: effects of serum protein binding on concentration and teratogenicity of valproic acid and 2-en-valproic acid in whole embryos in culture. *Arch. Toxicol.* **58**:239–242.

Mirkes, P. E. 1987. Molecular and metabolic aspects of cyclophosphamide teratogenesis, in: *Approaches to Elucidate Mechanisms in Teratogenesis,* F. Welsch, ed. Hemisphere, Washington, D.C., pp. 123–147.

Neubert, D. 1981. On the predictability of developmental toxicity—especially prenatal toxicity—on the basis of culture experiments, in: *Culture Techniques,* D. Neubert and H.-J. Merker, eds. de Gruyter, Berlin, pp. 567–583.

Neubert, D. 1982. The use of culture techniques in studies on prenatal toxicity. *Pharmacol. Ther.* **18**:397–434.

Neubert, D. 1985a. Benefits and limits of model systems in developmental biology and toxicology (in vitro techniques), in: *Prevention of Physical and Mental Congenital Defects,*

Part A: The Scope of the Problem, Volume 163A, M. Marois, ed. Liss, New York, pp. 91–96.

Neubert, D. 1985b. Toxicity studies with cellular models of differentiation. *Xenobiotica* **15:** 649–660.

Neubert, D. 1987. Panel discussion, in: *Approaches to Elucidate Mechanisms in Teratogenesis,* F. Welsch, ed. Hemisphere, Washington, D.C., p. 279.

Neubert, D. 1989. In vitro techniques for assessing teratogenic potential, in: *Advances in Applied Toxicology,* A. D. Dayan and A. J. Paine, eds. Taylor and Francis, London, pp. 191–211.

Neubert, D., and Barrach, H. J. 1983. Effect of environmental agents on embryonic development and the applicability of in vitro techniques for teratological testing, in: *In Vitro Toxicity Testing of Environmental Agents,* A. R. Kolber, T. K. Wong, L. D. Grant, and R. S. DeWoskin, eds. Plenum Press, New York, pp. 147–172.

Neubert, D., Barrach, H. J., and Merker, H.-J. 1980. Drug-induced damage to the embryo or fetus, in: *Drug-induced Pathology,* E. Grundmann, ed. Springer, Berlin, pp. 242–331.

Neubert, D., Blankenburg, G., Lewandowski, C., and Klug, S. 1985. Misinterpretations of results and creation of "artifacts" in studies on developmental toxicity using systems simpler than in vivo systems, in: *Developmental Mechanisms: Normal and Abnormal,* Volume 171, J. W. Lash and L. Saxen, eds. Liss, New York, pp. 241–266.

Newall, D. R., and Tesh, J. M. 1986. Embryo culture as an early screen for teratogenic potential. II. Thalidomide. *Food Chem. Toxicol.* **24:**635–636.

NTP. 1986. Final report. Evaluation of Two in Vitro Teratology Test Systems. Research Triangle Park, N.C., NTP-86-372.

Oglesby, L. A., Ebron, M. T., Beyer, P. E., Carver, B. D., and Kavlock, R. J. 1986. Coculture of rat embryos and hepatocytes: in vitro detection of a proteratogen. *Teratog. Carcinog. Mutag.* **6:**129–138.

Palmer, A. K. 1976. Assessment of current test procedures. *Environ. Health Perspect.* **18:**97–104.

Palmer, A. K. 1981. Regulatory requirements for reproductive toxicology: theory and practice, in: *Developmental Toxicology,* C. A. Kimmel and J. Buelke-Sam, eds. Raven Press, New York, pp. 259–288.

Palmer, A. K. 1987. An indirect assessment of the Chernoff/Kavlock assay. *Teratog. Carcinog. Mutag.* **7:**95–106.

Perraud, J. 1982. L'utilisation des méthodes in vitro pour l'étude de l'embryotoxicité: avantages et limites. *Sci. Tech. Anim. Lab* **7:**301–309.

Pool, B. L., and Schmahl, D. 1987. What's new in mutagenicity and carcinogenicity—status of short term assay systems as tools in genetic toxicology and carcinogenesis. *Pathol. Res. Pract.* **182:**704–712.

Pratt, R. M., and Willis, W. D. 1985. In vitro screening assay for teratogens using growth inhibition of human embryonic cells. *Proc. Natl. Acad. Sci. USA* **82:**5791–5794.

Pratt, R. M., Grove, R. K., and Willis, W. D. 1982. Prescreening for environmental teratogens using cultured mesenchymal cells from the human embryonic palate. *Teratog. Carcinog. Mutag.* **2:**313–318.

Price, C. J., Kimmel, C. A., George, J. D., and Marr, M. C. 1987. The developmental toxicity of diethylene glycol dimethyl ether in mice. *Fundam. Appl. Toxicol.* **8:**115–126.

Ritter, E. J., Scott, W. J., Randall, J. L., and Ritter, J. M. 1985. Teratogenicity of dimethoxyethyl phthalate and its metabolites methoxyethanol and methoxyacetic acid in the rat. *Teratology* **32:**25–31.

Rogers, J. M. 1987. Comparison of maternal and fetal toxic dose responses in mammals. *Teratog. Carcinog. Mutag.* **7:**297–306.

Rogers, J. M., Gray, L. E., Carver, B., and Kavlock, R. J. 1986. Comparison of maternal and fetal toxicity in mice, rats and hamsters treated with the fungicide dinocap. *Teratology* **33**:86C.

Rosa, F. W., Wilk, A. L., and Kelsey, F. O. 1986. Teratogen update: vitamin A congeners. *Teratology* **33**:355–364.

Rowan, A. N., and Goldberg, A. M. 1985. Perspectives on alternative to current animal testing techniques in preclinical toxicology. *Annu. Rev. Pharmacol. Toxicol.* **25**:225–247.

Sabourin, T. D., and Faulk, R. T. 1987. Comparative evaluation of a short-term test for developmental effects using frog embryos, in: *Developmental Toxicology: Mechanisms and Risk, Banbury Report 26,* J. M. McLachlan, R. M. Pratt, and C. L. Markert, eds. Cold Spring Harbor Laboratory, Cold Spring Harbor, New York, pp. 203–223.

Sabourin, T. D., Faulk, R. T., and Gross, L. B. 1985. The efficacy of three non-mammalian test systems in the identification of chemical teratogens. *J. Appl. Toxicol.* **5**:227–233.

Sadler, T. W. 1985. The role of mammalian embryo culture in developmental biology and teratology, in: *Issues and Reviews in Teratology,* Volume 3, H. Kalter, ed. Plenum Press, New York, pp. 273–294.

Sadler, T. W., and Horton, W. E. 1983. Effects of maternal diabetes on early embryogenesis: the role of insulin and insulin therapy. *Diabetes* **32**:1070–1074.

Sadler, T. W., and Warner, C. W. 1984. Use of whole embryo culture for evaluating toxicity and teratogenicity. *Pharmacol. Rev.* **36**:145S–150S.

Sadler, T. W., Horton, W. E., and Warner, C. W. 1982. Whole embryo culture: a screening technique for teratogens. *Teratog. Carcinog. Mutag.* **2**:243–253.

Schardein, J. L. 1983. Teratogenic risk assessment—past, present, and future, in: *Issues and Reviews in Teratology,* Volume 1, H. Kalter, ed. Plenum Press, New York, pp. 181–214.

Schardein, J. L. 1985. *Chemically Induced Birth Defects.* Dekker, New York.

Schardein, J. L. 1987. Approaches to defining the relationship of maternal and developmental toxicity. *Teratog. Carcinog. Mutag.* **7**: 255–271.

Schmid, B. P. 1984. Monitoring of organ formation in rat embryos after in vitro exposure to azathioprine, mercaptopurine, methotrexate or cyclosporin A. *Toxicology* **31**:9–21.

Schmid, B. P. 1985a. Action sites of known in vivo teratogens in extracorporeally exposed rat embryos, in: *Concepts in Toxicology,* Volume 3, F. Homburger, ed. Karger, Basel, pp. 74–85.

Schmid, B. P. 1985b. Xenobiotic influences on embryonic differentiation, growth and morphology in vitro. *Xenobiotica* **15**:719–726.

Schmid, B. P. 1987. Old and new concepts in teratogenicity testing. *Trends Pharmacol. Sci.* **8**: 133–137.

Schmid, B. P., and Cicurel, L. 1986. Application of the post-implantation rat embryo culture system to in vitro teratogenicity testing. *Food Chem. Toxicol.* **24**:623–626.

Schmid, B. P., and Cicurel, L. 1987. Development and validation of an alternative test for the detection of embryonic malformations: the postimplantation embryo culture test (PECT), in: *Alternative Methods in Toxicology,* Volume 5, A. M. Goldberg, ed. Liebert, New York, pp. 179–187.

Schmid, B. P., Trippmacher, A., and Bianchi, A. 1982. Teratogenicity induced in cultured rat embryos by the serum of procarbazine treated rats. *Toxicology* **25**:53–60.

Schmid, B. P., Hauser, R. E., and Donatsch, P. 1985. Effects of cyproheptadine on the rat yolk sac membrane and embryonic development in vitro. *Xenobiotica* **15**:695–699.

Schmid, B. P., Cicurel, L., and Marazzi, A. 1987. Correlation between drug concentrations and teratogenicity in vivo and in vitro, in: *Pharmacokinetics in Teratogenesis,* Volume II, H. Nau and W. J. Scott, eds. CRC, Boca Raton, Fla., pp. 209–216.

Schuler, R. L., Hardin, B. D., and Niemeier, R. W. 1982. Drosophila as a tool for the rapid assessment of chemicals for teratogenicity. *Teratog. Carcinog. Mutag.* **2**:293–301.

Schuler, R. L., Hardin, B. D., Niemeier, R. W., Booth, G., Hazelden, K., Piccirillo, V., and

Smith, K. 1984. Results of testing fifteen glycol ethers in a short-term in vivo reproductive toxicity assay. *Environ. Health Perspect.* **57**:141–146.

Schuler, R. L., Radike, M. A., Hardin, B. D., and Niemeier, R. W. 1985. Pattern of response of intact Drosophila to known teratogens. *J. Am. Coll. Toxicol.* **4**:291–303.

Schwetz, B. A., and Moorman, M. P. 1987. Assessment of adult toxicity in development versus prechronic toxicology studies. *Teratog. Carcinog. Mutag.* **7**:211–233.

Scott, W. J., Jr., Nau, H., Wittfoht, W., and Merker, H.-J. 1987. Ventral duplication of the autopod: chemical induction by methoxyacetic acid in rat embryos. *Development* **99**: 127–136.

Seidenberg, J. M., and Becker, R. A. 1987. A summary of the results of 55 chemicals screened for developmental toxicity in mice. *Teratog. Carcinog. Mutag.* **7**:17–28.

Seidenberg, J. M., Anderson, D. G., and Becker, R. A. 1986. Validation of an in vivo developmental screen in the mouse. *Teratog. Carcinog. Mutag.* **6**:361–374.

Shepard, T. H. 1986. *Catalog of Teratogenic Agents*, 5th ed. Johns Hopkins Press, Baltimore.

Shepard, T. H., Fantel, A. G., Mirkes, P. E., Greenaway, J. C., Faustman-Watts, E., Campbell, M., and Juchau, M. R. 1983. Teratology testing: I. Development and status of short-term prescreens. II. Biotransformation of teratogens as studied in whole embryo culture, in: *Developmental Pharmacology*, Volume 135, S. M. MacLeod, A. B. Okey, and S. P. Spielberg, eds. Liss, New York, pp. 147–164.

Sleet, R. B., and Greene, J. A., and Welsch, F. 1987. Teratogenicity and disposition of the glycol ether 2-methoxyethanol and their relationship in CD-1 mice, in: *Approaches to Elucidate Mechanisms in Teratogenesis*, F. Welsch, ed. Hemisphere, Washington, D.C., pp. 33–57.

Sleet, R. B., Greene, J. A., and Welsch, F. 1988. The relationship of embryotoxicity to disposition of 2-methoxyethanol in mice. *Toxicol. Appl. Pharmacol.* **93**:195–207.

Smith, M. K., Kimmel, G. L., Kochhar, D. M., Shepard, T. H., Spielberg, S. P., and Wilson, J. G. 1983. A selection of candidate compounds for in vitro teratogenesis test validation. *Teratog. Carcinog. Mutag.* **3**:461–480.

Smith, M. K., George, E. L., Zenick, H., Manson, J. M., and Stober, J. A. 1987. Developmental toxicity of halogenated acetonitriles: drinking water byproducts of chlorine disinfection. *Toxicology* **46**:83–93.

Solursh, M. 1983. Cell–cell interactions and chondrogenesis, in: *Development, Differentiation, and Growth*, Volume 2, B. K. Hall, ed. Academic Press, New York, pp. 121–141.

Solursh, M. 1984. Cell and matrix interactions during limb chondrogenesis in vitro, in: *The Role of Extracellular Matrix in Development*, R. L. Trelstad, ed. Liss, New York, pp. 277–303.

Solursh, M. 1986. Environmental regulation of limb chondrogenesis, in: *Articular Cartilage Biochemistry*, K. Kuettner, ed. Raven Press, New York, pp. 145–161.

Starr, T. B., and Gibson, J. E. 1984. Understanding formaldehyde toxicity with the delivered dose concept. *Trends Pharmacol. Sci.* **5**:477–480.

Steele, C. E., and Marlow, R. 1986. Teratological studies using whole-embryo culture. *Food Chem. Toxicol.* **24**:644.

Tennant, R. W., Margolin, B. H., Shelby, M. D., Zeiger, E., Haseman, J. K., Spaulding, J., Caspary, W., Resnick, M. 1987. Prediction of chemical carcinogenicity in rodents from in vitro genetic toxicity assays. *Science* **23**:933–941.

Warner, C. W., Sadler, T. W., Shockey, J., and Smith, M. K. 1983. A comparison of the in vivo and in vitro response of mammalian embryos to a teratogenic insult. *Toxicology* **28**: 271–282.

Warner, C. W., Sadler, T. W., Tulis, S. A., and Smith, M. K. 1984. Zinc amelioration of cadmium-induced teratogenesis in vitro. *Teratology* **30**:47–53.

Weaver, T. E., and Scott, W. J., Jr. 1984. Acetazolamide teratogenesis: association of maternal respiratory acidosis and ectrodactyly in C57BL/6J mice. *Teratology* **30:**187–193.

Welsch, F. 1987. Panel discussions, in: *Approaches to Elucidate Mechanisms in Teratogenesis*, F. Welsch, ed. Hemisphere, Washington, D.C., pp. 269–279.

Welsch, F., Stedman, D. B., Willis, W. D., and Pratt, R. M. 1986. Karyotype, growth, and cell cycle analysis of human embryonic palatal mesenchymal cells: relevance to the use of these cells in an in vitro teratogenicity screening assay. *Teratog. Carcinog. Mutag.* **6:** 383–392.

Welsh, J. J. 1987. Teratological research using in vitro systems. IV. Cells in culture. *Environ. Health Perspect.* **72:**225–235.

Whitby, K. E. 1987. Teratological research using in vitro systems. III. Embryonic organs in culture. *Environ. Health Perspect.* **72:**221–223.

Whitby, K. E., and Flynn, T. J. 1987. Introduction. *Environ. Health Perspect.* **72:**201.

Wickramaratne, G. A. de S. 1986. The teratogenic potential and dose-response of dermally administered ethylene glycol monomethyl ether (EGME) estimated in rats with the Chernoff–Kavlock assay. *J. Appl. Toxicol.* **6:**165–166.

Wickramaratne, G. A. de S. 1987. The Chernoff–Kavlock assay: its validation and application in rats. *Teratog. Carcinog. Mutag.* **7:**73–83.

Wiger, R., and Stottum, A. 1985. In vitro testing for developmental toxicity using the *Hydra attenuata* assay. *NIPH Ann.* **8:**43–47.

Wiger, R., and Stottum, A. 1986. Testing for the potential developmental toxicity of methoxyethanol, ethoxyethanol and butoxyethanol in three in vitro test systems. *Teratology* **34:**423.

Wilby, O. K., and Tesh, J. M. 1987. A Hydra assay as a pre-screen for teratogenic potential. *Food Chem. Toxicol.* **25:**637–638.

Wilby, O. K., Newall, D. R., and Tesh, J. M. 1986. A Hydra assay as a pre-screen for teratogenic potential. *Food Chem. Toxicol.* **24:**651–652.

Wilk, A. L., Greenberg, J. H., Horigan, E. A., Pratt, R. M., and Martin, G. R. 1980. Detection of teratogenic compounds using differentiating embryonic cells in culture. *In Vitro* **16:**269–276.

Wilson, J. G. 1973. *Environment and Birth Defects.* Academic Press, New York.

Wilson, J. G. 1977. Current status of teratology: general principles and mechanisms derived from animal studies, in: *Handbook of Teratology*, Volume 1, J. G. Wilson and F. C. Fraser, eds. Plenum Press, New York, pp. 47–74.

Wilson, J. G. 1978a. Survey of in vitro systems: their potential use in teratogenicity screening, in: *Handbook of Teratology*, Volume 4, J. G. Wilson and F. C. Fraser, eds. Plenum Press, New York, pp. 135–150.

Wilson, J. G. 1978b. Review of in vitro systems with potential for use in teratogenicity screening. *J. Environ. Pathol. Toxicol.* **2:**149–167.

WHO. 1984. Principles for Evaluating Health Risks to Progeny Associated with Exposure to Chemicals During Pregnancy. Health Criteria 30. WHO, Geneva.

Yonemoto, J., Brown, N. A., and Webb, M. 1984. Effects of dimethoxyethyl phthalate, monomethoxyethyl phthalate, and 2-methoxyacetic acid on postimplantation rat embryos in culture. *Toxicol. Lett.* **21:**97–102.

Issues and Reviews in Teratology **5**:155–180
Plenum Press, New York, 1990, 978-1-4612-7847-4

Twinning in Spontaneous Abortions and Developmental Abnormalities

4

IRENE A. UCHIDA

1. INTRODUCTION

Twins have been prominent in folklore from earliest times. In Greek mythology, Zeus, lord of the universe, fathered the twins Apollo, god of the sun, and Artemis, goddess of the moon. Castor and Pollux, guardians of mariners, were immortalized in the constellation Gemini, third sign of the zodiac. In Roman mythology, Romulus and Remus, twin sons of Mars, were raised by a wolf. Remus was eventually killed by Romulus who became the founder of Rome. In Taoist philosophy, the twin forces of Yin, female and negative, and Yang, masculine and positive, maintained cosmic harmony. In more recent times, twins have appeared in poetry and drama to tantalize and entertain us.

The early twins were mainly unlike pairs who maintained the world in balance by opposing forces. In the present-day, more practical world, interest is drawn to identical twins because of their greater value to scientific knowledge. Two people who look so much alike as to be almost indistinguishable from each other hold a special fascination for the more common single-born. Their identity stems from the separation of cells that originally were meant to form a single individual. It is readily conceivable that unique properties would be associated with this method of forming two bodies. Unequal division of cells or the sharing of a single gestational sac can lead to malformations unique to monozygotic twins and to more frequent development of abnormalities than in singletons or even dizygotic twins. The main impediment for the latter is crowding in a space intended for one infant.

IRENE A. UCHIDA • Departments of Pediatrics and Pathology, McMaster University, Hamilton, Ontario L8N 3Z5, Canada.

Twins arising from a single egg fertilized by a single sperm have been called identical, but with improved technical advances, some of these twins have been shown to develop into genetically different individuals and the term "monozygotic twins" is preferable. Much of the following discussion of abnormal development is logically focused upon monozygotic twinning.

2. TYPES OF TWINS

Two types of twins are well documented: monozygotic (MZ) and dizygotic (DZ) twins. Since MZ twins are formed from one fertilized ovum, they have identical genes, barring of course any post-conception mutations. Timing and uniformity of division of the original cell mass into two embryos are crucial to subsequent development. The division into two occurs most often after implantation and a single placenta is formed, but with earlier separation each twin will form its own placenta. In the latter case the implantation site will determine the formation of either two completely separate placentas or one fused dichorionic placenta. For this reason, cursory examination of an apparent "single" placenta will not provide an accurate diagnosis of zygosity. Since early and equal division of a single cell mass will produce MZ twins each with its own sac and placenta, their prenatal environment will be similar to that of DZ twins.

2.1. Monozygotic Twins

Two-thirds of MZ twins are formed after differentiation of a single chorion (Bulmer, 1970) and each twin is usually enclosed within its own amnion. Anastomosis between the blood vessels of these twins is common (Fig. 1). A disadvantage of a shared blood circulation is evident when unequal division of the cell mass leads to twins of considerable difference in size. It is apparent that the larger one may deprive the other of its food supply.

Greater delay in the twinning process will result in twins enclosed within a single amnion (Fig. 2). Entanglement of the umbilical cords within the shared cavity may tie off the blood supply and lead to fetal death. A complication of still later division of the cell mass is the formation of conjoined twins (Bergsma, 1967). In extreme cases, one twin may become a parasite upon the other, an instance of which is the acardiac monster.

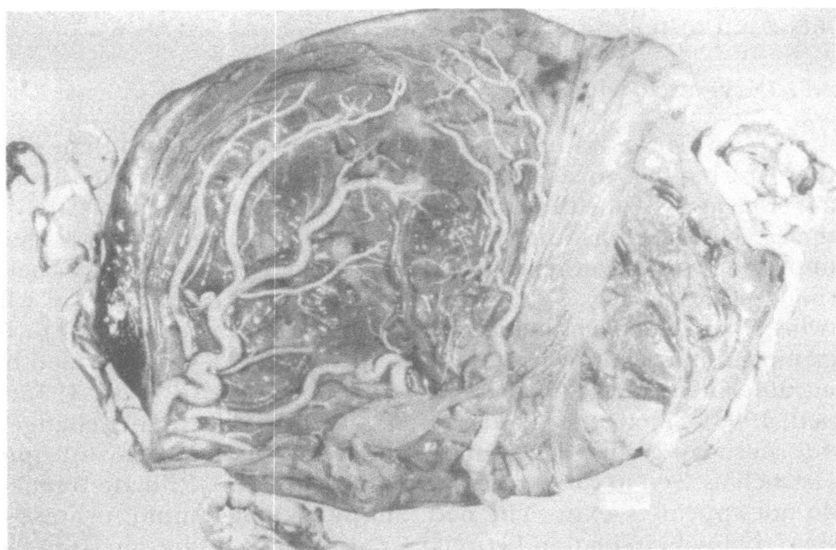

Figure 1. Dichorionic placenta of dizygotic triplets. Monochrionic-diamniotic placenta of MZ twins on the left. Rubber latex injected into left cord spread into arterial system of middle cord, at base of which can be seen remnants of amnion. Placenta of third infant on right fused with placenta on left but separated with thick chorion. Latex did not penetrate into this side. Genetic markers indicated this infant to be different from the other two triplets.

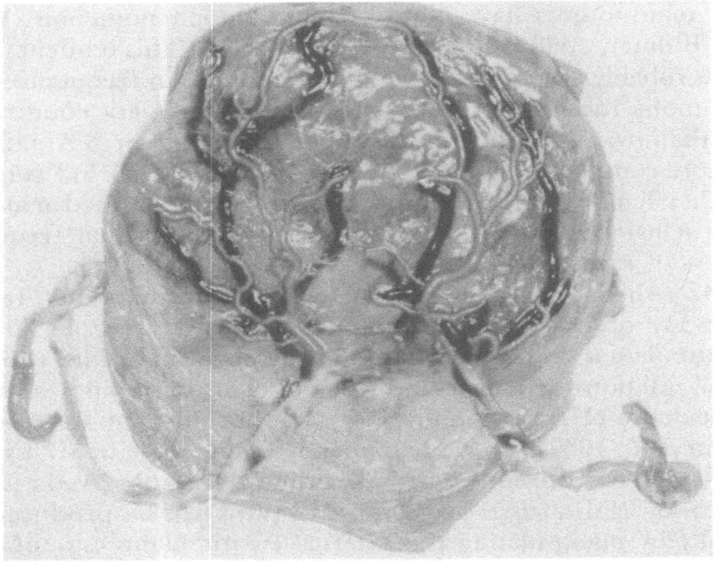

Figure 2. Monoamniotic placenta with shared arterial blood circulation.

2.2. Dizygotic Twins

DZ twins are no more alike in their genetic makeup than siblings; they just happen to share the same intrauterine environment and thus are born at the same time. Originating as separate individuals, they form their own placentas and extraembryonic membranes. Those that are implanted close together may give the misleading impression of having a single placenta (Fig. 1). Only under exceptional circumstances will DZ twins share blood circulations. This rare phenomenon leads to blood group chimeras (Chown *et al.*, 1963; Uchida *et al.*, 1964) formed by the intrauterine grafting of blood precursors (Dunsford *et al.*, 1953; Tippett, 1983). Since other tissues are not involved, vascular exchange does not cause congenital malformations nor infertility in twins of opposite sex such as seen in cattle freemartins (Owen, 1945). Primate freemartins do not appear to exist. The best known example among primates of a shared blood circulation between DZ twins with no freemartin effect in the female is the marmoset. This small New World monkey consistently bears DZ twins with a monochorionic placenta and vascular anastomosis and yet female offspring are not sterile.

Aside from genetic variation, other important differences between MZ and DZ twins have been observed. Increased maternal age and parity are factors relevant to DZ twinning (Bulmer, 1959). Since Weinberg's original report of "repeat frequency" of twinning among mothers of DZ twins, many others have observed the same phenomenon (Weinberg, 1901; Bulmer, 1958; White and Wyshak, 1964). This tendency to recurrence probably accounts for the large variations in frequencies of twinning among racial groups. DZ twins are rare in Asiatic countries, occur more than twice as frequently among Caucasians than in Asians, and are still more common in Negroid peoples (Bulmer, 1970). MZ twinning, on the other hand, has little if any correlation with elevated maternal age nor is it associated with parity, and occurs with equal frequency regardless of race.

MZ twins appear sporadically to the extent that when familial instances are observed they attract special attention (Fig. 3). The factors that stimulate a single cell mass to divide abnormally into two embryos are still unknown. Artificial induction of ovulation is known to increase the incidence of DZ twinning (Lamont, 1982) but a more recent study (Derom *et al.*, 1987) revealed an unexpected increase in MZ twinning as well. An increased frequency of MZ twins was also observed after *in vitro* fertilization (Edwards *et al.*, 1986). MZ twins can be produced experimentally by manipulating the external environment, e.g., of fish eggs (Stockard, 1921). In some species of armadillos the MZ quadruplets that

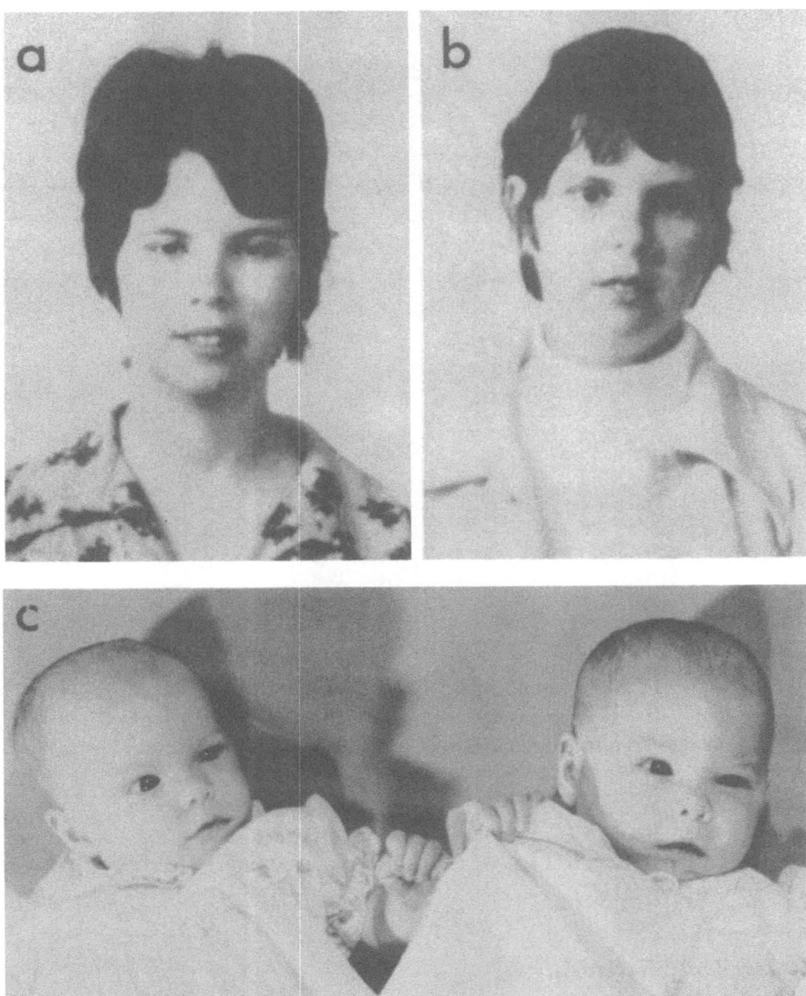

Figure 3. Two pairs of MZ twins in one kindred. (a, b) MZ twins who had a dichorionic placenta. (c) MZ twin daughters of twin a. Monozygosity of both sets was determined by blood grouping, red cell enzymes, serum proteins, and chromosome variants.

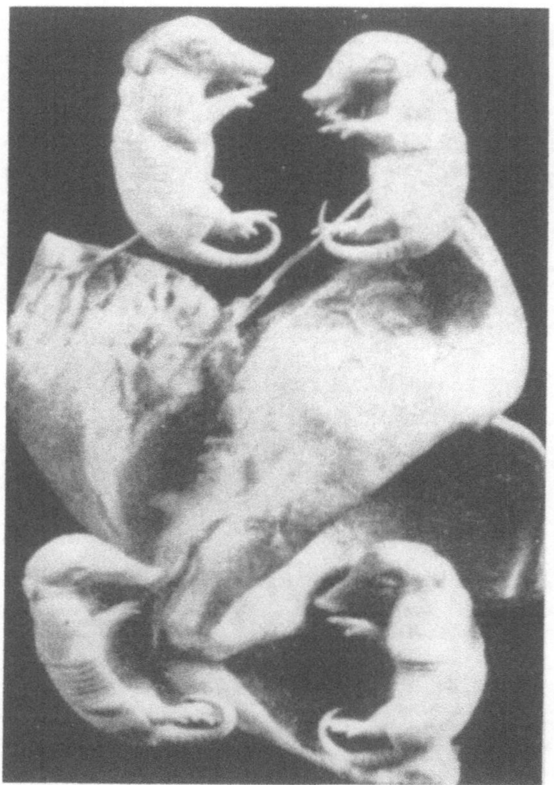

Figure 4. MZ armadillo quadruplet embryos attached to a single placenta. From Winchester (1951).

are regularly produced (Fig. 4) are said to be caused by developmental arrest (Newman, 1917).

2.3. Polar Body Twinning

A third type of twinning, sometimes referred to as polar body twinning, may result when two sperm fertilize the two meiotic cells that would normally form an ovum and a polar body. These twins are DZ in origin but may be more alike than DZ twins because some maternal genes may be held in common, particularly if the second polar body is fertilized.

Evidence for first polar body twinning was provided by Bieber *et al.*

(1981), in a report of a normal infant and its acardiac triploid twin, both enclosed within a single chorion. Fertilization of the first polar body, which has a haploid set of chromosomes each with two chromatids, was suggested to explain the formation of the acardiac triploid.

In second polar body twinning the two maternal components, ovum and polar body, have identical genes except for any crossover segments. Each of the components can be independently fertilized before or after extrusion of the polar body. Chimeric subjects with two different clones originating from double fertilization of the two meiotic components have been described (Corey *et al.*, 1967; Dewald *et al.*, 1980; Uchida *et al.*, 1985). Had the products separated into two independent zygotes, twins would have been formed.

3. SPONTANEOUS ABORTIONS

Twins among spontaneous abortions are readily confirmed by the presence of two gestational sacs or fetuses or if more than one sac has been detected ultrasonographically. There is evidence that relatively more twins occur among abortions than among live births. Kajii *et al.* (1973, 1980) and Creasy *et al.* (1976) found 50 twin pairs among a total of 2594 abortions, an incidence of 1/52 compared with the generally quoted frequency of 1/85 among live births.

In the absence of morphological evidence, additional twins can still be detected by cytogenetic analyses of aborted tissues. DZ twins with different karyotypes or of different sex are readily identified by the admixture of different clones of cells with the obvious proviso that mosaicism and maternal cell contamination have been excluded. By combining morphology and cytogenetics it was possible to detect a total of 15 twin pairs among 661 abortions, an incidence of 1/44 (Uchida *et al.*, 1983b). Even then, chromosomally normal DZ twins of the same sex can be missed unless striking differences in chromosomal variations happen to draw attention to the presence of more than one clone of cells. It is virtually impossible to detect the presence of MZ twins in the absence of two discrete specimens.

Some 50% of spontaneous abortions have abnormal chromosomes (Hassold *et al.*, 1980; Warburton *et al.*, 1980; Carr, 1983), but among abortions a much lower frequency of twins are chromosomally abnormal (Creasy *et al.*, 1976). In a follow-up study of spontaneously aborted twins (Uchida *et al.*, 1983b), 20 pairs with normal chromosomes (11 MZ and 9 DZ) were found among 30 twin pairs. The remainder consisted of eight pairs that were discordant normal/abnormal and two sets (1 MZ, 1 DZ) that were concordant for abnormal chromosomes. This excessive loss of twins with normal karyotypes may be due solely to the fact that the products of conception were twins although it is difficult to explain the loss of normal twins in the first trimester.

4. DEVELOPMENTAL ABNORMALITIES

Many structural anomalies are unique to twins. They range from interesting curiosities of no clinical relevance to severe lethal defects. The phenomenon of mirror-imaging, as demonstrated in handedness, dermatoglyphics, and direction of hair whorls, occurs commonly in MZ twins. Another form of lateral inversion of more clinical importance is situs inversus viscerum which may not be a serious handicap unless accompanied by other defects.

Minor abnormalities, such as malrotation of feet, may result from overcrowding during the later stages of prenatal life. This anomaly is found more often in DZ twins not only because they occur more frequently but also because they are usually larger than are MZ twins. Other more debilitating defects, such as congenital heart disease and neural tube defects, when they occur, are usually discordant among MZ twins.

Fortunately, many severe anomalies cause early fetal death. Some have their roots in the MZ twinning process itself. Early developmental problems can arise from the unequal distribution of blood from a vascular system shared between monochorionic twins. Severe circulatory imbalance can lead to the transfusion syndrome or to the formation of an acardiac monster.

4.1. Conjoined Twins

From earliest times, conjoined twins have attracted attention as curiosities. A very early example is traced back to the Stone Age, ca. 6500 B.C. (Fig. 5). The best known example of conjoined twins are Chang and Eng, born in Siam in 1814. Their fame has been preserved in the eponym "Siamese twins" in popular usage for conjoined twins.

Incompletely separated twins are the result of very late division of the inner cell mass into two embryos. They may be joined at any part of the body, the most common being ventral union. In extreme cases a conjoined twin may be found merely as an appendage to its cotwin (Fig. 6). A fetus can be completely embedded within the cotwin, known as fetus-in-fetu, and may be found in almost any organ of the body.

By origin, conjoined twins are MZ, excluding the unlikely possibility

Figure 5. Double-headed twin goddes (*sic*) carved in the Stone Age, c. 6500 B.C. From Kunze and Nippert (1986).

of the fusion of two DZ embryos. An instance of conjoined twins reported as being of different sex was noted by Milham (1966), but no documentation was provided. Confirmation of sex discordance is shown in Fig. 7 in a pair of aborted conjoined twins preserved from the early days of a study of twin placentas in Toronto (Walker, 1957), but no background information remains. It is suspected that these instances of disparity in sex could have resulted from sex chromosome mosaicism, a phenomenon that has been demonstrated in live-born MZ twins of different sex (see below).

Incomplete separation of twins occurs so rarely that it is difficult to

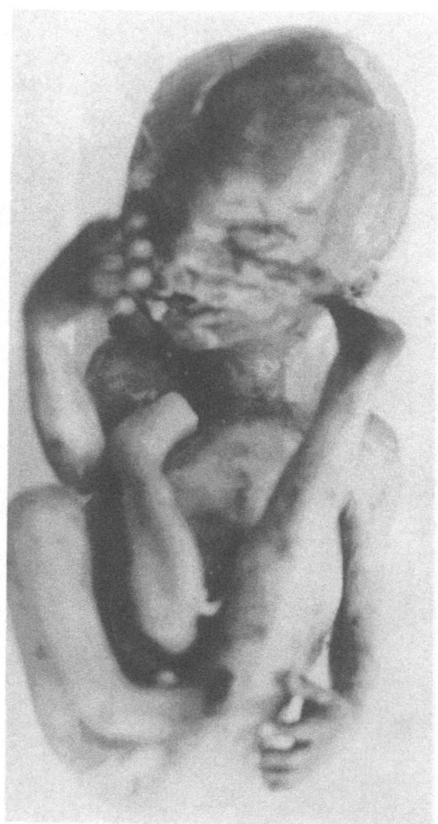

Figure 6. Male fetus with leg of incompletely separated twin attached to lower abdomen.

provide even a rough estimate of frequency. Potter (1973) found two sets among more than 100,000 births. A strikingly high frequency of females was found by Milham (1966) who postulated that more males may be lost in early pregnancy to account for the unusual sex ratio. In a series of 2000 spontaneous abortions there were two pairs of conjoined twins (Fig. 8), a frequency of one in 1000 abortuses. Both sets were males with normal chromosomes (Uchida *et al.*, 1983b).

4.2. Severe Discordant Anomalies

Serious defects may occur in one of a pair of MZ twins. If a twin should die in midgestation, when abortion is less likely to occur, it may be

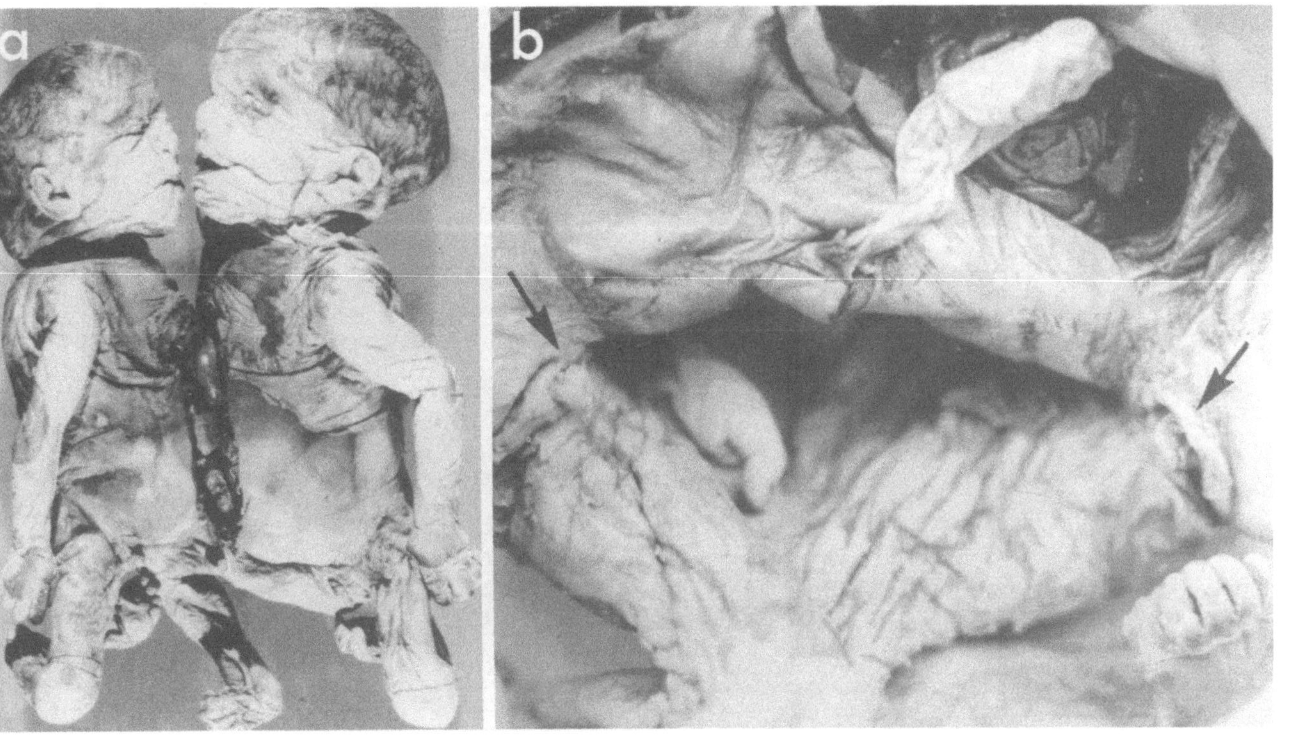

Figure 7. Conjoined twins of unlike sex found preserved in formalin circa 1930. (a) Two fetuses united in abdominal region. Abdominal cut made after recent rediscovery. (b) Genitalia indicated by arrows, male on left and female on right.

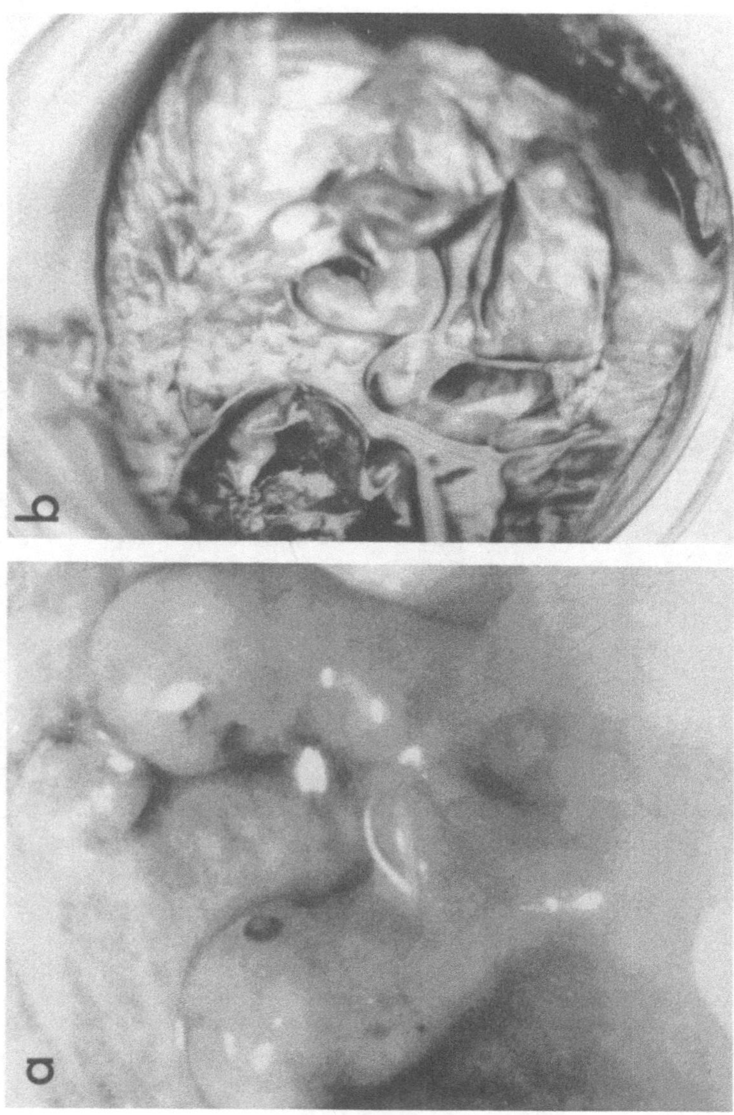

Figure 8. Conjoined fetuses in spontaneous abortions. (a) Twins united in umbilical region. (b) two-headed fetus. Both sets were male with normal chromosomes.

compressed and flattened by the normal twin, hence the descriptive term fetus papyraceous (Fig. 9).

Most monochorionic twins share blood circulations, the connection usually occurring between the two arterial systems. When there is an arteriovenous connection, the blood of one twin may be drained into the other. The resulting transfusion syndrome causes one twin to be plethoric while the other will be anemic. Consequently, the heart and kidneys of the former may be enlarged and a considerable difference in size of the twins may ensue. Antenatal death of one twin may result in the death of the cotwin but in many instances the surviving twin is normal even though prematurity is high (Enbom, 1985).

A severe form of malformation resulting from a shared blood circulation between MZ twins is acardiac anomaly. The twin with this anomaly becomes a parasite of the normal cotwin, who maintains the blood circulation for both. The holoacardiac amorphous twin shown in Fig. 10 was attached with a poorly defined umbilical cord to the cord of the normal twin. The globular mass showed good axial orientation but no cardiac structure was identifiable. The karyotype of the fibroblasts was that of a trisomy 9 female while the surviving twin's fibroblasts were normal. Neither twin gave any evidence of mosaicism in 100 fibroblasts

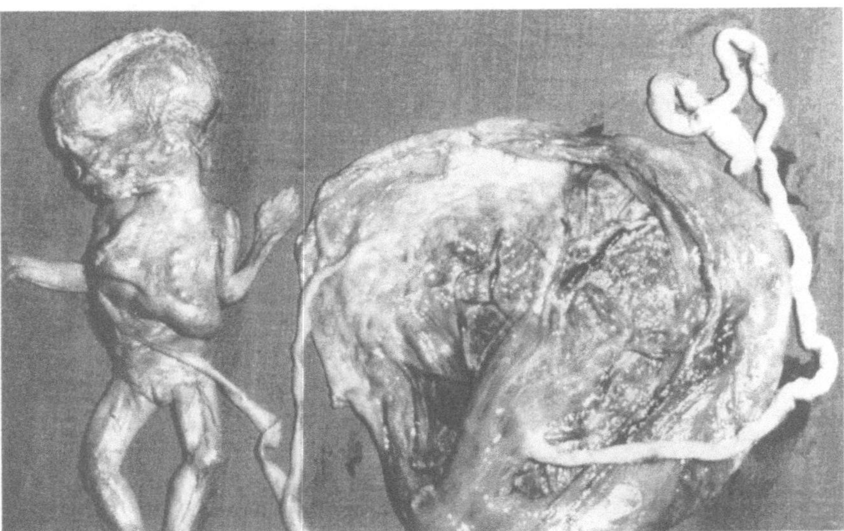

Figure 9. Fetus papyraceous. To the right is the cord of the normal live-born MZ twin. Placenta is monochorionic-diamniotic.

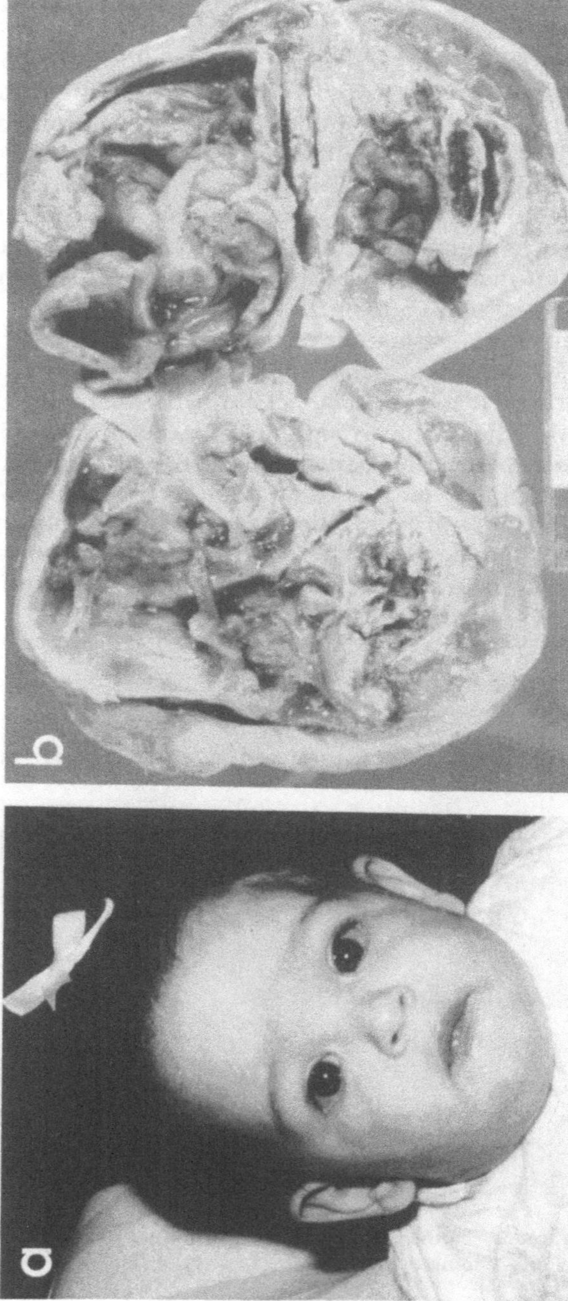

Figure 10. MZ twins, one normal and the other acardiac. (a) Normal twin has normal chromosomes. (b) Acardiac twin with trisomy 9 opened for pathological assessment. (Courtesy of Dr. Derek J. deSa, McMaster University.)

analyzed from each. All 200 lymphocytes analyzed in the surviving twin were normal. No blood was available from the acardiac twin but there is no doubt that its lymphocytes were also normal since both twins shared blood circulation.

Other instances of aneuploidy in acardiac twins with normal karyotypes in the cotwins have been described (Kerr and Rashad, 1966; Scott and Ferguson-Smith, 1973; Bieber *et al.*, 1981); usually, however, both twins have been reported to have normal chromosomes. The pathogenesis of the acardiac twin is still obscure (Benirschke and Harper, 1977; Kaplan and Benirschke, 1979; Nance, 1981). Be it placental anastomosis, chromosomal defect, cytoplasmic deficiency, or combinations of these, it is interesting to speculate that if the normal twin had not been capable of maintaining the blood flow for the acardius, these instances could well have resulted in the vanishing twin phenomenon described below.

4.3. Mosaics and Chimeras

Chromosomal mosaicism, the presence of more than one cell line in a single subject, is not a rare phenomenon. Chimerism, a special type of mosaicism, is relatively rare and is categorized separately. The origins of these two types of cell admixture are distinctly different. Mosaics arise from a single zygote whereas the cell lines of a chimera originate from different zygotes. It follows therefore that chromosome variants in mosaic subjects are identical but unlikely to be the same in chimeras. Although mosaicism for chromosomal rearrangements is known, most mosaics involve diploid and aneuploid clones. Autosomal mosaics usually originate as trisomies and lose the extra chromosome during mitotic division.

Mosaicism can be the cause of MZ twins having grossly different phenotypes (Fig. 11) which in such cases belies the term "identical twins." For example, they may be of different sex, one a normal male and the other a female 45,X Turner syndrome, resulting from loss of the Y chromosome (Turpin *et al.*, 1961; Pedersen *et al.*, 1980). Similarly, loss of an X chromosome from one twin of an MZ female pair will result in discordant twins of the same sex (Mikkelsen *et al.*, 1963; Uchida *et al.*, 1983a). Instances of MZ twins, one with Down syndrome and the other normal, are also well known (deWolff *et al.*, 1962; Fanconi, 1962).

In aborted tissue the presence of two clones of cells arouses suspicion of mosaicism particularly if all chromosome variants are identical in the two cell lines. If any differences are detected, other possibilities, such as maternal cell contamination, chimerism, or twinning, must be consid-

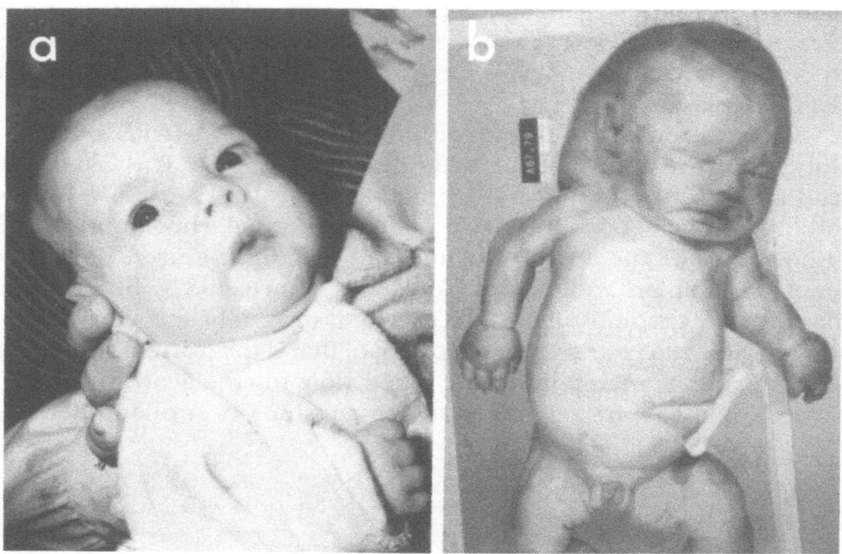

Figure 11. Discordant MZ twins. (a) Normal twin girl, 46,XX. (b) Twin with 45,X Turner syndrome. Placenta was monochorionic-diamniotic with common vascular circulation.

ered. It is not always possible to tell the difference between twins and a chimeric singleton if the aborted material consists only of fragmented tissue.

Chimerism confined to the blood cells of DZ twins has been well documented. This phenomenon is usually detected by chance during blood-grouping studies of normal twins. It has been postulated that at some time during early embryogenesis there must have been transient vascular communication between DZ twins with grafting of blood-forming tissue. One instance of blood group chimerism found in a girl was puzzling until it was later discovered that she had had a stillborn twin brother (Valez-Orozco, 1961). Another unusual example of blood grafting involved DZ male twins one of whom had Down syndrome and the other was normal (Bias and Migeon, 1967). Chimerism was first suspected because of isoagglutinin deficiency in the Down syndrome twin. The normal twin was blood group B while the abnormal one was A with no anti-B in his serum. Five and a half percent of the trisomic infant's cells had a normal karyotype which was interpreted to represent a graft from the normal twin; proof through chromosome banding was not available in those days. The normal twin had no evidence of chimerism.

5. THE VANISHING TWIN

Long before the advent of ultrasonography, it was sometimes suspected that a newborn singleton could have had a twin. What prompted the suspicion was the unexpected presence of a fetus papyraceous at delivery of a normal infant. Though such fetuses are formed during the second trimester, and occasionally the third (Livnat *et al.*, 1978), earlier demise, it was thought, might lead a blighted fetus or an empty sac to disappear completely. Unexplained early vaginal bleeding during an otherwise normal pregnancy also aroused suspicion that a twin may have been aborted. However, pregnancies suspected of being multiple were difficult to verify by examination of the placenta after delivery of a singleton (Robinson and Caines, 1977; Finberg and Birnholz, 1979; Landy *et al.*, 1982).

Recently, visualization of more than one gestational sac during the early weeks of pregnancy has been made possible by the technical advances in diagnostic ultrasound and the skill of sonographers in interpreting the tracings. Attention was drawn to the efficiency of this technique when a quintuplet pregnancy was identified by ultrasonic examination at 9 weeks of gestation and five infants were subsequently delivered (Campbell and Dewhurst, 1970). It soon became evident that the number of twins at delivery was much lower than the number detected by ultrasound (Levi, 1976). Sonographic evidence of a twin preg-

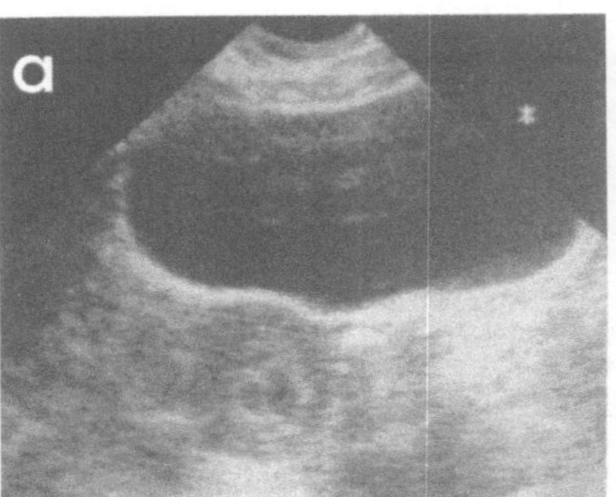

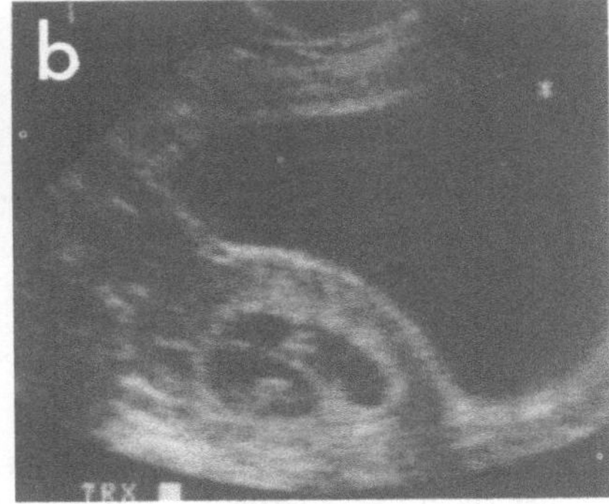

Figure 12. Pelvic ultrasound. (a) Two hypoechoic areas are seen within the uterus which was thought to represent an early pregnancy with an implantation bleed. A fetal pole was not seen at this time. (b) Two gestational sacs each with a fetal pole were seen 2 weeks later. Fetal cardiac pulsations were demonstrated at real time examination. (Courtesy of Dr. C. Caco, McMaster University.)

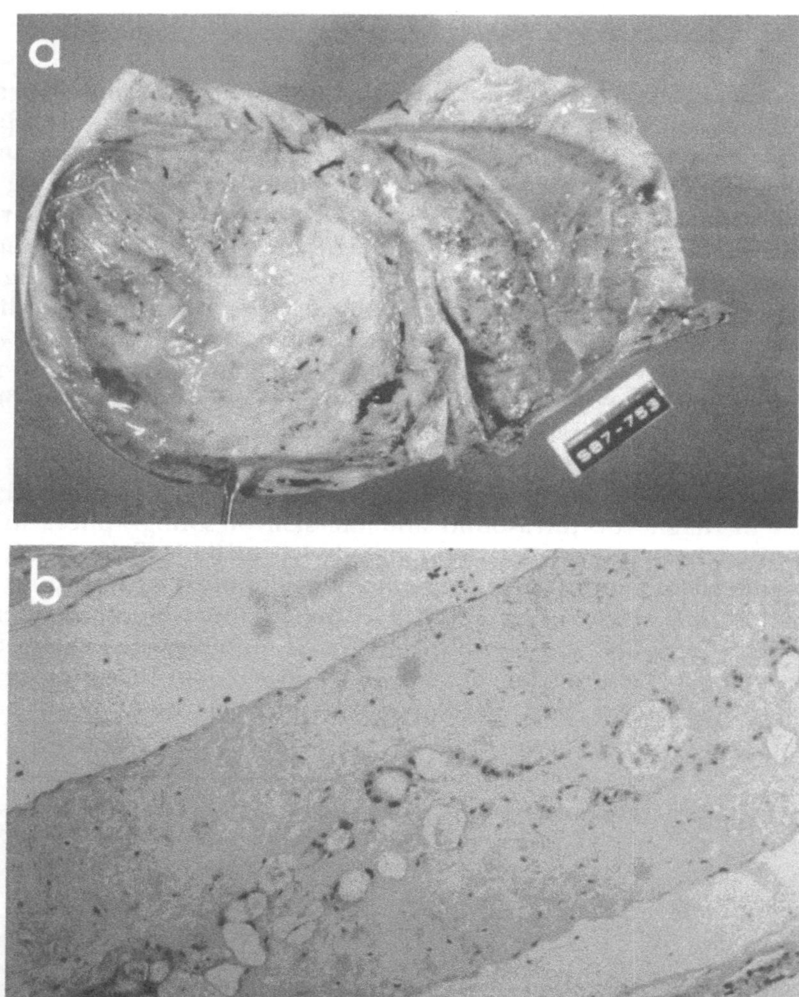

Figure 13. (a) Dense plaque in fetal membranes away from main placental disk. Needle indicates small dense structure resembling old yolk sac (between amnion and chorion). Separate yolk sac was seen on the fetal aspect of the main placental disk. (b) From thickened plaque—tissue resembling reduplicated amniotic epithelial tubule seen. (c) Low-power view of plaque showing placental cotyledons with degeneration and collapsed cystic structure lined by squamous epithelium. (d) Plaque with occasional stratified squamous-lined ductular structures. (Courtesy of Dr. Derek J. deSa, McMaster University.)

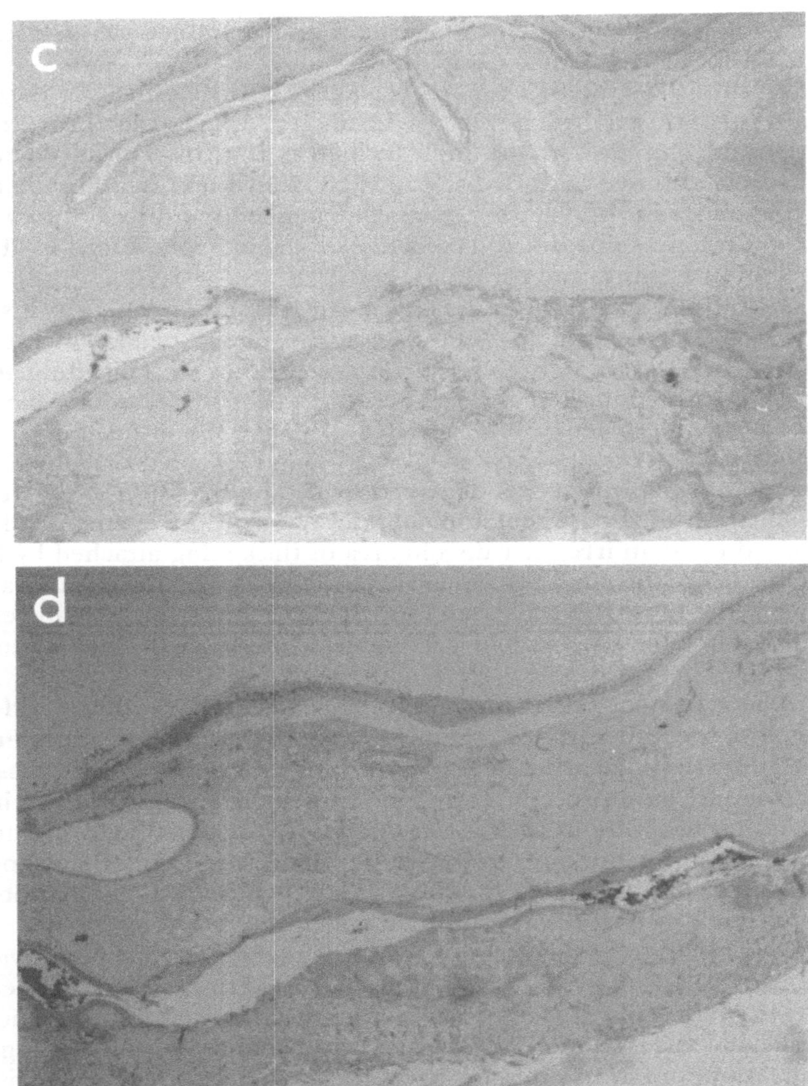

Figure 13. (*Continued*)

nancy followed eventually by the birth of a singleton led to the fascinating concept of "the vanishing twin."

In the 1970s the first reports began to appear confirming the phenomenon of the vanishing twin (see Landy *et al.*, 1982). Data provided by ultrasound emphasized the high incidence of twins conceived when compared with live-born twins. The blighted ovum or anembryonic sac predominates in the vanishing twin phenomenon and loss of a twin is thought usually to be due to resorption or spontaneous abortion of one sac without harm to the cotwin.

Ultrasonograms and pathological evidence of vanishing twins can be seen in Figs. 12 and 13. A double sac was first visualized with ultrasound which was thought to be an early pregnancy with bleeding at the site of implantation. Two weeks later at 8 weeks of gestation, two sacs each containing a fetus were detected. Two fetal heartbeats were still present at 14 weeks but by 22 weeks ultrasound revealed only one fetus. A single normal infant was delivered by cesarean section at 38 weeks. Examination of the placental membrane revealed the remnants of the resorbed twin: an irregular discoid area of thickening attached by fetal membranes to the main placenta. There was no vascular communication between the normal placenta and the thickened area but a small yellow focus compatible with an old yolk sac was found on the membranous mass.

The reported fetal twinning incidence detected by ultrasound has ranged between 0 and 78% (Landy *et al.*, 1982), the higher incidences probably reflecting earlier detection of twins with sonograms or possible errors in technical procedures and interpretation. Examination with ultrasound is not without problems. Objectivity is essential to avoid misinterpretation of echograms to give false positives. On the other hand, a special interest in twinning may be required to alert the sonographer to the presence of twins during routine imaging.

Confirmation of the vanishing twin phenomenon may be provided by chromosomal analysis. If cells from an aborted twin are maintained in the extraembryonic tissue of the surviving twin, detection of mosaicism in placental tissue would suggest a twin pregnancy. An interesting interpretation for trisomy mosaicism confined to placental chorionic tissue of infants with intrauterine growth retardation was proposed by Kalousek and Dill (1983), who suggested that the mosaicism may be associated with placental insufficiency leading to retardation of chromosomally normal infants. The same discrepancy between infant and placenta has been observed by others (Verjaal *et al.*, 1987). However, some instances of apparent confined mosaicism may be examples of the vanishing twin phenomenon. This alternative interpretation should be considered and the chromosomes examined for possible differences in inherited chromosome variants.

Because of the possibility that first-trimester bleeding may indicate loss of a twin, early documentation of a twin pregnancy with ultrasonography becomes clinically important. Dilatation and curettage should be avoided to prevent loss of a surviving twin (Finberg and Birnholz, 1979; Landy *et al.,* 1982). Identification of a papyraceous fetus with ultrasound will also alert the obstetrician to provide specialized prenatal care for the viable twin (Landy *et al.,* 1986). Once a twin pregnancy has been identified, serial sonographic examinations may provide information on the timing of twin loss, should it occur.

The beneficial role played by ultrasound during pregnancy is without question. However, what is not yet known is whether ultrasonography is hazardous to the fetus. There is still disagreement and much confusion about the possible genetic damage on rapidly dividing cells (Martin, 1984; Nesheim *et al.,* 1987). Damage in the form of chromosome or chromatid breaks was noted by Macintosh and Davey (1970, 1972) but not supported by the majority of investigators (Bobrow *et al.,* 1971; Abdulla *et al.,* 1971). Sister chromatid exchanges in cultured lymphocytes did not appear to be increased (Morris *et al.,* 1978; Miller *et al.,* 1983). High experimental doses and timing of exposure appear to have harmful effects on animals as well as Drosophila (Pizzarello *et al.,* 1978) but the low doses used in diagnostic ultrasound are probably inconsequential to humans (Hellman *et al.,* 1970). However, indiscriminate sonographic examinations may not be justifiable until the effects on the growing fetus are proven to be harmless.

6. INCIDENCE OF TWINNING

The well-known Hardy–Weinberg law (Hardy, 1908; Weinberg, 1908) states that in a randomly mating population the proportions of a gene and its allele occur with the frequencies $p^2 + 2pq + q^2$. Similarly, among sibling or twin pairs the sexes occur in the proportions 1 : 2 : 1, i.e., one male pair to two male/female pairs to one female pair, assuming that the two sexes occur with similar frequencies. The same principle, called the differential method, was applied by Weinberg (1901) to estimate the frequency of MZ twins. Thus, in a random sample of twins the total number of DZ twins is twice the number of unlike-sex pairs and all other twins must be MZ. The proportion of MZ twins among live births in Caucasoid populations has been reported to be one-third of all twins (Bulmer, 1970).

In a study of 661 spontaneous abortions there were 15 pairs of detectable twins, a frequency of 1/44 (2.3%), seven of which were of unlike sex (Uchida *et al.*, 1983b). Under the expectation that one third were MZ twins, an additional six like-sex pairs can be said to have been undetected. Hence, the frequency of aborted twins is estimated to be 1/30 (3.3%). However, since proportionately more MZ twins are found among stillbirths and neonatal deaths (Naeye *et al.*, 1966), even a frequency as high as one twin pair among 30 spontaneous abortions is no doubt an underestimate.

The incidence of twins in the live-born Caucasoid population has been generally accepted to be on the order of 1/85. More recently, however, a gradual reduction in twin frequencies has been noted (Elwood, 1973, 1985). The *Vital Statistics of Canada* (1978–1980) reported the incidence of live-born twins in Ontario to be 0.9% (1/107). This rate, however, may rise again with the popularity of *in vitro* fertilization. From the Canadian statistics the frequency of stillbirths was 0.8% and of these 5.7% (1/17) were twin pairs, one or both of which were stillborn.

Among clinically detectable pregnancies 15% result in spontaneous abortions (Warburton and Fraser, 1964), 1% are stillborn, and 84% are live-born. The proportions of twins, estimated above, for these three groups are 1/30, 1/17, and 1/107, respectively. Hence, the frequency of recognizable twin conceptions is 1.3% or 1/75. This rate is similar to the twinning rate of 1/76 for British Columbia estimated by Livingston and Poland (1980).

From the many reports of vanishing twins an average rate of disappearance is difficult to estimate. Because of possible sources of error in the detection of twin pregnancies with ultrasound, reported frequencies have ranged widely. For the present it can only be said that there appear to be many instances of vanishing twins as well as undetectable aborted MZ twins. Thus, the overall incidence of twinning is no doubt in excess of 1/75.

ACKNOWLEDGMENTS. All photographs were prepared and contributed by Mrs. Viola C. P. Freeman, Cytogenetics Laboratory; Dr. Derek J. deSa, Department of Pathology; and Dr. C. Caco, Department of Radiology; McMaster University, Hamilton, Ontario, Canada. Permission to publish the photographs of the normal subjects is gratefully acknowledged.

REFERENCES

Abdulla, U., Campbell, S., Dewhurst, C. J., Talbert, D., Lucas, M., and Mullarkey, M. 1971. Effect of diagnostic ultrasound on maternal and fetal chromosomes. *Lancet* **2**:829–831.

Benirschke, K., and Harper, V. D. R. 1977. The acardiac anomaly. *Teratology* **15**:311–316.

Bergsma, D., ed. 1967. Conjoined twins. *Birth Defects* **3**:1–147.

Bias, W. B., and Migeon, B. R. 1967. Blood-group chimaerism with Down's syndrome. *Lancet* **2**:257.

Bieber, F. R., Nance, W. E., Morton, C. C., Brown, J. A., and Redwine, F. O. 1981. Genetic studies of an acardiac monster: evidence of polar body twinning in man. *Science* **213**:775–777.

Bobrow, M., Blackwell, N., Unrau, A. E., and Bleaney, B. 1971. Absence of any observed effect of ultrasonic irradiation on human chromosomes. *J. Obstet. Gynaecol. Br. Commonw.* **78**:730–736.

Bulmer, M. G. 1958. The repeat frequency of twinning. *Ann. Hum. Genet.* **23**:31–35.

Bulmer, M. G. 1959. The effect of parental age, parity and duration of marriage on the twinning rate. *Ann. Hum. Genet.* **23**:454–458.

Bulmer, M. G. 1970. *The Biology of Twinning in Man.* Oxford University Press (Clarendon), London.

Campbell, S., and Dewhurst, C. J. 1970. Quintuplet pregnancy diagnosed and assessed by ultrasonic compound scanning. *Lancet* **1**:101–103.

Carr, D. H. 1983. Cytogenetics of human reproductive wastage, in: *Issues and Reviews in Teratology,* Volume 1, H. Kalter, ed. Plenum Press, New York, pp. 33–72.

Chown, B., Lewis, M., and Bowman, J. M. 1963. A pair of newborn human blood chimeric twins. *Transfusion* **3**:494–495.

Corey, M. J., Miller, J. R., MacLean, J. R., and Chown, B. 1967. A case of XX/XY mosaicism. *Am. J. Hum. Genet.* **19**:378–387.

Creasy, M. R., Crolla, J. A., and Alberman, E. D. 1976. A cytogenetic study of human spontaneous abortions using banding techniques. *Hum. Genet.* **31**:177–196.

Derom, C., Derom, R., Vlietinck, R., Berghe, H. V., and Thiery, M. 1987. Increased monozygotic twinning rate after ovulation induction. *Lancet* **1**:1236–1238.

Dewald, G., Haymond, M. W., Spurbeck, J. L., and Moore, S. B. 1980. Origin of chi,46,XX/46,XY chimerism in a human true hermaphrodite. *Science* **207**:321–323.

deWolff, E., Scharer, K., and Lejeune, J. 1962. Contribution a l'étude des jumeaux mongoliens: un cas de monozygotisme heterocaryote. *Helv. Paediatr. Acta* **17**:301–328.

Dunsford, I., Bowley, C. C., Hutchison, A. M., Thompson, J. S., Sanger, R., and Race, R. R. 1953. A human blood-group chimera. *Br. Med. J.* **2**:81.

Edwards, R. G., Mettler, L., and Walters, D. E. 1986. Identical twins and in vitro fertilization. *J. In Vitro Fertil. Embryo Transf.* **3**:114–117.

Elwood, J. M. 1973. Changes in the twinning rate in Canada. 1926–70. *Br. J. Prev. Soc. Med.* **27**:236–241.

Elwood, J. M. 1985. Temporal trends in twinning, in: *Issues and Reviews in Teratology,* Volume 3, H. Kalter, ed. Plenum Press, New York, pp. 65–93.

Enbom, J. A. 1985. Twin pregnancy with intrauterine death of one twin. *Am. J. Obstet. Gynecol.* **152**:424–428.

Fanconi, G. 1962. Weitere Fälle von wahrscheinlich eineiigen Zwillingen, von denen der eine gesund ist, der andere einen Mongolismus zeigt. *Helv. Paediatr. Acta* **17**:490–491.

Finberg, H. J., and Birnholz, J. C. 1979. Ultrasound observations in multiple gestation with first trimester bleeding: the blighted twin. *Radiology* **132:**137–142.

Hardy, G. H. 1908. Mendelian proportions in a mixed population. *Science* **28:**49–50.

Hassold, T., Chen, N., Funkhouser, J., Jooss, T., Manuel, B., Matsuura, J., Matsuyama, A., Wilson, C., Yamane, J. A., and Jacobs, P. A. 1980. A cytogenetic study of 1000 spontaneous abortions. *Ann. Hum. Genet.* **44:**151–178.

Hellman, L. M., Duffus, G., Donald, I., and Sunden, B. 1970. Safety of diagnostic ultrasound in obstetrics. *Lancet* **2:**1133–1134.

Kajii, T., Ohama, K., Niikawa, N., Ferrier, A., and Avirachan, S. 1973. Banding analysis of abnormal karyotypes in spontaneous abortion. *Am. J. Hum. Genet.* **25:**539–547.

Kajii, T., Ferrier, A., Niikawa, N., Takahara, H., Ohama, K., and Avirachan, S. 1980. Anatomic and chromosomal anomalies in 639 spontaneous abortuses. *Hum. Genet.* **55:**87–98.

Kalousek, D. K., and Dill, F. J. 1983. Chromosomal mosaicism confined to the placenta in human conceptions. *Science* **221:**665–667.

Kaplan, C., and Benirschke, K. 1979. The acardiac anomaly: new case reports and current status. *Acta Genet. Med. Gemellol.* **28:**51–59.

Kerr, M. G., and Rashad, M. N. 1966. Autosomal trisomy in a discordant monozygotic twin. *Nature* **212:**726–727.

Kunze, J., and Nippert, I. 1986. *Genetics and Malformations in Art.* Grosse, Berlin.

Lamont, J. A. 1982. Twin pregnancies following induction of ovulation: a literature review. *Acta Genet. Med. Gemellol.* **31:**247–253.

Landy, H. J., Keith, L., and Keith, D. 1982. The vanishing twin. *Acta Genet. Med Gemellol.* **31:**179–194.

Landy, H. J., Weiner, S., Corson, S. L., Batzer, F. R., and Bolognese, R. J. 1986. The "vanishing twin": ultrasonographic assessment of fetal disappearance in the first trimester. *Am. J. Obstet. Gynecol.* **155:**14–19.

Levi, S. 1976. Ultrasonic assessment of the high rate of human multiple pregnancy in the first trimester. *J. Clin. Ultrasound* **4:**3–5.

Livingston, J. E., and Poland, B. J. 1980. A study of spontaneously aborted twins. *Teratology* **21:**139–148.

Livnat, E. J., Burd, L., Cadkin, A., Keh, P., and Ward, A. B. 1978. Fetus papyraceus in twin pregnancy. *Obstet. Gynecol.* Suppl. **51:**41–45.

Macintosh, I. J. C., and Davey, D. A. 1970. Chromosome aberrations induced by an ultrasound fetal pulse detector. *Br. Med. J.* **4:**92–93.

Macintosh, I. J. C., and Davey, D. A. 1972. Relationship between intensity of ultrasound and induction of chromosome aberrations. *Br. J. Radiol.* **45:**320–327.

Martin. A. O. 1984. Can ultrasound cause genetic damage? *J. Clin. Ultrasound* **12:**11–20.

Mikkelsen, M., Froland, A., and Ellebjerg, J. 1963. XO/XX mosaicism in a pair of presumably monozygotic twins with different phenotypes. *Cytogenetics* **2:**86–98.

Milham, S. 1966. Symmetrical conjoined twins: An analysis of the birth records of twenty-two sets. *J. Pediatr.* **69:**643–647.

Miller, M. W., Wolff, S., Filly, R., Cox, C., and Carstensen, E. L. 1983. Absence of an effect of diagnostic ultrasound on sister-chromatid exchange induction in human lymphocytes in vitro. *Mutat. Res.* **120:**261–268.

Morris, S. M., Palmer, C. G., Fry, F. J., and Johnson, L. K. 1978. Effect of ultrasound on human leucocytes. Sister chromatid exchange analysis. *Ultrasound Med. Biol.* **4:**253–258.

Naeye, R. L., Benirschke, K., Hagstrom, J. W. C., and Marcus, C. C. 1966. Intrauterine growth of twins as estimated from liveborn birth-weight data. *Pediatrics* **37**:409–416.

Nance, E. W. 1981. Malformations unique to the twinning process, in: *Twin Research 3: Biology and Epidemiology*. Liss, New York, pp. 123–133.

Nesheim, B. I., Benson, I., Braekken, A., Evensen, A. R., Forde, O. H., Hjort, P. F., Lonning, I., Maltau, J., Sjoli, S. I., and Thesen, J. 1987. Ultrasound in pregnancy: consensus statement, 1986. *Int. J. Techn. Assess. Health Care* **3**(3):463–470.

Newman, H. H. 1917. *The Biology of Twins*. University of Chicago Press, Chicago.

Owen, R. D. 1945. Immunogenetic consequences of vascular anastomoses between bovine twins. *Science* **102**:400–401.

Pedersen, I. K., Philip, J., Sele, V., and Starup, J. 1980. Monozygotic twins with dissimilar phenotypes and chromosome complements. *Acta Obstet. Gynecol. Scand.* **59**:459–462.

Pizzarello, D. J., Vivino, A., Madden, B., Wolsky, A., Keegan, A. F., and Becker, M. 1978. Effect of pulsed low-power ultrasound on growing tissues. *Exp. Cell Biol.* **46**:179–191.

Potter, E. L. 1973. *Pathology of the Fetus and Infant*. Year Book Medical, Chicago.

Robinson, H. P., and Caines, J. S. 1977. Sonar evidence of early pregnancy failure in patients with twin conceptions. *Br. J. Obstet. Gynaecol.* **84**:22–25.

Scott, J. M., and Ferguson-Smith, M. A. 1973. Heterokaryotypic monozygotic twins and the acardiac monster. *J. Obstet. Gynaecol. Br. Commonw.* **80**:52–59.

Stockard, C. R. 1921. Developmental rate and structural expression: an experimental study of twins, 'double monsters' and single deformities, and the interaction among embryonic organs during their origin and development. *Am. J. Anat.* **28**:115–277.

Tippett, P. 1983. Blood group chimeras. *Vox Sang.* **44**:333–359.

Turpin, R., Lejeune, J., Lafourcade, J., Chigot, P. L., and Salmon, C. 1961. Présomption de monozygotisme en dépit d'un dimorphisme sexuel: sujet masculin XY et sujet neutre haplo X. *C.R. Acad. Sci.* **252**:2945.

Uchida, I. A., Wang, H. C., and Ray, M. 1964. Dizygotic twins with XX/XY chimerism. *Nature* **204**:191.

Uchida, I. A., deSa, D. J., and Whelan, D. T. 1983a. 45,X/46,XX mosaicism in discordant monozygotic twins. *Pediatrics* **71**:413–417.

Uchida, I. A., Freeman, V. C. P., Gedeon, M., and Goldmaker, J. 1983b. Twinning rate in spontaneous abortions. *Am. J. Hum. Genet.* **35**:987–993.

Uchida, I. A., Freeman, V. C. P., and Chen, P.-L. 1985. Detection and interpretation of two different cell lines in triploid abortions.*Clin. Genet.* **28**:489–494.

Valez-Orozco, A. C. 1961. Estudio de una quimera. *Bol. Inst. Estud. Med. Biol. Univ. Nac. Auton. Mex.* **19**:41–50.

Verjaal, M., Leschot, N. J., Wolf, H., and Treffers, P. E. 1987. Karyotypic differences between cells from placenta and other fetal tissues. *Prenat. Diagn.* **7**:343–348.

Vital Statistics Annual Reports (1978–1980). Statistics Canada, Ottawa, 1980, 1981, 1982.

Walker, N. F. 1957. Determination of the zygosity of twins. *Acta Genet.* **7**:33–38.

Warburton, D., and Fraser, F. C. 1964. Spontaneous abortion risks in man: data from reproductive histories collected in a Medical Genetics Unit. *Am. J. Hum. Genet.* **16**:1–27.

Warburton, D., Stein, Z., Kline, J., and Susser, M. 1980. Chromosome abnormalities in spontaneous abortion: data from the New York City study, in: *Reproductive Loss*, E. B. Hook and I. H. Porter, eds. Academic Press, New York, pp. 261–287.

Weinberg, W. 1901. Beiträge zur Physiologie und Pathologie der Mehrlingsgeburten beim Menschen. *Arch. Ges. Physiol.* **88**:346–430.

Weinberg, W. 1908. Über den Nachweis dere Vererbung beim Menschen. *Jahresh. Ver. Vaterl. Naturkd. Württemb.* **64**:368–382.

White, C., and Wyshak, G. 1964. Inheritance in human dizygotic twinning. *N. Engl. J. Med.* **271**:1003–1005.

Winchester, A. M. 1951. *Genetics*. Riverside Press, Cambridge, Mass.

Issues and Reviews in Teratology **5:**181–253
Plenum Press, New York, 1990, 978-1-4612-7847-4

Experimental Induction of Dominant Mutations in Mammals by Ionizing Radiations and Chemicals

5

PAUL B. SELBY

1. INTRODUCTION

Dominant mutations cause their effects in heterozygotes, and if they have complete penetrance, all heterozygotes are affected. Many mutations discussed in this review have incomplete dominance (semidominance), which means that the heterozygote is intermediate in effect between the homozygotes. Often the mutation is homozygous lethal. Some dominant mutations are X-linked; most are autosomal. This review covers dominant mutations with effects detected in late pregnancy or postnatally, and it omits results of tests for dominant lethals, chromosomal abnormalities, histocompatibility mutations, and isozyme variants detected by electrophoresis.

Since many human disorders have dominant inheritance (McKusick, 1983), it is important to understand the process by which new dominant mutations occur. Many irregularly inherited disorders, which make up most of the genetic load in human populations (BEIR III, 1980), may result from dominant mutations with incomplete penetrance. If the mutation frequency in humans increased, most of the

The submitted manuscript has been authored by a contractor of the U.S. Government under contract No. DE-AC05-84OR21400. Accordingly, the U.S. Government retains a nonexclusive, royalty-free license to publish or reproduce the published form of this contribution, or allow others to do so, for U.S. Government purposes.

PAUL B. SELBY • Biology Division, Oak Ridge National Laboratory, Oak Ridge, Tennessee 37831-8077.

impact on health, at least for many generations, would result from dominant mutations.

In spite of the great need to understand induction of dominant mutations in mammals, most (Selby, 1981) of what is known about induction of gene mutations and small chromosomal deficiencies in mammals has been learned using a method that measures induction of recessive mutations, namely the mouse visible specific-locus test, developed by W. L. Russell (1951). That test is ideally suited for providing information about the physical and biological factors affecting mutation rate (e.g., dose rate, dose fractionation, germ-cell stage, and sex) and, compared with the methods discussed in this review, is simple to apply. It usually requires large sample sizes, but only a few seconds of observation is needed to see whether a mouse carries a mutation. Some specific-locus mutations also have associated dominant effects, usually on body size (W. L. Russell, 1951; L. B. Russell, 1971).

Studies of mutagenesis in mammals have great bearing on genetic risk estimation (Section 4). Specific-locus data can be used to estimate relative risk, in which induced and spontaneous mutation frequencies are compared, but specific-locus data are not useful for estimating overall phenotypic damage from induced mutations. In fact, W. L. Russell (1974) has stressed the need to improve estimates of induction of dominant mutations, and pointed out that "... probably the most serious lack in information necessary for the adequate estimation of genetic hazards of radiation in man ... is the nature, extent, and persistence of the actual anatomical and physiological damage expressed in the descendants of irradiated populations."

Several important reasons for studying induction of dominant mutations in mammals follow, and the individual tests to be described in this review usually are useful for only some of the reasons listed. Dominant mutation tests can be used to determine whether a particular agent can induce dominant mutations in mammalian germ cells, to quantify the mutational response, to study the physical and biological factors affecting mutation rate (e.g., dose rate and germ-cell stage), and to provide data useful for risk estimation. Some of the methods also yield mutations that are useful for genetic research, including some valuable models of human disorders.

2. GAMETOGENESIS

Interpretation of mutation studies requires knowledge of which germ-cell stages were exposed. For genetic risk estimation, the most important stages are the spermatogonial stem cell and the immature arrested primary oocyte because mutations induced in these specific cells persist for long times.

In male mice, germ cells are gonocytes (primordial germ cells) from midpregnancy until sometime in the first 3 d after birth, when spermatogonia first appear (Mintz, 1960; Widmaier, 1963). Germ cells in adults range from stem-cell spermatogonia to mature sperm. An adult male exposed to a large dose of acute radiation or certain chemicals remains fertile for several weeks, becomes sterile for several weeks or months, and then regains fertility. This delayed transitory sterility results from the killing of differentiating germ cells at stages of development between the rather radioresistant stem cells and the radioresistant spermatids and sperm (Oakberg, 1969, 1975). Long delays in the return to fertility result when there is extensive killing of stem-cell spermatogonia. Oakberg (1956a,b) timed the duration of the various stages of spermatogenesis, which as a whole lasts about 35 d. From his findings and the time required for sperm to reach the ejaculate, it is expected that matings during the weeks after treatment shown below would involve sperm that had been exposed when in the germ-cell stages indicated: week 1, spermatozoa; weeks 2 and 3, spermatids; weeks 4 and 5, primary spermatocytes; week 6, spermatogonia; later, stem-cell spermatogonia (Oakberg, 1969). It is also important for this review to know that spermatogenesis lasts about 48 d, or about 2 weeks longer, in rats than in mice (Clermont *et al.*, 1959).

In female mice, most germ cells have entered the leptotene or zygotene stage of meiosis by d 14 postconception (pc) (Borum, 1961). By 4 d after birth, all oocytes are in diffuse diplotene (Brambell, 1927; Borum, 1961), in which most remain until they start to mature, which may be many months later. Although the correspondence between germ-cell stages in male mice and men seems close, it is unclear how well arrested primary oocytes in mice correspond to those in women because the diffuse diplotene chromosomal state is apparently unique to mice and a few other rodents (Baker, 1973). Oocytes in adult female mice are sometimes subdivided into (1) immature arrested oocytes and (2) mature and maturing oocytes, the latter being those within 6 weeks of ovulation. The second grouping of oocytes is based upon W. L. Russell's (1977) discovery that the mutation frequency in adult females falls abruptly to zero, or almost zero, at the end of the 6th week following irradiation and upon timing of oocyte development by Oakberg (1979).

3. REVIEW OF RESULTS OBTAINED FOR DIFFERENT ENDPOINTS

Major findings will be described and discussed for many of the types of phenotypic damage reported in F_1 offspring of mice and rats exposed to mutagens. Possible explanations are given for some of the conflicting results reported. The categories covered include skeletal abnormalities, cataracts, visibles (i.e., externally visible), litter-size reduction, congenital malformations, stunted growth, shortened life span, tumors, and effects on behavior.

Units of exposure for radiation will be listed as reported by authors, without any attempt (except in tables) to convert them to a standard unit. For the purposes of this review, 1 R, 1 r, and 1 cGy are essentially equal.

3.1. Skeletal Abnormalities

Because the rodent skeleton is an intricate structure formed over several weeks, many genes presumably play a role in its development. The skeleton can easily be prepared for detailed examination (Selby, 1987), and the many similarities between mouse and human skeletons make it relatively easy to predict whether an abnormality in a mouse corresponds to what would be a serious handicap in a person.

3.1.1. Ehling's Experiments

At the suggestion of W. L. Russell, Ehling in 1960 collaborated with W. L. and L. B. Russell to attempt to detect skeletal variants in the offspring of male mice exposed to radiation (W. L. Russell, 1981). The Russells had considerable experience studying mouse skeletons, and they thought the skeleton might be a body system in which overall induced dominant damage could be measured. The approach used by Ehling to separate mutational damage from within-strain variation hinged on the idea that abnormalities caused by mutations would probably occur no more than once in an experiment involving no more than a few thousand mice. In several experiments (Ehling and Randolph, 1962; Ehling, 1965, 1966), male mice of the 101 inbred strain were exposed to acute X radiation or neutrons in various doses. F_1 offspring, from matings with females of the C3H inbred strain, were killed for skeletal preparation at 26–28 d of age, and alizarin-stained and cleared intact preparations were examined. Experimental and control groups were coded before examination.

The following criteria were developed in an attempt to distinguish between mutational and nonmutational variation. Skeletal abnormalities were classified as Class 1 if they occurred only one time in the experiment (experimental and control groups combined) and Class 2 if they occurred more than once. Animals having Class 1 abnormalities were further classified as Class 1 multiple if they had one or more additional abnormalities of Class 1 or Class 2, but not counting a small number of extremely variable abnormalities. Class 1 animals with single abnormalities of the appendicular skeleton were subdivided according to whether the anomaly was bilateral or unilateral. Mice were concluded to be presumed dominant mutants if they were either Class 1 multiple or Class 1 bilateral appendicular. Both categories of presumed mutants were of statistically significantly higher frequency in the combined exposed groups, compared with the control. Although these criteria are not presently used (Section 3.1.2), they were an important step in the development of ways to distinguish mutational damage and nonmutational variation.

In experiments with high-dose-rate X irradiation, the presumed mutation frequencies were as follows: combined controls of three experiments, 1/1739 (0.06%); 600 R to postspermatogonial stages, 10/569 (1.8%); various doses to stem cells: 600 R, 5/754 (0.7%); 100 + 500 R, 24-h interval, 5/277 (1.8%); and 500 + 500 R, 10-week interval, 2/131 (1.5%). Including the few data on neutrons, 23 presumed mutations were observed in 1968 offspring derived from irradiated males.

The results from different radiation conditions paralleled those for specific-locus mutations (W. L. Russell *et al.*, 1958; W. L. Russell, 1964) in that (1) X irradiation of postspermatogonial stages yielded about twice the mutation rate obtained in stem cells and (2) there were about twice as many mutations when 600 R was administered as 100 + 500 R (with 24 h between exposures) instead of as a single exposure. For both comparisons, differences in the frequencies of skeletal mutations approach statistical significance, $P = 0.06$ and $P = 0.10$, respectively, in a 1-tailed Fisher's exact test. Because of these similarities, and because there were statistically significantly more mutations in most experimental groups than in the control, Ehling (1966) concluded that a sizable proportion of the presumed dominant skeletal mutations probably resulted from gene mutations or small deficiencies.

Ehling published descriptions of 14 of his 24 presumed mutants. Almost all of the presumed mutants were killed before breeding age. Ehling (1967, 1970) did report that "3 dominant mutations affecting the skeleton were found to be transmitted to the second and later generations," but he presented no details on the transmissibility of those mutations. Skepticism by others over whether Ehling's presumed mutants were really mutants was one reason why his data were not used by risk estimation committees until 1977, when Selby and Selby (1977a) reported a high frequency of transmitted radiation-induced dominant skeletal mutations (Section 3.1.2a). Most of Ehling's mutations manifested effects similar to malformations for which transmission has since been demonstrated by Selby and Selby (1978b).

3.1.2. Selby's Experiments

3.1.2a. Breeding-Test Experiment. The Selby and Selby (1977a, 1978a,b) experiment was designed to use a breeding test to judge the correctness of Ehling's conclusion that radiation induces a high frequency of dominant skeletal mutations. Male mice of the 102 inbred strain were exposed to 100 + 500 R (24-h interval) of acute ^{137}Cs γ radiation and, after the sterile period, mated to C3H females to study induction of mutations in stem-cell spermatogonia.

Of 37 dominant skeletal mutations found in 2646 F_1 offspring, 31 were confirmed by breeding tests (Selby and Selby, 1978b) and 6 with no offspring were classified as mutants based on presumed-mutation criteria strongly supported by breeding-test results (Selby and Selby, 1978a). There was no control group because Ehling's results suggested that very few presumed mutations would be found in a control, and the main concern was the number of transmissible effects in the experimental group. It was assumed that the induced mutation frequency was roughly equal to the observed one. This assumption was further supported by examinations of parents and sibs of mutants, from which Selby and Selby (1977a) concluded that few, if any, of the 37 mutations had preexisted in the stocks. Lüning and Eiche (1977) stressed the possible importance of preexisting mutations in skeletal experiments, and they felt that the absence of a control might have kept us from recognizing an important contribution of preexisting spontaneous mutations. Although their concern is valid for many experiments (Section 3.1.2b), we have pointed out (Selby and Selby, 1977a,b) that the data rule out any important component of preexisting mutations in our breeding-test experiment.

Twenty-five of the 31 proven mutations caused multiple effects, often in widely separated parts of the skeleton. Most abnormalities involved fusions of bones or other changes in the numbers of separate bones, gross changes in the shapes of bones, or changes in their relative positions. Because at least 9 of the 31 proven mutations had incomplete penetrance and essentially all had variable expressivity, it seems clear that many of the malformations are analogous to irregularly inherited disorders in humans. Because of the common incomplete penetrance and variable expressivity, Selby and Selby (1977a) suggested that many of the skeletal malformations are probably threshold traits, similar to the situation described for mutation *Px* in guinea pigs (Wright, 1968). Many of the individual skeletal abnormalities resemble malformations found in humans, and at least one mutation, namely *Ccd* (cleidocranial dysplasia), is strikingly homologous to a whole syndrome in humans (Sillence *et al.*, 1987).

Eight of the 31 proven mutations were tested for effects in homozygotes and all were homozygous lethal (Selby, 1979; Selby and Lee, 1980). For seven of them, homozygotes were found as resorbing moles, and for the eighth, homozygotes, which manifested skeletal malformations more extreme than those in heterozygotes, died at about 2 weeks of age. Three mutations are heritable chromosomal translocations (Selby, 1979). Twenty-four specific skeletal malformations were found to be caused by two or more of the 31 proven mutations (Selby and Selby, 1977a). Thus, either several different mutations can cause each of these malformations or multiple mutations occurred at many gene loci. The former seems more likely.

3.1.2b. Non-Breeding-Test (NBT) Experiments. Low penetrance, sterility, or death before breeding age can prevent the detection of mutations by methods that require a demonstration of transmissibility before counting variants as mutants (Selby, 1982). To overcome these disadvantages of strict breeding-test methods, and to make the study of dominant skeletal mutations less labor intensive, most of the more recent effort has been devoted to development of NBT methods, in which presumed mutations are identified by phenotype alone. These methods have been described in detail elsewhere, and the specific abnormalities that are considered sensitive indicators and index abnormalities (described below) have been listed (Selby, 1983; Selby and Niemann, 1984). The NBT methods are based mainly on the understanding gained from the large breeding-test experiment (Section 3.1.2a).

Sensitive indicators are abnormalities that are thought not to occur (or at least hardly ever), in F_1 progeny unless a mutation is present, and to result from mutations at any one of many different genes. A mouse is considered a presumed mutant if any one of the following three criteria is met: (1) the presence of any sensitive indicator, (2) the presence of two or more rare (i.e., less than 1 in 400 in the control) and major (i.e., easily seen using a dissecting microscope) defects that do not seem to result from one accident of development, and (3) the presence bilaterally of a single extremely rare (i.e., less than 1 in 1000 in control) major abnormality (Selby, 1983). The induced mutation frequency is calculated by subtracting the presumed mutation frequency in the concurrent control from that of the experimental group. Both the experimental and control groups are assumed to contain approximately equal numbers of spontaneous mutations (both new and preexisting) and nonmutant variants.

The results of the breeding-test experiment (Section 3.1.2a) strongly suggest that these criteria would result in relatively few errors of falsely classifying nonmutant mice as mutants, because all 15 offspring that met the presumed-mutation criteria (criteria 2 and 3 above, except that the degree of rarity was based on the experimental group) and had offspring were proved to be mutants (Selby and Selby, 1978a,b). At the same time, the breeding-test results show that many mutations are not identified by criteria 2 and 3 alone. The addition of criterion 1 substantially increases the number of mutations found, and most mutations with serious effects should be identified using these three criteria. These criteria are similar to Ehling's (Section 3.1.1) in being based on the idea that malformations caused by mutations are rare in F_1 progeny, but they differ (1) in allowing for the certainty that some individual abnormalities are caused by different mutations and (2) by providing more guidance as to which combinations of abnormalities are sufficiently unusual to suggest a mutational basis.

In NBT experiments, parents are randomized between experimental and control groups and F_1 offspring are killed for skeletal preparation at 3 or 6–8 weeks of age. They are coded, prepared (stained with alizarin and cleared), examined, classified, and uncoded. Usually only the skeleton anterior to the pelvis is prepared. Table I and Fig. 1 show the presumed mutation frequencies obtained in all NBT experiments. Preliminary or final results have been published for most of these experiments (Selby, 1983; Selby and Niemann, 1984). All experiments had a concurrent control and, except for the fractionated ethylnitrosourea (ENU) experiments, parents were randomized between the experimental and control groups. The results show clearly that both ENU and X rays induce dominant skeletal mutations in stem-cell spermatogonia. The induced presumed-mutation frequency of 0.49%/gamete, at the acute 600-R exposure, is not statistically significantly different from the expectation of 0.74%/gamete for a single acute exposure, as estimated from the earlier breeding-test data using a correction for the fractionation effect (Section 4.2.1).

Preexisting mutations accounted for a large fraction of the mutations found in all experiments (Fig. 1). The preexisting mutations consisted mainly of two widespread mutations with low penetrance, which

Table I. Dominant Skeletal Mutation Frequencies Corrected (and Uncorrected) for Preexisting Mutations[a]

Agent	Dose (cGy or mg/kg)	Dose rate (cGy/min)	Germ-cell stages	Strains used[b]	Number of newly arising presumed mutations[c]	Number of offspring	Mutations/gamete ($\times 10^2$)
X rays	600	85–93	Stem cells[d]	1	6(12)	1022	0.59[j]
	100 + 500[e]	85–93	Stem cells	1	3(7)	452	0.66[j]
	300	85–93	Stem cells	1	2(9)	1307	0.15
γ rays	600	0.005	Stem cells	1	1(2)	205	0.49
ENU	150	—	Stem cells	1	4(7)	331	1.21[k]
Control for above experiments				1	2(8)	2060	0.10
Dichlorvos	See below[f]		All in adult male and female	1	0(4)	653	0
	See below[f]		Stem cells	1	1(2)	308	0.32
Control for dichlorvos experiment				1	1(7)	845	0.12
^{239}Pu[g]	58	—	All in adult male	1	6(20)	3353	0.18
Control for plutonium experiment				1	3(29)	1987	0.15
ENU	3 × 100[h]	—	Stem cells	2	10(10)	243	4.12[k]
	4 × 100[h]	—	Stem cells	2	10(10)[i]	180	5.56[l]
Control for fractionated ENU experiments				2	2(2)	374	0.53

[a] All mutations identified using non-breeding-test methods.
[b] 1: 102/Sl strain males are mated to C3H/Sl females; 2: (101/Rl × C3H/Rl)F_1 males are mated to random-bred T-stock females.
[c] Numbers in parentheses are numbers of mutations before correcting for preexisting ones. Preexisting mutations consisted mainly of two widespread mutations with low penetrance and one mutation with high penetrance, which was found in only one sibship.
[d] Spermatogonial stem cells.
[e] 24-h interval between exposures.
[f] Both sexes were exposed for 80 d to one-third of a resin strip impregnated with dichlorvos (Johnson Wax BOLT brand) that was placed on top of pen; mated immediately afterward; when exposed females died they were replaced with unexposed females who produced offspring listed as being derived from exposed stem cells.
[g] 58 rad of α radiation from monomeric ^{239}Pu-citrate incorporated into the testes.
[h] 1-week interval between exposures.
[i] Two of these mutations are a cluster of 2; the cluster is corrected for in the statistical comparison.
[j] Significantly higher than concurrent control, $P < 0.05$ in a 1-tailed Fisher's exact test.
[k] Significantly higher than concurrent control, $P < 0.01$ in a 1-tailed Fisher's exact test.
[l] Significantly higher than concurrent control, $P < 0.001$ in a 1-tailed Fisher's exact test.

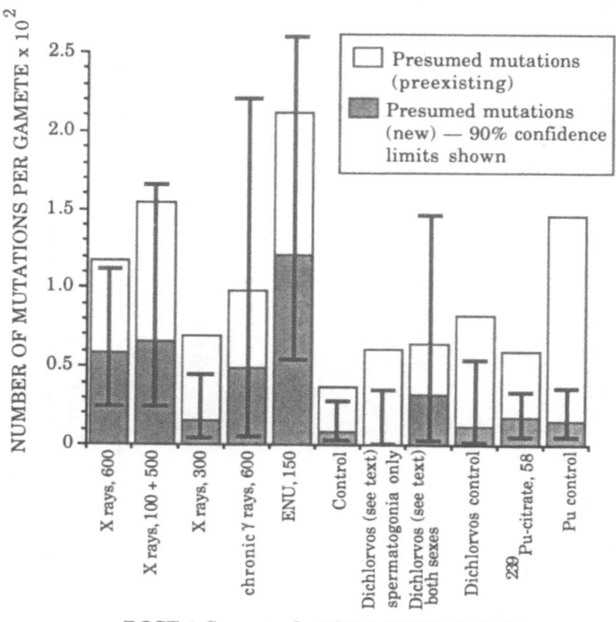

Figure 1. Frequencies of presumed dominant skeletal mutations. The full height of each bar shows the presumed mutation frequency observed for the specified treatment. The shaded part of each bar shows the presumed mutation frequency after a correction was made for preexisting mutations. Exposure conditions are shown in Table I.

were described elsewhere (Selby, 1983), and one mutation with high penetrance, which was found in only one sibship. Although the estimates of the numbers of preexisting mutations are preliminary, they are being reported here so that investigators studying dominant mutations will realize the great impact that preexisting mutations, especially those with incomplete penetrance, can have on their results. It has been recognized for years that inbred lines are not entirely genetically homogeneous owing to the presence of spontaneous mutations and the residual heterozygosity (Haldane, 1936; Deol *et al.*, 1957; Bailey, 1978); however, it was still surprising to see how many of the skeletal mutations were preexisting mutations in these experiments.

From theoretical considerations, Haldane (1936) estimated that the heterozygosity resulting from mutations might be 9–13 times the spontaneous mutation frequency. The control data shown in Fig. 1 appear to be in good agreement with his estimate if it is assumed that all of the preexisting mutations in our experiments have been identified as such. Of course, preexisting mutations in the stocks could only complicate the analysis of a mutation experiment if they happened to affect the endpoint under study. Thus, e.g., a preexisting mutation causing a belly spot would not complicate a skeletal experiment. Selby (1983) proposed a method of using multiple sublines of inbred strains to deal more effectively with this problem. It is important to note that the mutation frequencies for the 600-R and 100 + 500-R X-ray treatments and for the 150-mg/kg ENU exposure were statistically significantly higher than the control both before and after the correction for preexisting mutations.

The mutational-index method (Selby, 1983), which is another NBT method, is a first attempt to use many additional types of dominant skeletal damage to determine whether mutations are being induced. The skeletal abnormalities that are used in calculating the index of mutation are termed index abnormalities, and they are anomalies that are known, or strongly suspected, to be caused by dominant mutations. However, except for the sensitive indicators, which are included with them, they probably all occur at much too high a frequency as nonmutant variants to be used by themselves to classify a particular mouse as a presumed mutant. The more index abnormalities present in a given mouse, the more likely it is to be mutant. For this reason, as many as three index abnormalities are counted per mouse in calculating the index of mutation (Selby, 1983). As a result, the index of mutation is not a mutation frequency, but it is useful for deciding whether the treatment in question induces dominant mutations. Results obtained on two different genetic backgrounds (Selby, 1983; Selby and Niemann, 1984) suggest that this approach may be useful.

Selby and Niemann (1984) applied NBT methods to offspring collected in fractionated ENU experiments (Table I). For both treatments, the experimental presumed mutation frequencies and the indices of mutation were statistically significantly higher than those of the concurrent control. Of seven different comparisons made between the experimental groups and the control, experimental frequencies or indices of mutation were always statistically significantly higher than the control and the point estimate at 4 × 100 mg/kg was always higher than that at 3 × 100 mg/kg. Because the F_1 offspring were taken from an ongoing specific-locus experiment (Hitotsumachi *et al.*, 1985), this experiment had the potentially serious drawback of not having the parents randomized. However, the consistent results of comparisons, the uniqueness of most syndromes, and the results of analyses of sibships suggested that preexisting mutations did not seriously influence the conclusions.

Selby *et al.* (1984a,b) carried out a small breeding-test experiment on offspring drawn from the same 4 × 100-mg/kg ENU experiment. Of 114 experimental F_1 males, 5 were presumed mutants and 23 had at least one index anomaly. There were no index abnormalities or presumed mutants among 27 control males. Two of the five presumed mutations were confirmed, with both having low penetrance; two were sterile. These results are consistent with the effect seen for ENU in NBT experiments, but because the control was so small, it is impossible to rule out that preexisting mutations might have accounted for an important fraction of the index abnormalities that were transmitted. A much larger breeding-test experiment in progress with ENU should provide a stronger foundation for NBT experiments carried out on mice from specific-locus experiments. That experiment should also provide more information on the transmissibility of some of the more common index abnormalities in both the experimental and control groups.

All of Selby's skeletal experiments were said to have used the 102 mouse strain even though previous papers indicated that the 101 strain was used. A single name for this strain was used because it seems likely that all of Selby's experiments were carried out long after the 101 inbred strain in Ehling's colony became genetically contaminated, possibly by a cross to the C3H strain. The name 102 was recommended by West *et al.* (1984, 1985c) who discovered the contamination. Selby derived his 101 strain from Ehling's.

3.1.3. Importance of Randomizing Parents and of Coding Offspring

The purpose of randomizing parents between experimental and control groups is to distribute any preexisting mutations between the groups as evenly as possible. To accomplish this, littermates are distributed at random, with the exception that they are divided as evenly as possible. To illustrate, if four C3H females from one litter are to be distributed between an experimental and a control group, each female has an equal chance of being put into either group until one group has two of them, at which time any remaining females are put into the other group.

The finding of so many preexisting skeletal mutations has implications for many, if not all, areas of research to be discussed. This finding becomes much more important because, as far as can be judged from publications, only in very few experiments are parents randomized between the experimental and control groups. There is thus the danger that an experiment might be set up with many more of the parents carrying a preexisting mutation in the experimental group than the control group. If this happened, an observer might be misled into concluding that he had found a high induced mutation frequency when few if any mutations were actually induced. The converse of this situation would also be important because a chemical might not be identified as a mutagen owing to an unusually high control mutation frequency. It is important to identify and correct for preexisting mutations if at all possible. Because the skeleton is a large complex structure, different syndromes can be distinguished more readily than for most, if not all, other endpoints; hence, the skeletal methods offer greater potential for recognizing preexisting mutations.

For the many experiments in which subjectivity is involved in observing effects and in applying criteria, coding of offspring is important to avoid unintentionally biasing results. Although coding is always done in skeletal experiments, the literature suggests that it is a rather uncommon practice in other experiments. In this review, experiments in which parents were randomized or offspring were coded will be noted.

3.1.4. Earlier Skeletal Experiments

Searle (1964) attempted to use the skeleton to show the effects of induced mutations by examining 23 threshold traits and 4 metrical characters. In contrast with the emphasis put on rare abnormalities by Ehling and Selby, the threshold traits studied by Searle were common variants, the least common of which occurred in 1.2% of control offspring. Many were found in 10–40% of control animals. Ancestors of the experimental group had been exposed to low-level γ radiation for many generations. The rate of divergence between two experimental sublines was expected to be higher than that between two control sublines, but just the opposite happened, probably because a major mutation became fixed in a control subline. This experiment suggests that common variants are of little value for demonstrating mutation induction in mammals. Grüneberg *et al.* (1966) applied a rather similar approach to natural populations of rats exposed to natural radioactivity in south India and found no effect of the radiation.

Röhrborn and Vogel (1969) unsuccessfully attempted to detect induction of dominant mutations, including skeletal mutations, in F_1 progeny of male mice injected intraperitoneally with 0.125 mg/kg triethyleneiminobenzoquinone-1,4 (Trenimone) and mated for eight weekly intervals. They concluded that Ehling's method, which they applied to coded samples, "cannot be recommended for screening of chemicals as to their mutagenic activity." However, they may have failed to detect dominant skeletal mutations because their samples were small, because they examined skeletons in less detail using "pocket-lens magnifications," or because the damage induced may have been gross chromosomal damage incompatible with life.

3.1.5. Lovell's Experiment

Lovell *et al.* (1985) used detailed morphometric measures of six large bones in the mouse in an attempt to identify dominant mutations induced in stem-cell spermatogonia by 250 mg/kg ENU. Their study involved 225 experimental and 253 control offspring. Detailed statistical analyses of as many as 15 measurements per bone were made in hopes of identifying subtle changes in bone shape to provide a more sensitive test for dominant mutations (CCEM, 1983). Because skeletons of F_1s were prepared by papain digestion, they were disarticulated, with the result that many parts of the skeleton important in NBT experiments were discarded without study. An analysis was also made of a series of common variants like those Searle studied unsuccessfully (Section 3.1.4).

The authors concluded that it was unlikely that any of the variants that they found resulted from mutations, and they (Johnson and Lovell, 1983) stated accordingly that "dominant skeletal mutations are not induced by ENU in mouse spermatogonia." On the basis of their results, they argued against the application of our skeletal methods for studying chemical mutagenesis. Lovell *et al.* (1985) were especially critical of the NBT methods, which had shown ENU to be a powerful mutagen in three separate experiments (Section 3.1.2b).

The results of Lovell *et al.* provide no support for such strong conclusions. They implied that they would have found many mutations if our conclusion that ENU induces a high frequency of dominant skeletal mutations is correct. However, based on our experiments with ENU and on the relative numbers of specific-locus mutations (W. L. Russell *et al.*, 1982; Hitotsumachi *et al.*, 1985) found for those same treatments and the 250-mg/kg treatment (W. L. Russell *et al.*, 1979, 1982), the expected number of presumed dominant skeletal mutations in the Lovell *et al.* sample of 225 offspring is only from 3 to 5. According to the binomial distribution, the finding of no, 1, 2, or 3 mutations in 225 offspring rules out, at the 5% significance level, presumed mutation frequencies (expressed per gamete) higher than 1.32, 2.09, 2.77, and 3.41%, respectively. The numbers of mutants to which these frequencies correspond in the sample of 225 are 2.97, 4.70, 6.23, and 7.67. Because their data do not rule out a mutation frequency of over 1% per gamete, it seems unsupportable for them to claim absence of induction of dominant skeletal mutations. Their methods are so different from ours that it seems likely that they could have overlooked several presumed mutations.

Thus, there seems to be no justification for their conclusion that the skeletal NBT methods are unreliable.

Some of the experimental procedures used by Lovell *et al.* (1985) decrease the likelihood that they could have found the 3–5 mutations predicted from our experiments. Because they killed their mice for examination at 88–104 days of age, they could not have seen any gross abnormalities associated with syndromes severe enough to cause death before that age. In two of our ENU experiments (Selby and Niemann, 1984), mice were killed for study at 3 weeks of age. Some of the presumed mutants were extremely small then and, accordingly, would have been much more likely to die in the few weeks following weaning. Most of the presumed mutations we have found could not be identified using only the bones Lovell *et al.* examined. Another factor that might have changed the expectation from our work was that they used entirely different strains (reciprocal hybrids of the C57BL/6J and DBA/2J strains).

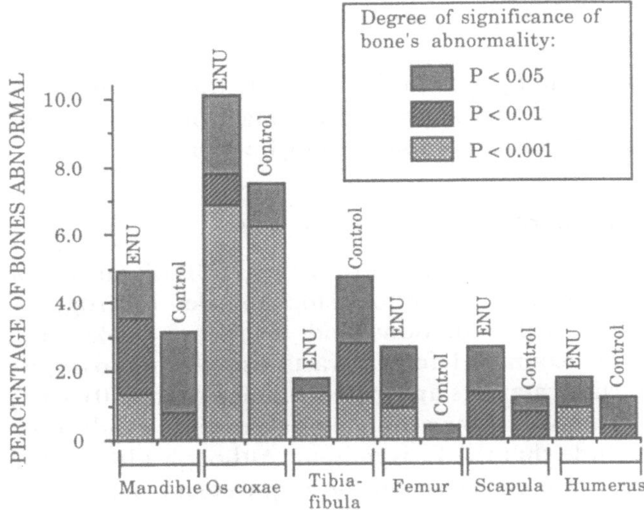

Figure 2. Morphometric data of Lovell *et al.* (1985) showing effects on bone shape of mutations induced by exposure of stem-cell spermatogonia to 250 mg/kg ethylnitrosourea. Total height of each bar indicates the percentage of bones abnormal at the 5% significance level. Shading patterns indicate percentages of bones abnormal at other significance levels. For example, the top of the diagonally striped section represents those bones judged to be outliers at the 1% significance level.

Although Lovell *et al.* did not think so, their results suggest that ENU does induce dominant mutations with effects that can be detected by detailed morphometric analysis. Figure 2 shows their morphometric data. They found a higher percentage of abnormal femurs in the experimental group ($P = 0.04$ in a 1-tailed Fisher's exact test). For the bones that were the most extreme outliers at a significance level of $P < 0.001$, in all five cases in which the experimental and control groups differed, the experimental group had a higher percentage of variants. If there were no real difference between the experimental and control groups, the probability of finding five out of five higher in the ENU group would be 0.031. It thus appears that significantly more variants occurred in the experimental group. From our skeletal studies, it appears that relatively few dominant mutations cause gross shape changes in any of the six bones they studied; however, five of these bones are known to be affected by at least one mutation. The frequencies of "gross abnormalities" were 4/225 and 2/253 in their experimental and control groups, respectively. Again, the point estimate for the ENU group is higher.

3.2. Cataracts

A cataract is an opacity in the normally transparent lens of the eye. The term "cataract" can apply to anything from a small asymptomatic opacity to opacities that completely block vision.

3.2.1. Description of Methods and Results

In most cataract experiments, $(101 \times C3H)F_1$ male mice are exposed to a mutagen and mated to T-stock females. F_1 progeny are examined for specific-locus mutations, and, when 4–6 weeks old, their eyes are dilated and examined for cataracts using a microscope and a slit lamp. Those with cataracts are mated to mice of the 101 strain to see if enough offspring will show a phenotype similar enough to that seen in the F_1 to conclude that the F_1 is mutant. Although Ehling (1984a, 1985) has placed extreme emphasis on the requirement that transmission be demonstrated before considering a variant a mutant, exceptions have been made by his group.

Tables II and III show results following treatment of stem-cell spermatogonia and postspermatogonial stages, respectively. ENU and γ radiation induce dominant cataract mutations in stem cells and postspermatogonial stages, although the conclusion for ENU in postspermatogonial stages is based on only two mutations, one of which belongs to a class of variants often discarded without testing. Kratochvilova *et al.*

Table II. Dominant Cataract Mutation-Rate Data in Stem-Cell Spermatogonia

Agent	Gy or mg/kg	Dose rate (Gy/min)	Interval (h)	Strains used[a]	Number of mutations	Number of offspring	Mutations/ gamete (× 10^4)	Mutations/ gamete/cGy or mg/kg (× 10^7)	Reference
γ rays	4.55 + 4.55	0.55	24	1	6	5,231	11[g]	13	Ehling *et al.* (1982)
	5.34	0.53	—	1	3[b]	10,212	2.9	5.5	Ehling *et al.* (1982)
	6.0	0.53	—	1	3	11,095	2.7	4.5	Ehling *et al.* (1982)
X rays	3.0 + 3.0	0.75	24	1	3[c]	15,551	1.9	3.2	Graw *et al.* (1986)
	5.1 + 5.1	0.75	24	1	3[d]	11,205	2.7	2.6	Graw *et al.* (1986)
	3 + 3	0.75	24	2	5	14,132	3.5[h]	5.9	Favor *et al.* (1987)
	3 + 3	0.75	24	3	12[i]	15,931	7.5[h]	13	Favor *et al.* (1987)
Procarbazine	5 × 200	—	168	1	1	2,249	4.4	4.4	Kratochvilova *et al.* (1988)
	600	—	—	1	2	25,090	0.8	1.4	Kratochvilova *et al.* (1988)
ENU	80	—	—	1	8	5,090	16[g]	196	Favor (1986)
	160	—	—	1	14	6,435	22[g]	136	Favor (1986)
	250	—	—	1	17	9,352	18[g]	73	Favor (1983)
	80 + 80	—	24	1	8[i]	17,665	4.5[g]	28	Favor *et al.* (1988)
Untreated historical control[e]				1	1	22,594	0.4	—	Favor *et al.* (1987)
Solvent control[f]				1	0	1,780	0	—	Ehling *et al.* (1982)
ENU	250	—	—	4	6[c]	923	65	260	West and Fisher (1986)
Control of West and Fisher				4	0	538	0	—	West and Fisher (1986)

[a] 1: $(101/El \times C3H/El)F_1$ males mated to untreated T-stock females; 2: BALB/c males mated to untreated T-stock females; 3: DBA/2 males mated to untreated T-stock females; 4: C3H/HeH males mated to untreated PT-stock females.

[b] Because one of these mice was sterile, it was assumed to be mutant based on its phenotype.

[c] Two of these mutations are a cluster of 2.

[d] One of these mutants was a Class III variant, which means that in some experiments it would have been discarded without testing.

[e] It consists of the following parts (No. of mutations/No. of offspring): 0/8174 in control for 4.55 + 4.55-Gy and 5.34-Gy experiments (Kratochvilova, 1981); 0/2517 in control for 250-mg/kg ENU experiment (Favor, 1983); 0/11,036 in control for X-ray experiments of Graw *et al.* (1986); and 1/867 in control for 80- and 160-mg/kg ENU experiments (Favor, 1986).

[f] Received equal volume of 0.03 M phosphate buffer (pH 6.0) as that used for solvent for ENU in the 250-mg/kg experiment.

[g] Frequency is statistically signficantly higher than that of the untreated historical control. $P \leq 0.008$ in a 1-tailed Fisher's exact test.

[h] No control is available for comparison on this genetic background.

[i] Two of these mice were sterile, so they were assumed to be mutant based on phenotype

Table III. Dominant Cataract Mutation-Rate Data in Postspermatogonial Stages

Agent	Gy or mg/kg	Dose rate (Gy/min)	Interval (h)	Strains used[a]	Number of mutations	Number of offspring	Mutations/ gamete ($\times 10^4$)	Mutations/ gamete/cGy or mg/kg ($\times 10^7$)	Reference
γ rays	4.55 + 4.55	0.55	24	1	1	272	37[d]	40	Ehling *et al.* (1982)
	5.34	0.53	—	1	1	1721	5.8	11	Ehling *et al.* (1982)
	6.0	0.53	—	1	1[b]	865	12	19	Ehling *et al.* (1982)
X rays	3.0 + 3.0	0.75	24	1	1	1120	8.9	15	Graw *et al.* (1986)
	5.1 + 5.1	0.75	24	1	0	425	0	0	Graw *et al.* (1986)
ENU	250	—	—	1	2[c]	3360	6.0[d]	24	Favor (1983)
	250	—	—	2	0	86	0	0	West and Fisher (1986)
Procarbazine	600	—	—	1	0	413	0	0	Kratochvilova *et al.* (1988)

[a] 1: (101/El × C3H/El)F_1 males mated to untreated T-stock females; 2: C3H/HeH males mated to untreated PT-stock females.
[b] Because this mouse was sterile, it was assumed to be mutant based on its phenotype.
[c] One of these mutants was a Class III variant. In some experiments such variants are discarded without testing.
[d] The frequency is statistically significantly higher than that of the untreated historical control, $P < 0.05$ in a 1-tailed Fisher's exact test.

(1988) claimed that procarbazine, in the fractionated exposure, significantly induced cataract mutations in stem cells, but my calculation of the P value is 0.17 in a 1-tailed Fisher's exact test.

Approximately half of the offspring had cataracts, when heterozygotes were mated to normal mice, for 7 of 10 γ-ray-induced cataract mutations (Kratochvilova, 1981), for 2 of 3 procarbazine-induced cataract mutations (Kratochvilova *et al.*, 1988), and for 12 of 23 ENU-induced ones (Favor, 1984; Favor *et al.*, 1988). This is, of course, as expected for autosomal dominant mutations with full penetrance and normal viability. For some mutations, fewer affected offspring were found because of viability effects, but a significant number of cataract mutations had incomplete penetrance. Some cataract mutations had variable expressivity (Ehling, 1980a; Favor, 1984; Graw *et al.*, 1986), and some manifested additional phenotypes, e.g., white belly spots, abnormal teeth, or stunted growth (Kratochvilova, 1981; Favor, 1984). Kratochvilova (1981) concluded that 7 of 10 transmissible γ-ray-induced dominant cataract mutations were homozygous lethal. Of three mutations recovered in procarbazine experiments, one was homozygous lethal, one was homozygous viable, and the test of the third was inconclusive (Kratochvilova *et al.*, 1988). Because Favor (1984) recovered at least some homozygotes from 10 of 11 ENU-induced cataract mutations for which he had definitive data, he (Favor, 1986) concluded that there is a substantial difference between radiation- and ENU-induced dominant cataract mutations in the likelihood of being homozygous lethal, probably because radiation-induced mutations are mainly small deletions while ENU-induced ones are mainly intragenic changes. Favor (1984) also noted the difficulty of assessing homozygous lethality for mutations with incomplete penetrance.

Many cataract mutations are being made congenic on the inbred 101 strain, and many are being tested for allelism. The fully penetrant and homozygous-viable mutation found in the procarbazine experiment is one of five phenotypically different mutations that are alleles, with the others coming from radiation experiments or breeding stocks (Kratochvilova *et al.*, 1988). The cataract test is clearly a rich source of material for learning more about this type of abnormal development. These experiments have yielded a large fraction of the lens abnormalities known in the mouse (West and Fisher, 1985).

Although the mutation frequency in West and Fisher's ENU experiment is statistically significantly higher than that in Favor's experiment at the same dose, these studies may not have dealt with comparable mutations. West and Fisher (1986) and Kratochvilova *et al.* (1988) stressed that cataracts in the West–Fisher study had milder effects. Kratochvilova *et al.*

(1988) suggested that an additional reason for the difference might be that West and Fisher counted all types of lens opacities in determining transmission instead of just those of a defined phenotype.

3.2.2. Possible Reasons for Poor Repeatability of Some Results

The cataract method has the major advantage of permitting easy detection of cataracts in living F_1s, but the method has serious drawbacks, one being the low and not consistently repeatable mutation frequencies found even for powerful mutagens. For example, Graw *et al.* (1986) found no significant increase in the mutation frequency in either stem cells or postspermatogonial stages despite the relatively high mutation frequency found by Ehling *et al.* (1982) and the well-known enhancement of radiation-induced mutation frequencies by a 24-h fractionation interval with large fractions (W. L. Russell, 1963). Figure 3 compares the results of Graw *et al.* (1986) with Ehling and colleagues' (1982) 4.55 + 4.55-Gy experiment. Even for the combined stem-cell data in Graw's experiments, the P value for the comparison with the untreated historical control only reaches 0.10 in a 1-tailed Fisher's exact test. The specific-locus data collected in the same experiments (Figure 4) show that the exposures were clearly effective. The cataract mutation frequency per Gy in the 5.1 + 5.1-Gy experiment is statistically significantly lower ($P < 0.05$) than that found in the earlier 4.55 + 4.55-Gy experiment of Ehling *et al.* (1982). The authors (Graw *et al.*, 1986) attributed their unexpected results to sampling error, but, whatever the reason, the failure to demonstrate clear-cut effects in large experiments following massive exposures to such a strong mutagen raises the question of how useful the method would be for defining the genetic risk for a less effective mutagen.

The recent 2 × 80-mg/kg ENU experiment of Favor *et al.* (1988) provides another example of poor repeatability because it yielded a mutation frequency significantly lower than that of the earlier 80-mg/kg experiment ($P = 0.03$ in a 2-tailed Fisher's exact test), even though the eight mutations claimed by the authors in the fractionated experiment included two sterile, and thus unconfirmed, Class 1 variants. The specific-locus mutation frequencies indicated that the effects of the doses in the 2 × 80-mg/kg experiment were additive. Therefore, on an equal-dose basis, the cataract mutation frequency in the fractionation experiment was only 14% of that in the earlier experiment. The poor agreement was attributed to sampling error, but the significant difference suggests more serious problems.

Graw *et al.* (1986) ruled out scoring and exposure errors as possible

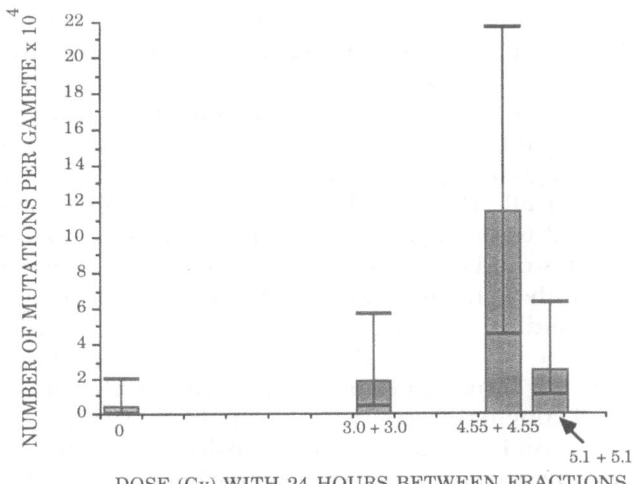

Figure 3. Dominant cataract mutation frequencies following exposure of stem-cell spermatogonia to acute X rays or γ rays (details in Table II). Fractionation interval was 24 h. Ninety percent confidence limits are shown. Data of Graw *et al.* (1986) and Ehling *et al.* (1982).

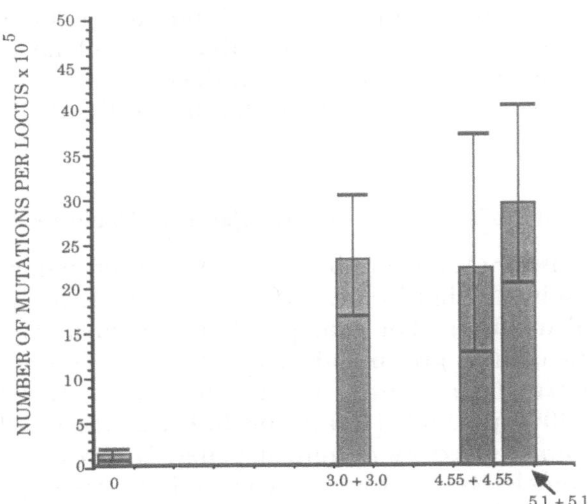

Figure 4. Specific-locus mutation frequencies in the same experiments shown in Fig. 3. Ninety percent confidence limits are shown. Data of Graw *et al.* (1986) and Ehling *et al.* (1982).

reasons for the poor consistency of radiation results. It is worth considering other possible reasons for the frequent inconsistency in both radiation and ENU cataract experiments. One possibility is that some experiments were contaminated with preexisting mutations having incomplete penetrance. A standard precaution taken in cataract experiments, presumably as a safeguard against preexisting mutations, has been to examine the lenses of all (101 × C3H)F_1 and T-stock animals feeding into experiments and to discard any with a lens opacity (Graw *et al.*, 1986). This precaution would eliminate the problem of preexisting mutations if the test dealt only with mutations having complete penetrance, but it does not. The finding of only one control mutation to date suggests that there may be little problem from preexisting mutations, but it should be kept in mind that some experiments, particularly the ones with ENU, had only small concurrent controls (Favor, 1983, 1986).

Another possible cause for the inconsistency in the γ-ray experiments may be that the test cross in the experiment that yielded the high mutation frequency was made to (101 × C3H)F_1 and T-stock animals (Kratochvilova and Ehling, 1979), whereas in more recent experiments it was made to 101 strain animals. The change in the stock used in the breeding test could influence the chances of confirming transmissibility, perhaps considerably, because some cataract (Favor, 1984) and skeletal mutations differ in their expressivity on different genetic backgrounds. Unlike the breeding test for cataracts, that for skeletal mutations attempts to keep the genetic background of the offspring examined in the breeding test as similar as possible to that in the F_1s (Selby and Selby, 1977a).

3.2.3. Favor's Criteria for Classifying Cataract Variants

West and Fisher (1986) emphasized that another major drawback of the cataract test is the high frequency of F_1s with cataracts thought to be nonmutational in origin. For example, Kratochvilova (1981) reported that 10% of the control group and 11% of the experimental group had small lens abnormalities, most of which were small opacities. Breeding tests on about 300 carriers led her to conclude that the small lens abnormalities were not caused by dominant genes. In order to reduce the number of animals chosen for time-consuming breeding tests, Favor (1983) developed criteria for classifying cataract variants based on phenotype. Definitions used in his criteria are: *severe,* which means that the opacity affects the entire lens and permits little if any light to pass through it, and *unique,* meaning (1) that the opacity affects more than one region, or (2) that it has never been seen before, or (3) that it was

Table IV. Classification of Variants by Severity Class in Mice from ENU and X-ray Experiments, with Number of Variants Proved to Be Mutant in Parentheses

Agent	Dose (mg/kg or Gy)	Germ-cell stage	Strain of males	Number of offspring	Number of severity Class I[a]	Number of severity Class II[b]	Number of severity Class III[c]	Reference
ENU	80	Stem-cell spermatogonia	(101 × C3H)F_1	5,090	5 (4)	20 (4)	NT[d]	Favor (1986)
	160	Stem-cell spermatogonia	(101 × C3H)F_1	6,435	12 (12)	25 (2)	NT	Favor (1986)
Control	0	—	(101 × C3H)F_1	867	—[e]	4 (1)	1 (0)	Favor (1986)
ENU	250	Stem-cell spermatogonia	(101 × C3H)F_1	9,352[f]	10 (7)	199 (10)	143 (0)	Favor (1983)
	250	Postspermatogonial stages	(101 × C3H)F_1	3,360[f]	0	56 (1)	23 (1)	Favor (1983)
Control	0	—	(101 × C3H)F_1	2,517[f]	0	11	15	Favor (1983)
X rays	3.0 + 3.0	Stem-cell spermatogonia	(101 × C3H)F_1	15,551	3 (1)	84 (2)	28	Graw *et al.* (1986)
	3.0 + 3.0	Postspermatogonial stages	(101 × C3H)F_1	1,120	1 (1)	6	—	Graw *et al.* (1986)
	5.1 + 5.1	Stem-cell spermatogonia	(101 × C3H)F_1	11,205	2 (2)	25	25 (1)	Graw *et al.* (1986)
	5.1 + 5.1	Postspermatogonial stages	(101 × C3H)F_1	425	—	5	2	Graw *et al.* (1986)
Control	0	—	(101 × C3H)F_1	11,036	3	14	1	Graw *et al.* (1986)
X rays	3 + 3	Stem-cell spermatogonia	BALB/c	14,132	4 (4)	37 (1)	2 (0)	Favor *et al.* (1987)
	3 + 3	Stem-cell spermatogonia	DBA/2	15,931	6 (4)	49 (8)	6 (0)	Favor *et al.* (1987)
ENU	80 + 80	Stem-cell spermatogonia	(101 × C3H)F_1	17,665	6 (3)[g]	31 (3)	2	Favor *et al.* (1988)

[a]Bilateral severe or unique.

[b]Unilateral severe or unique; or bilateral neither severe nor unique.

[c]Unilateral neither severe nor unique.

[d]NT, neither reported nor tested.

[e]"—" means assumed to be 0, but publications listed as "—."

[f]All variants found in this group were accounted for. Some groups in this table may list only those variants thought worthwhile to test.

[g]Even though two of the six offspring were sterile, they were counted as mutants. Only the four tested ones were included in calculating the fraction of the Class I variants that were found to be transmitted.

seen in a previously confirmed mutant. Animals with cataracts are placed into one of the following categories—Class I: severe or unique lens opacities that are bilateral; Class II: either (1) unilateral opacities that are unique or severe or (2) bilateral opacities that are neither severe nor unique; or Class III: unilateral opacities that are neither severe nor unique. Each class was further subdivided into one of six subgroups based on the location of a cataract in the lens, with the hope that some of these subcategories could be ignored in future work.

Table IV shows the numbers in each severity class in all experiments where the classification scheme has been applied, along with the numbers of confirmed mutants. Only a small fraction of F_1 offspring with Class II or Class III cataracts have been confirmed as mutants, but 38 of 50 completely tested F_1 offspring with Class I cataracts have been confirmed.

3.2.4. Reanalysis Suggests Many Mutations with Low Penetrance Are Induced

Ehling's (1984a, 1985) emphasis in the cataract test is on mutations confirmed by breeding tests, but it is important to consider whether mutations might account for some of the many cataracts for which there is no evidence of transmission. I have checked this by comparing the control and experimental frequencies of the cataracts classified by the authors as presumably nonmutational in the various experiments. This is shown in Table V and in Figs. 5 and 6 for all groups having a concurrent control in which Favor's classification scheme was applied. The figures also include proven cataract mutations. Several interesting conclusions can be drawn. In the 250-mg/kg ENU experiment (Fig. 5), for which all cataracts in all three classes are definitely listed (Favor, 1983), the frequency of Class II cataracts is significantly higher than control for both stem-cell spermatogonia and postspermatogonial stages. For Class III cataracts, the frequency is significantly higher than control for stem-cell spermatogonia. In all three cases, $P < 0.0001$. Figure 6 relates to the experiment of Graw *et al.* (1986), which surprisingly showed no effect of radiation. The much lower control values for Classes II and III in Fig. 6 compared with Fig. 5, and examination of Table V, indicate that Graw *et al.* (1986) probably excluded many of the subclasses of cataracts included by Favor. In sharp contrast with the conclusion of no effect based on confirmed mutations, the comparisons shown in Fig. 6 and Table V provide much evidence for induction of rather high frequencies of dominant cataract mutations by both radiation treatments in both stem-cell spermatogonia and postspermatogonial stages. It thus appears that a

Table V. Percentages of Animals in Each Experiment That Were in Each Severity Class and Were Not Proved to be Mutant

Agent	Dose	Germ-cell stage	Strain of males	Severity Class I[a] (%)	Severity Class II[b] (%)	Severity Class III[c] (%)
ENU	80	Stem-cell spermatogonia	$(101 \times C3H)F_1$	0.02	0.31	NT[d]
	160	Stem-cell spermatogonia	$(101 \times C3H)F_1$	0	0.36	NT
Control	0	—	$(101 \times C3H)F_1$	0	0.35	0.12
ENU	250[e]	Stem-cell spermatogonia	$(101 \times C3H)F_1$	0.03	2.02[g]	1.53[g]
	250[e]	Postspermatogonial stages	$(101 \times C3H)F_1$	0	1.64[g]	0.65
Control	0[e]	—	$(101 \times C3H)F_1$	0	0.44	0.56
X rays	3.0 + 3.0	Stem-cell spermatogonia	$(101 \times C3H)F_1$	0.01	0.53[k]	0.18[h,k]
	3.0 + 3.0	Postspermatogonial stages	$(101 \times C3H)F_1$	0	0.54[j]	0[h]
	5.1 + 5.1	Stem-cell spermatogonia	$(101 \times C3H)F_1$	0	0.22[l]	0.21[h,k]
	5.1 + 5.1	Postspermatogonial stages	$(101 \times C3H)F_1$	0	1.18[j]	0.47[j]
Control	0	—	$(101 \times C3H)F_1$	0.03	0.13[h]	0.00[i]
X rays	3 + 3[f]	Stem-cell spermatogonia	BALB/c	0	0.25	0.01
	3 + 3[f]	Stem-cell spermatogonia	DBA/2	0.01	0.26	0.04

[a]Bilateral severe or unique.
[b]Unilateral severe or unique; or bilateral neither severe nor unique.
[c]Unilateral neither severe nor unique.
[d]NT, neither reported nor tested.
[e]All variants found in this group are accounted for. Some groups in this table may list only those variants thought worthwhile to test.
[f]These two experiments are only compared to each other.
[g]Significantly higher than value for same class in control of 250-mg/kg ENU experiment, $P < 0.0001$ (1-tailed Fisher's exact test).
[h]Significantly lower than value for same class in control of 250-mg/kg ENU experiment, $P < 0.01$ (2-tailed test).
[i]Significantly lower than value for same class in control of 250-mg/kg ENU experiment, $P < 0.0001$ (2-tailed test).
[j]Significantly higher than the control for its severity class in the experiment of Graw *et al.* (third row from bottom), $P < 0.01$ (1-tailed Fisher's exact test).
[k]Significantly higher than the control for its severity class in the experiment of Graw *et al.*, $P < 0.0001$ (1-tailed Fisher's exact test).
[l]On borderline of being significantly higher than the control, $P = 0.07$ in a 1-tailed Fisher's exact test.

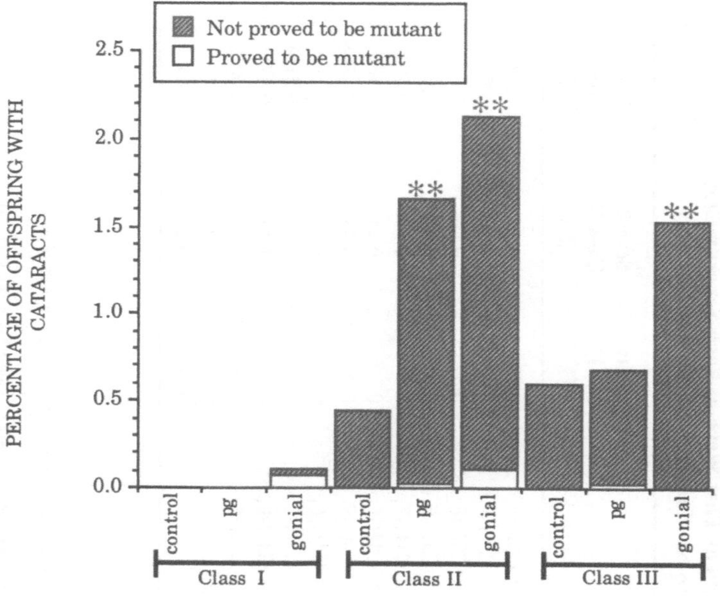

Figure 5. Graphical presentation of the reanalysis of the cataract data of Favor (1983). The experimental treatment was 250 mg/kg ethylnitrosourea. Bars show the percentages of offspring with cataracts of each of the three classes described by Favor (1983), which are described in Section 3.2.3, for exposed postspermatogonial stages (pg) and exposed stem-cell spermatogonia (gonial). Asterisks indicate that the experimental frequency is statistically significantly higher than that of the control for its class, $P < 0.0001$. Notice that very few of the cataracts were confirmed to be mutants by breeding tests.

cataract test could be applied to compare frequencies of certain classes of cataracts (or the sums of all classes) without even testing animals for transmission.

3.2.5. Limitations of Breeding Test for Cataracts of Classes II and III

My reanalysis of the cataract data suggests that, for some of the treatments, upwards of 1% of the F_1s had cataracts caused by mutations for which no proof of transmission was found. This indicates that the breeding test may be missing a high proportion of the mutations, probably those with low penetrance. Favor and Ehling claim that, by their procedures, no mutations would be missed with penetrance greater than

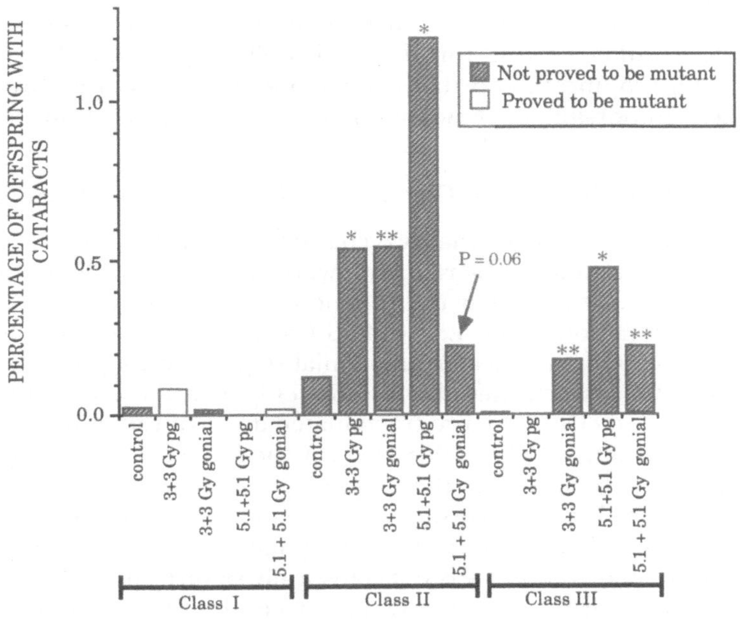

Figure 6. Graphical presentation of the reanalysis of the cataract data of Graw *et al.* (1986). The experimental treatments were acute exposures to X radiation with 24-h intervals between exposures. Bars show the percentages of offspring with cataracts of each of the three classes described by Favor (1983), which are described in Section 3.2.3, for exposed postspermatogonial stages (pg) and exposed stem-cell spermatogonia (gonial). **, experimental frequency statistically significantly higher than that of the control for its class, $P < 0.0001$; *, statistical significance at the 5% significance level. Notice that very few of the cataracts were confirmed to be mutants by breeding tests.

0.32 (Favor, 1983; appendix in Ehling *et al.*, 1982), and Favor (1986) stated that cataract mutations with "penetrance values as low as 0.10 have a surprisingly high probability of being confirmed." Favor's claim that the cataract breeding test, using at least 20 offspring, fares so well in confirming mutations with rather low penetrance is based on a mathematical model in which one affected offspring is sufficient to confirm transmission (appendix in Ehling *et al.*, 1982; Favor, 1986). Although this model is applied to all three classes of cataracts, Favor (1983) made statements that make it clear that, at least for Class II and Class III cataracts, the model is inadequate because more than one affected offspring must be found to confirm transmission. The model may be valid for Class I cataracts, but these comprise few of the total cataracts found.

Therefore, it remains likely that many mutations with low to moderate penetrance could account for the statistically significant differences shown in Figs. 5 and 6, and that complexities in the cataract breeding test make it impossible to show that many of them are mutants.

3.2.6. Comparisons between Cataract and Specific-Locus Mutations

Ehling (1983) reported that two characteristic features of induction of specific-locus mutations by radiation were paralleled by the induction of dominant cataracts, namely the enhanced mutation frequency following dose fractionation in the 4.55 + 4.55-Gy experiment and the higher mutation frequency in postspermatogonial stages. Although these conclusions may be correct, they should not yet be considered established. The finding of no effect of large fractionated doses of radiation by Graw *et al.* (1986) certainly weakens the basis for the first conclusion. Regarding Ehling's other conclusion, the mutation frequencies induced by single 534- and 600-R γ-ray exposures of postspermatogonial stages were 2–4 times higher than those in spermatogonia (Ehling *et al.*, 1982), but even the combined frequency is not statistically significantly higher ($P = 0.21$ in a 1-tailed Fisher's exact test). The strength of the second conclusion is further weakened because one of the two mutants counted for postspermatogonial stages was sterile and thus unconfirmed. If data from the fractionated radiation experiments are included, the mutation frequency in postspermatogonial stages of 4/4403 seems high and in line with Ehling's conclusion. However, because of the different dose regimens used and the poor repeatability of results in spermatogonia, any conclusion on this point should be tempered with caution. Interestingly, in my reanalysis of cataract data shown in Fig. 6, for the 5.1 + 5.1-Gy exposure, the frequency of Class II cataracts is significantly higher in postspermatogonial stages than in stem-cell spermatogonia ($P = 0.004$ in a 1-tailed Fisher's exact test).

Ehling *et al.* (1982) and Favor (1983) noted that the mutation frequency in postspermatogonial stages for ENU is higher, but not significantly higher at the 5% level, than the untreated historical control frequency. Now that the sample size of the historical control is larger, this difference has become statistically significant ($P = 0.046$), but it is unclear how meaningful it is because one of the two mutations found was a Class III cataract, and those cataracts are confirmed so rarely that they are sometimes not even reported and tested. It is interesting to note that my reanalysis of the cataract data (Fig. 5) supports the conclusion that ENU may induce a relatively high frequency of cataract mutations in postspermatogonial stages. If this is true, it would conflict with the expectation from specific-locus experiments, in which few mutations have been found in offspring derived from postspermatogonial stages (Ehling *et al.*, 1982; W. L. Russell and Hunsicker, 1984).

3.3. Dominant Visible Mutations

Dominant visible mutations are defined as those whose effects can be seen in a living mouse with the naked eye. Searle (1974) reviewed the information available on spontaneous and induced frequencies of dominant visibles. In the control groups of several radiation experiments, mostly carried out at the Radiobiology Unit at Harwell, the spontaneous dominant visible mutation frequency was 3/187,612, or 0.5×10^{-5} per gamete, with two of the three mutations being X-linked. The radiation-induced mutation frequencies for dominant visible mutations are very low, similar to those for cataract mutations confirmed by breeding tests. Visibles represent only the tip of the iceberg of induced dominant mutations, presumably because only a few parts of the body are externally visible. Other laboratories have put less emphasis on dominant visibles largely because of the subjectivity involved. Owing to the latter, comparisons between different treatments have more validity when restricted to one laboratory. For this reason, Table VI shows only results collected at Harwell. (Offspring were coded in at least some of these experiments.) The results support Searle's (1974) claim "that most of the phenomena described with respect to specific locus mutations apply also to dominant visible mutations."

From the skeletal breeding-test experiment (Selby and Selby, 1978b), it appears that dominants with externally visible effects often have those effects in just a small proportion of the offspring. Among 31 proven dominant skeletal mutations, none had an externally visible effect on the skeleton, other than stunted growth, in any more than a small fraction of the carriers of the mutation. There were seven skeletal mutations that had externally visible effects on other body systems, but because their non-skeletal effects were subtle or present in only a small fraction of the carriers of the mutation, it is unclear whether such mutations would have been considered dominant visibles in Harwell experiments. Lyon *et al.* (1979) reported that nine dominant visibles were found in one experiment and then added that "six additional mutants with irregular inheritance [also] occurred," leaving the impression that visible mutations with incomplete penetrance might not have been considered dominant visibles in their studies.

Searle (1974) noted the similarity of the spectrum of spontaneous and X ray-induced dominant visibles and the dissimilarity of those spec-

Table VI. Dominant Visible Mutation Frequencies Obtained in Experiments at Harwell (Autosomal and X-Linked)

Type of radiation	Dose or exposure	Germ-cell stage	Dose rate	Number of mutations	Number of offspring	Mutations/ gamete ($\times 10^5$)	Reference
X rays	600 rad	Stem cells[a]	60–70 rad/min	2	11,138	18	Phillips (1961), Lyon *et al.* (1972)
	600 rad	Stem cells	88 rad/min	9	24,834	36	Lyon and Morris (1966, 1969)
	600 R	Stem cells	68 R/min	0	838	0	Carter and Lyon (1961)
	12 × 50 rad[b]	Stem cells	60–70 rad/min	2	18,119	11	Lyon *et al.* (1972)
	500 + 500 rad[c]	Stem cells	88 rad/min	24	32,763	73	Lyon and Morris (1969)
	600 + 600 R[d]	Stem cells	217 R/min	2	3,612	55	Lyon *et al.* (1964)
γ rays	37.5 rad	Stem cells	0.001 rad/min	1	63,322	2	Carter *et al.* (1958)
	606 rad	Stem cells	0.005 rad/min	6	58,795	10	Batchelor *et al.* (1966)
	600 rad	Stem cells	0.008 rad/min	0	22,682	0	Lyon *et al.* (1972)
	600 rad	Stem cells	17 rad/min	1	12,021	8	Lyon *et al.* (1972)
	60 × 10 rad[e]	Stem cells	17 rad/min	0	23,982	0	Lyon *et al.* (1972)
Fission neutrons	62 rad	Stem cells	0.001 rad/min	7	39,083	18	Batchelor *et al.* (1967)
	214 rad	Stem cells	0.002 rad/min	24	41,875	57	Batchelor *et al.* (1966)
	188 rad	Stem cells	57 rad/min	2	39,028	5	Batchelor *et al.* (1967)
X rays	200 rad	Oocytes[f]	52–72 rad/min	2	18,867	11	Lyon *et al.* (1979)
	400 rad	Oocytes[f]	52–72 rad/min	3	7,501	40	Lyon *et al.* (1979)
	600 rad	Oocytes[f]	52–72 rad/min	6	9,875	61	Lyon *et al.* (1979)

[a]Spermatogonial stem cells.
[b]Exposures 1 week apart.
[c]Exposures 24 h apart.
[d]Exposures 8 weeks apart.
[e]Exposures daily on weekdays.
[f]Mature oocytes, all offspring were conceived in the 1st week after treatment.

tra to the one induced by fission neutrons. He also reported that a "very high proportion" of induced dominant visibles were homozygous lethal and that many had deleterious effects in heterozygotes. Similar to what was reported for some skeletal mutations (Section 3.1.2a), some dominant visibles, in particular some Steeloids (Cacheiro and Russell, 1975) and Diver (Rutledge *et al.*, 1986), are probably balanced translocations.

Dominant visibles are of interest because they are the dominant mutations easiest to spot initially. However, it remains to be seen how well their mutation frequencies will correlate with those of dominant mutations in general after exposure to different mutagens.

3.4. Litter-Size Reduction

The litter-size reduction (LSR) considered here is the difference in mean litter size between experimental and control groups at weaning age, and it includes all dominant mutations that cause death between conception and weaning age. It is important to know how much induced death occurs before weaning age because methods for estimating overall genetic damage (Section 4) often overlook mutations causing death during that period. Estimates of LSR can be combined with estimates of induced dominant damage observed in living mice at weaning, or soon thereafter, to make a more inclusive estimate of the total dominant damage resulting from a given exposure (Selby and Russell, 1985). Various attempts to measure LSR have been described elsewhere (W. L. Russell, 1954; Roderick, 1964; Green, 1968), but they were complicated by such factors as small sample size, exposure over successive generations, and exposure to protracted irradiation during pre- and postnatal life.

Selby and Russell (1985) estimated the frequency of induction of mutations causing death between conception and weaning (approximately 3 weeks postpartum) by comparing litter sizes of experimental and concurrent control animals in 14 experiments, involving 158,490 F_1 litters, in which males were exposed to X or γ rays. All data came from exposed stem-cell spermatogonia. To decrease variability, comparisons between experimental and control were made only between groups of litters having mothers of approximately the same age. Weighted mean LSRs at 300, 600, and 1000 R of 90-R/min X irradiation were 3.90, 4.49, and 1.55%, paralleling the humped dose–response curve seen for specific-locus mutations. The result at 300 R agreed well with a much earlier preliminary estimate of 3.8% at that dose (W. L. Russell and L. B. Russell, 1959).

Selby and Russell (1985) showed that there is a dose-rate effect for

LSR, with the mutational responses being much lower at dose rates of 0.009 and 0.001 R/min than at 90 R/min. The slope of the LSR for the low-dose-rate experiments was 0.00194 ± 0.000760% (S.E.), which is significantly above zero ($P < 0.006$). Selby and Russell estimated from the slope that if all men were exposed to 1 R of γ or X radiation, the number of deaths caused by induced dominant mutations among their conceptuses would be about 19 per million children born, with about half of the deaths occurring in early pregnancy. As is discussed in detail elsewhere (UNSCEAR, 1986), independent calculations based on much smaller data sets of Lüning (1972) and Searle and Papworth (1986) are in close agreement with this risk estimate.

Recently, Russell and Hunsicker (1988), using the same strains of mice, showed that four weekly 100-mg/kg i.p. injections of ENU induced a statistically significant LSR of approximately 11%, which is more than twice that obtained with 600 R of acute X radiation. This result is in rough agreement with expectation when the much higher effectiveness of ENU, compared with X rays for specific-locus mutations (Russell *et al.*, 1979), is balanced against the fact that a large proportion of the LSR with X rays is due to translocations (Selby and Russell, 1985), a type of damage not induced by ENU in spermatogonia. This experiment had a concurrent control, and parents were randomized between groups.

3.5. Congenital Malformations

Congenital malformations have been defined as "structural abnormalities of prenatal origin that are present at birth and that seriously interfere with viability or physical well-being" (Kalter and Warkany, 1983). Mutation studies directed specifically at anomalies detectable at birth in mice and rats often examine fetuses late in pregnancy, instead of soon after birth, to avoid the error introduced by mothers eating grossly abnormal offspring.

3.5.1. Lyon's Experiments

Figures 7 and 8 summarize the experiments on males conducted by Lyon and coworkers using (C3H/HeH × 101/H)F_1 hybrid mice. Uterine contents were examined 18 days pc, and matings were coded to eliminate observer bias. In the experiment shown in Fig. 7, and in controls 1 and 2 of Fig. 8, all fetuses were dissected to search for internal abnormalities, of which relatively few were found. In other experiments, only those fetuses showing some external malformation, small size, or cleft palate were dissected.

Figure 7 shows the results of the first series of experiments by Kirk and Lyon (1984) on X-irradiated males. The separate control group for the spermatogonial part of the experiment, here referred to as control 2, is almost an order of magnitude higher than the control for the post-spermatogonial part, here called control 1. With such a high control frequency, it is not surprising that no evidence was found for induction of congenital-malformation mutations in stem-cell spermatogonia. In contrast, the combined results on postspermatogonial stages were statistically significantly above their much lower control at every dose level. Figure 7 shows, however, that only the 15- to 21-d interval showed the dose–response relationship expected if radiation induced the mutations.

Figure 8 shows the results of Kirk and Lyon (1984) and Lyon and Renshaw (1986). All five control groups in experiments on males are shown for comparison, including controls 1 and 2. The radiation and ENU experiments in Fig. 8 dealt with stem-cell spermatogonia; the EMS (ethylmethanesulfonate) experiment involved postspermatogonial stages. The frequencies of abnormal fetuses after acute exposures to 500 or 500 + 500 cGy of X rays were statistically significantly higher than the concurrent control, but not nearly as different from each other as expected from specific-locus results. Neither X-ray experiment would have been significantly higher than control 2. Statistical comparisons of ENU and EMS results with their respective concurrent controls led to the conclusions that "there was no evidence of induction of congenital malformations" by ENU and that "EMS gave a clear induction of malformations." The authors suggested, accordingly, that whether or not mutagens induce congenital malformations depends on the type of genetic lesion caused. Unlike EMS, ENU primarily induces base-pair substitutions, and the authors suspected that such mutations might not cause dominant damage. However, because the observed mutation frequency is high in the ENU group, it seems possible that the lack of statistical significance may result merely from an unduly high concurrent control frequency, such as presumably happened in their first X-ray experiment on spermatogonia. (No experimental frequency in males is statistically significantly higher than that of the ENU experiment's concurrent control, with the smallest P value being 0.15.) This control frequency may have been so high as the result of an unequal distribution of preexisting mutations between the experimental and control groups. Another way of judging whether ENU induced congenital malformations is to compare the ENU frequency of 22/866 (2.5%) with the combined control frequency for all experiments on males of 36/3108 (1.2%). The ENU frequency is significantly higher in a 1-tailed Fisher's exact test, $P = 0.004$.

Kirk and Lyon (1982) also demonstrated X-ray induction of con-

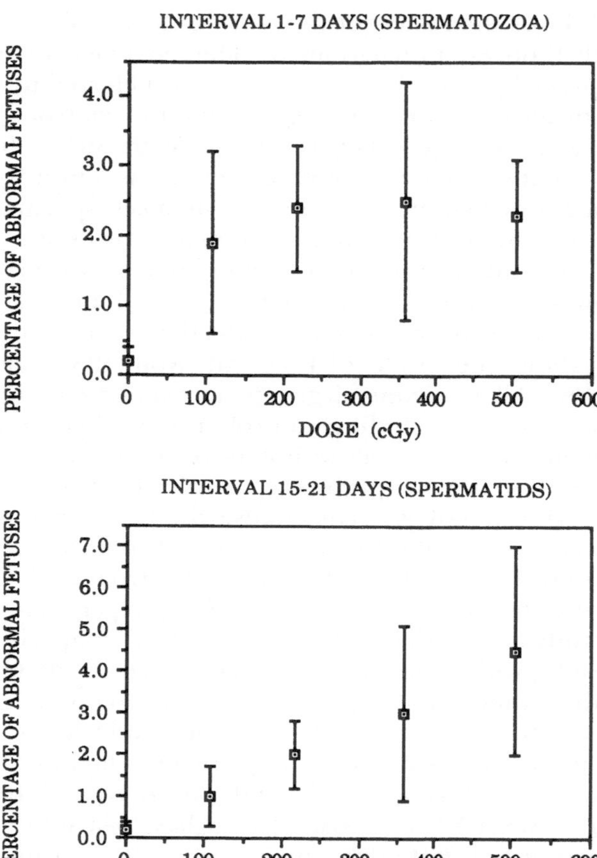

Figure 7. Percentages of abnormal fetuses in first series of experiments of Kirk and Lyon (1984). Male mice were exposed to acute X radiation and mated to untreated females at the intervals shown. Standard errors of the percentages are shown for four different intervals after treatment. Notice that the spermatogonial stem-cell group had a separate control.

genital-malformation mutations in mature and maturing oocytes during the first 4 weeks after treatment. In accordance with expectations for other types of mutations, the mutation frequency increased over the first 3 weeks after treatment. The control frequency was 1.1% and, for the 2nd- and 3rd-week samples, the slopes of the simple regression lines of dose versus frequency were statistically significantly above zero. Experi-

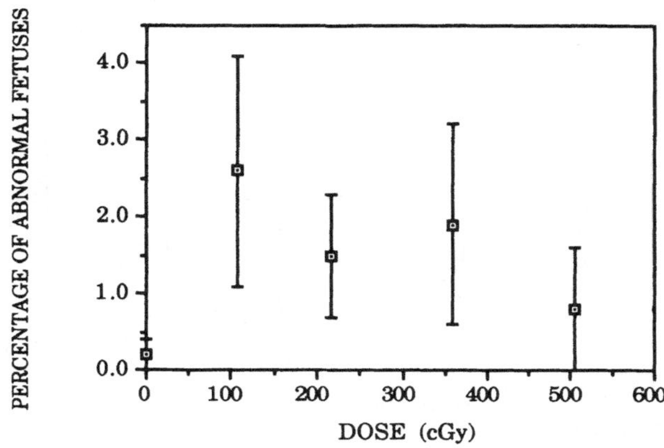

INTERVAL 64-80 DAYS (SPERMATOGONIAL STEM CELLS)

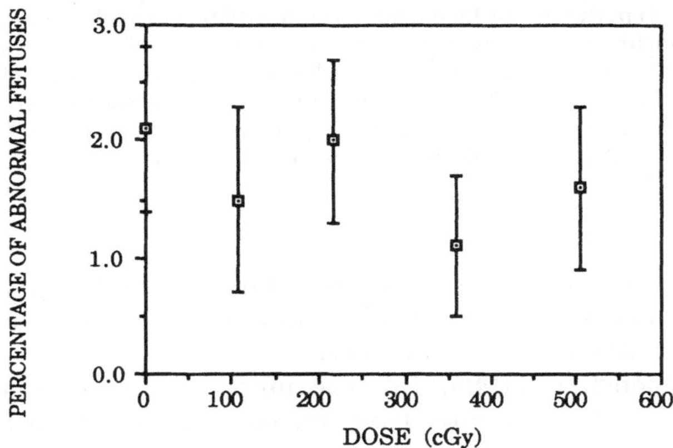

Figure 7. *(Continued)*

ments using embryo transfers and others using irradiation of externalized ovaries (West *et al.*, 1985a,b) increased the likelihood that this is a genetic effect rather than a maternal one.

A test of homogeneity of the eight separate controls reported in the above-described papers shows that the probability that they all came from the same population is only 0.025. The mean frequency of abnormal fetuses for the eight controls is 1.14%, and the means of four of them are outside the 95% confidence limits of the overall mean. Such

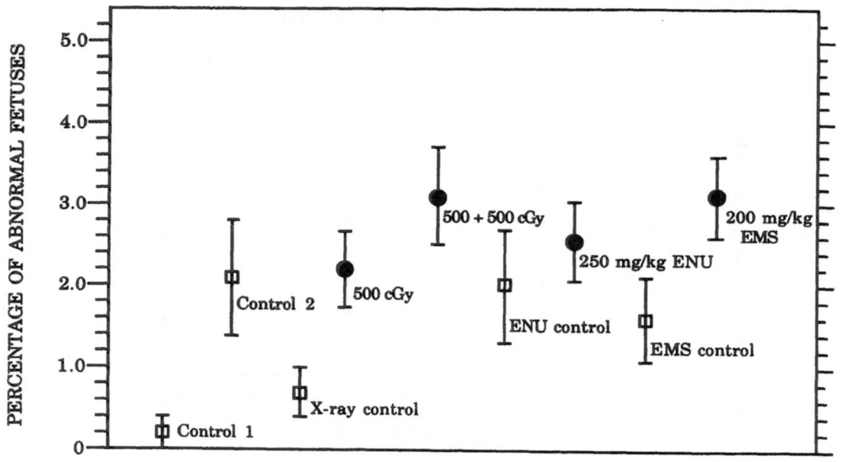

Figure 8. Percentages of abnormal fetuses in second series of experiments of Kirk and Lyon (1984) and in experiments of Lyon and Renshaw (1986). Controls 1 and 2 are those shown in Fig. 7. Concurrent controls are shown to the left of their respective experimental group or groups. The fractionation interval for the X-ray experiment was 24 h. The X radiation and ethylnitrosourea (ENU) experiments involved exposed stem-cell spermatogonia. The Ethylmethanesulfonate (EMS) experiment involved exposed postspermatogonial stages. Standard errors of the percentages are shown.

wide fluctuations in the control could possibly indicate the presence in the strain of preexisting mutations with low penetrance for some of the more common congenital abnormalities.

Lyon and Renshaw (1986, 1988) assumed that some congenital-malformation mutations might have incomplete penetrance, which would allow some carriers to transmit the mutations. They collected offspring from females exposed to either 360 (two groups) or 504 cGy of acute X radiation and found high frequencies of congenital malformations in the F_1 fetuses, namely 4.1 ± 1.8, 7.7 ± 1.9, and 12.5 ± 3.2%, respectively. They mated normal-appearing F_1s, both experimental and control, to untreated females. The frequencies of congenital malformations in fetuses with irradiated grandmothers were 1.6 ± 0.2% for 360 cGy and 2.6 ± 0.5% for 504 cGy, which compared to control frequencies of 1.3 ± 0.3% and 1.3 ± 0.2%, respectively. The frequency in the second generation was statistically significantly higher only at the higher dose. Although the authors argued that the marked drop in the frequency of congenital malformations seen between the first and second generations "would only be expected if the average penetrance of the genes concerned was high," the standard errors were wide enough to accommodate a fairly large component of mutations with rather low penetrance.

For the 360-cGy dose, Lyon and Renshaw (1986, 1988) attempted to identify F_1 carriers of mutations with low penetrance by using an extensive breeding test. Experimental and control males were mated to enough females to obtain 50 living fetuses, and experimental males with two or more abnormal fetuses were permitted to sire an additional 50 living fetuses. Males that produced an excess of malformed fetuses in both batches of 50 fetuses were tested further. In this way, two definite and two probable carriers of mutations with low penetrance for exencephaly, dwarfism, or both were identified. The authors considered this as further evidence that congenital abnormalities seen after irradiation of parental mice are of genetic origin. Although this conclusion is probably correct in part, there is no way to know from their data whether the transmitted mutations were induced or were preexisting. Indeed, no controls were tested beyond the first batch of 50 fetuses even though, for the first batch, the point estimate of the frequency of F_1s with two or more abnormal fetuses was slightly higher in the control than in the experimental group.

3.5.2. Nomura's Experiments

Nomura (1978) reported that irradiation of young adult male or female mice with acute X rays (single doses of 36, 108, 216, or 504 rad or the same doses administered as repeated 36-rad fractions at 2-h intervals), with matings at various intervals from 1 to 80 d after treatment, induced a statistically significant increase in congenital malformations in offspring examined at 18 d pc. He also reported a statistically significant increase in congenital malformations after single s.c. injections of 1500 or 2000 mg/kg urethan to male or female mice that were 21 or 63–65 d old, or at day 15 pc. Mating intervals in these experiments were similar to those in the X-ray experiments. All nine (seven urethan, two X ray) statistical comparisons of different age, sex, or dose groups with the single control group were statistically significant.

Recently, Nomura (1988) almost doubled the sizes of his X-ray and urethane samples and applied his method to four other chemicals. Table VII shows his conclusions, based on comparisons of these data with concurrent controls. His five control groups are homogeneous. All chemicals were injected subcutaneously, and in the more recent experiments he also collected data on X rays, urethan, and ENU from matings as late as 180 d after treatment. Either the overall data, or particular germ-cell stages, indicated a statistically significant effect for all agents

Table VII. Nomura's Conclusions about the Effectiveness of Various Agents in Causing Congenital Malformations following Treatment of Various Germ-Cell Stages in Mice[a]

		Germ-cell stages				
				Oocytes, time after treatment in adult		
Agent[b]	Doses[c] (rad or mg/kg)	Postspermatogonial stages	Stem-cell spermatogonia	≤6 weeks	>6 weeks	Oocytes in 21-d-old mouse
X rays	36, 108, 216, 360, 504	S	S	S	—[d]	S
Urethane	1500, 2000	S	S	S	NS	S
DMBA	20	S	NS	—	—	—
ENU	50, 100	S[e]	S	—	—	—
4NQO	12.5	S[g]	—	—	—	—
Furylfuramide	3 × 50[g]	NS[f,h]	—	—	—	—

[a]S, statistically significantly higher than the control, $P < 0.05$; NS, not statistically significant.
[b]Urethan is ethyl carbamate, DMBA is 7,12-dimethylbenz[*a*]anthracene, ENU is *N*-ethyl-*N*-nitrosourea, 4NQO is 4-nitroquinoline 1-oxide, and furylfuramide is 2-(2-furyl)-3-(5-nitro-2-furyl) acrylic acid amide. All injections were subcutaneous.
[c]These doses were tested, but they did not all necessarily yield statistically significant results.
[d]Not tested.
[e]All conceptions were 1–14 d after treatment.
[f]All conceptions were 8–21 d after treatment.
[g]Interval between injections was 1 d.
[h]Point estimate is the same as for DMBA, which is significant, but the sample is much smaller.

except furylfuramide. All mutagens for which Nomura claimed significant results have some mutation frequencies in the neighborhood of 3–4% or higher.

Nomura (1988) reported that frequencies of congenital malformations increased with doses of X rays, urethan, and ENU both in spermatozoa and in spermatogonia and that frequencies increased with doses in oocytes for X rays and urethan. However, in all of these experiments, the slope of the simple regression line of dose versus frequency was statistically significantly above zero only in irradiated mature and maturing oocytes. Nomura attached much importance to the clear linear relationship between dose and frequency in spermatogonia between 0 and 216 rad; however, he ignored the much lower frequency found at 504 rad. He also collected large samples of offspring first examined for external abnormalities at 7 d of age. Even though those offspring had some abnormalities that could not be seen in fetuses, their frequencies were usually lower than those for the prenatal anomalies on which his main conclusions were based.

Induction of congenital malformations by ENU doses of 50 and 100 mg/kg was tested in stem cells and postspermatogonial stages. Three of four data points were highly statistically significantly above the concurrent control, with the highest induced frequency being 3.1% in stem cells at the higher dose. Nomura (1988) pointed out that, in accord with specific-locus results (W. L. Russell *et al.*, 1979; W. L. Russell and Hunsicker, 1984), spermatogonia were more sensitive to mutation induction by ENU than were postspermatogonial stages. The effect in postspermatogonial stages seemed higher than expected from specific-locus results, which is interesting in view of the ENU results found for cataracts (Section 3.2.6).

Nomura's results for urethan conflict with the expectation from specific-locus experiments, in which urethan at a dose of 1750 mg/kg gave no hint of inducing mutations, in either postspermatogonial stages or stem-cell spermatogonia (L. B. Russell *et al.*, 1987). Section 5 discusses the implications of the lack of agreement between results on induction of dominant and specific-locus mutations. No specific-locus data for appropriate germ-cell stages exist with which to compare Nomura's findings for 7,12-dimethylbenz[*a*]anthracene (DMBA) and 4-nitroquinoline 1-oxide (4NQO), or for urethan in the female.

Altogether 42% of the anomalies found by Nomura were open eyelid, 25% were dwarfism, and most of the rest were tail anomalies or cleft palate. It is important to realize that most of the mice with open eyelid showed only a small unilateral gap and most were probably indistinguishable from normal mice by a few days after birth (Nomura, 1988).

Nomura (1988) studied the heritability of congenital malformations by collecting F_1 offspring in a urethan experiment (dose not stated), mating them to untreated mice, and then mating the second generation to collect an F_3 generation. (He did not indicate whether the last mating was made to untreated mice or whether the parents used to produce the last two generations were normal in appearance.) Because even the highest frequencies in the F_1 for urethan were approximately 4%, and because the frequency of anomalies, even if caused by mutations with low penetrance, would be expected to decrease considerably by the F_3 generation (especially if normal-appearing mice were mated to untreated mice), it is puzzling that Nomura found that 9.9% of the 274 F_3 offspring had anomalies (14 tail anomalies, 6 dwarfs, and 7 open eyelids). Of those offspring, he found that one of four tail anomalies tested was inherited and had 7% penetrance, and that three of six open eyelids tested were inherited and had penetrance values of 5, 11, and 50%. He also demonstrated transmissibility of six X-ray-induced dwarfs (penetrance of 7–40%) and of some open eyelids and tail anomalies from other experiments. No mention was made of testing any malformations in the control for transmissibility. He obtained many of the abnormal mice for testing by fostering 18-d-old fetuses obtained in the F_1 uterine analyses.

3.5.3. Other Studies

Rutledge *et al.* (1986) found that if males were (C3H × 101)F_1 hybrids instead of (SEC × C57BL)F_1 hybrids, there was significantly more induction of mutations causing congenital abnormalities when stem-cell spermatogonia were exposed to 4 × 500 R of acute X rays given 4 weeks apart. The induced frequencies for the two hybrids were 3.31 and 1.07%, respectively, and each experimental frequency was statistically significantly higher than its control. This finding paralleled a difference they found in X-ray induced embryonic lethality and in the frequency of induction of reciprocal translocations observed cytologically in the meiocytes of irradiated males. In this study, males were randomized between groups and mated to (C3H × C57BL)F_1 females; fetuses were coded and dissected.

Nagao (1987) reported that i.p. injections of ICR male mice with 5, 15, or 25 mg/kg methylnitrosourea (MNU) for 5 successive days induced congenital malformations in the offspring conceived 1–7, 8–21, or 64–80 d after treatment. Weighted regression analysis showed that there was a dose-dependent statistically significant increase in the frequency of congenital anomalies over the control level for each of the three intervals tested. The relative sensitivity of germ cells sampled at the three intervals was 1 : 1.6 : 2, respectively, with the highest induced frequency, at the total dose of 125 mg/kg to stem cells, being approximately 4.5%. These results conflict with the expectation from specific-locus experiments, in which stem-cell spermatogonia were extremely resistant to mutation induction by 75 mg/kg MNU and the induced mutation frequency was rather low in weeks 1–5 after treatment (W. L. Russell and Hunsicker, 1983). MNU is, however, extraordinarily mutagenic from 32 to 42 d after injection (W. L. Russell and Hunsicker, 1983), an interval not studied by Nagao.

In Nagao's study, live fetuses were coded, weighed, and examined for external abnormalities, including cleft palate, under a dissecting microscope. Skeletons were also stained with alizarin and cleared to permit examination of bones. There were 62 fetuses with external abnormalities and 17 with skeletal abnormalities among 3614 fetuses from treated germ cells. The frequency of skeletal abnormalities was statistically significantly higher in stem cells than in the concurrent control or postspermatogonial stages. The frequency of abnormal fetuses in the control (3/662 or 0.45%) agreed well with that in the historical control (25/4424 or 0.57%). The number of skeletal abnormalities reported was surprisingly low, and only slight indications were given as to how it was decided which skeletal malformations should be counted.

Bartsch-Sandhoff (1974) examined alizarin-stained and cleared skeletons of 19-d-old mouse fetuses from fathers exposed to 600 R of 60-R/min γ radiation. Offspring derived from irradiated postspermatogonial stages, from stem-cell spermatogonia, and from a concurrent control had frequencies of Class 1 abnormalities (i.e., those that occurred once) of 9/371 (2.4%), 4/343 (1.2%), and 5/1190 (0.4%), respectively. Only the frequency for postspermatogonial stages was statistically significantly higher than the control frequency. Several fetuses had such serious malformations that it is unlikely they could have lived until weaning. Skeletons were coded, and some serious Class 2 malformations were found. Because the fetal skeleton is incompletely developed, the chances of recognizing a mutation based on the pattern of anomalies are reduced.

Knudsen *et al.* (1977) provided limited evidence that 25–100 mg/kg cyclophosphamide (CP), given by gavage to male Wistar rats, induces malformations in offspring derived from germ cells exposed in postspermatogonial stages. Trasler *et al.* (1985) reported that a paternal low-dose chronic regimen of CP (5.1 mg/kg per d 6 d/week by gavage) was very effective in inducing congenital malformations in Sprague-Dawley rats. Offspring were collected in matings occurring during the following weeks after treatment: 1 + 2, 3 + 4, 5 + 6, and 7 + 9. The frequencies of congenital malformations were 3/93 (3%) for weeks 3 + 4, 4/57 (7%) for weeks 7 + 9, and 0% for the other two groups. The two groups with malformations were statistically significantly higher than the control frequency of 2/832 (0.2%), but not significantly different from each other.

Jenkinson *et al.* (1987) demonstrated that 75–100 mg/kg CP administered to male mice by gavage induced congenital malformations. The outbred CD1 parents were randomly distributed to experimental and control groups. Matings were 8–21 d posttreatment in the first experiment and 15–21 d posttreatment in the second. Uterine contents were examined 18 d pc. Combined results for both experiments showed a statistically significant increase in the frequency of congenital malformations at the higher dose level only. A few hundred alizarin-stained skeletons were examined and showed no significant effect of CP, which is not surprising in view of the sample sizes. These findings with CP are consistent with specific-locus results: Cumming and Walton (1971) found that 350 mg/kg CP induced specific-locus mutations in postspermatogonial stages.

Johnson and Lewis (1981), in two strains of mice, found no suggestion that 250 mg/kg ENU induces congenital-malformation mutations in stem-cell spermatogonia. For DBA males mated to C57BL females, the frequencies in the experimental and control groups were 4/2875 (0.1%) and 2/479 (0.4%), respectively; in the reciprocal cross the frequencies were 3/1128 (0.3%) and 0/154, respectively. They may have found so few abnormalities because they made their uterine examinations at 14–17 d of gestation, which is a few days earlier than in other studies.

3.5.4. General Comments on Congenital Malformation Experiments

Experience with preexisting skeletal mutations indicates that there is reason to suspect that preexisting mutations are present when many mice have the same syndrome. In the experiments of Lyon, Jenkinson, and Rutledge, small body size accounted for 50% or more of all abnormalities; neural tube defects were also quite common. In Nagao's experiments, 55% of the nonskeletal congenital abnormalities were cleft palate, and, as noted earlier, a few specific anomalies predominated in Nomura's experiments. If some of these predominant abnormalities are often caused by preexisting mutations with low penetrance, it would be likely for such experiments to give anomalous results, especially when there was no randomization of parents.

A complication in the classification of dwarfs can occur because a dwarf is defined as a fetus less than a certain percent of the mean weight of the rest of its litter. As West *et al.* (1985a) noted, classification problems can occur when the litter sizes are very small (as commonly occurs when a treatment induces many dominant lethals) because the mean is based on many fewer animals, and is thus much less reliably known.

Many authors (Lyon and Renshaw, 1986; Nomura, 1986; Rutledge *et al.,* 1986; Nagao, 1987) have suggested that chromosomal aberrations might be the cause of the congenital malformations, but, with the exception of an inversion near the centromere of chromosome 12 in a dwarf (Nomura, 1986), no consistent chromosomal abnormalities were found in a combined total of 79 abnormal fetuses (Lyon and Renshaw, 1986; Nomura, 1986; Jenkinson *et al.,* 1987). Duplication deficiencies may not have been seen, however. ENU produces primarily intragenic lesions, and since it now appears that ENU induces congenital malformations, it seems likely that many congenital malformation mutations could be gene mutations.

Study of induction of congenital malformations provides a way of identifying some of the types of damage included in estimates of the LSR. Some studies appear to have given clear-cut results; others have shown weak effects, or no effects, for strong mutagens. This inconsistency, and the sometimes high control levels, led Lyon and Renshaw (1986) to conclude that "the incidence of malformations is likely to be a relatively insensitive indicator of an increased mutation rate in man." However, if preexisting mutations are the explanation for these two drawbacks, randomization of parents would help to guard against misinterpretations. Coding of fetuses until after classification would also seem to be especially important in such experiments, and fortunately many of the studies reported in this section included this precaution.

3.5.5. Novel Effect Reported by Generoso

Generoso *et al.* (1987) found congenital malformations and fetal death when pregnant mice were exposed by inhalation to 1200 ppm ethylene oxide for 1.5 h when their embryos were in early zygotic stages. These effects differed both from genetic damage induced in germ cells just before mating, which leads mainly to death near the time of implantation, and from teratological damage, which leads to malformations primarily when embryonic exposure occurs during the period of major organogenesis. When the exposure started 6 h after mating, 37% of the living fetuses in the experimental group, versus 2% in the control, were grossly abnormal. (Growth retardation made up only a very small part of the percentage in the experimental group.) Experiments are in progress to see if this is a genetic effect.

3.6. Stunted Growth

Small body size has been associated with skeletal mutations (Selby and Selby, 1977a; Selby and Niemann, 1984), cataract mutations (Kratochvilova, 1981), and visibles (Lewis and Johnson, 1983; Searle and Beechey, 1986), and it was an effect sometimes seen in experiments designed to look for population effects of radiation exposures in successive generations (McGregor and Newcombe, 1961; Green, 1968). Selby and Selby (1977a, 1978a) noted that small body size is the most common external indicator of the presence of a dominant mutation in mice, and that if an F_1 offspring in a mutation experiment is very small and sterile, the chances are high that its skeleton will contain enough rare malformations to indicate the presence of a skeletal mutation.

Searle and Beechey (1986) studied induction of mutations causing stunted growth and dominant visibles in mice following exposure of stem-cell spermatogonia to 5 + 5 Gy of acute X radiation in an experiment with no control group. Out of 7309 offspring, of those 112 that were "markedly smaller than their litter-mates" at weaning, 46 were kept for testing, but 16 died or failed to breed before they were completely tested for dominant inheritance. Of 30 completely tested ones, 12 showed dominant inheritance for small size, but only 2 had complete penetrance. Mutations causing small size were termed "growth retardation" mutations. Although mice were not weighed to determine which ones were small, the weighing of some suggested that a small mouse is probably 41–80% of the mean weight of normal littermates.

Searle (1987) reported an experiment on maturing oocytes exposed to 6 Gy of X radiation (presumably acute) and a 5 + 5-Gy experiment on stem-cell spermatogonia (same as above with more data). There was also an untreated control, which was probably concurrent with the experiment on females. The frequencies of offspring recorded as small at weaning were 35/2951 (1.2%) in the control, 36/1881 (1.9%) from irradiated oocytes, and 112/7388 (1.5%) from irradiated stem cells. The frequency was statistically significantly higher than the control in oocytes ($P = 0.03$ in a 1-tailed Fisher's exact test) but not in stem cells ($P = 0.12$). In both the control and the experiment on males, matings were omitted in which preexisting mutations causing small size were segregating. Of those small mice saved for testing, many could not be fully tested; however, 19 were confirmed to be mutants, including 1 in the control. The percentage of the fully tested variants that showed dominant inheritance increased markedly after irradiation, as would be expected if there were many nonmutational variants in the control and experimental groups and many induced mutations in the experimental groups. The number of offspring required for a variant to be "fully tested for heritability" was not reported.

Selby *et al.* (1988) found a high frequency of induction of stunted growth following exposure of stem-cell spermatogonia to four 100-mg/kg doses of ENU given at weekly intervals. F_1 male mice, weighed at 79 ± 4 d of age, were considered to have stunted growth if their weights were two or more standard deviations below the mean for control litters of like number at weaning. Because there was a strong, and highly statistically significant, negative correlation between weight at 79 d and number of offspring in the litter at weaning, a correction for number in the litter was needed to prevent the 11% LSR reported by W. L. Russell and Hunsicker (1988), in this same experiment (Section 3.4), from partly obscuring the differences in weights between the groups. The frequencies of mice with stunted growth were 49/492 (10%) in the experimental group and 16/521 (3.1%) in the concurrent control group, with the experimental group being significantly higher, $P < 0.0001$. Parents were randomized between the groups.

3.7. Shortened Life Span

W. L. Russell (1957) reported that exposure of male mice to neutron radiation from a nuclear detonation shortened the life spans of F_1 offspring conceived 19–23 d after treatment. There were five dose groups, from 31 to 186 rep (Roentgen equivalent proton) and an unirradiated concurrent control. Death before weaning was not included, and only one mouse died between weaning and 1 year of age. Regression analysis by the method of weighted least squares yielded an intercept of 786 d and a slope of -0.609 ± 0.238 d, which differed significantly from zero.

Russell estimated that life shortening in offspring of exposed men would be 20 d/rep (95% confidence limits of 5 and 35). He assumed that shortening of life in humans, with a 70-year normal life span, would be proportional to that seen in mice. (This was probably the first application of a direct method for estimating genetic risk in humans from induced damage in mice.) Russell emphasized that X or γ irradiation of spermatogonia would almost certainly produce a smaller effect.

W. L. Russell (1981) did not pursue attempts to measure effects on vital statistics because he concluded that such endpoints have so much natural variability, and are so easily affected by numerous hard-to-control factors, that a small increment of damage caused by mutations would be hard to detect. He also felt that effects on vital statistics, e.g., longevity or body weight, would be difficult to translate into human detriment. Other investigators either found no effect (Spalding, 1964;

Frölen, 1965) or little effect (Kohn *et al.*, 1965) of irradiation on longevity of offspring, but as W. L. Russell (1981) noted, these experiments had several major differences from his. Attempts to study longevity in offspring of mutagenized animals might be easier to extrapolate to humans if they included pathological study to determine causes of death.

3.8. Tumors

Because some human dominant disorders predispose individuals to develop tumors, it seems likely that effective mutagens would induce dominant mutations that cause tumors in descendants. Many papers have reported such an effect, but the evidence is usually weak. Early work in this area was discussed by Tomatis (1965, 1979) and Tomatis *et al.* (1975), and this review will focus primarily on research of Tomatis and Nomura. The phrase *tumor mutations* will be used in this section to denote dominant mutations that cause tumors to develop.

3.8.1. Results of Tomatis and Closely Related Studies

Tomatis (1965) reported that 350 μg of DMBA administered to random-bred Swiss mice late in pregnancy caused their F_2 descendants to have a wide variety of tumors, similar to the spectrum seen in controls, with a latent period of at least 55 weeks. (In this experiment, and in others like it, involving in utero exposures of F_1 offspring to known carcinogens, any effects seen beyond the F_1 generation are thought to result from tumor mutations induced in the germ cells of the exposed F_1 fetuses.) Of 32 F_2 descendants checked for tumors after natural death, 84% had tumors. There was no concurrent control, but Tomatis thought this high frequency was significant because only 36% of the females and 21% of the males had tumors in the strain itself. However, when the spontaneous occurrence of tumors is so high, it is hard to rule out the alternative hypothesis that preexisting mutations segregating in the stocks account for the difference observed. There was no mention of randomization of parents in this experiment or in any experiments discussed in this section.

Tomatis and Goodall (1969) reported a similar result when they studied the occurrence of tumors in F_1, F_2, and F_3 descendants of inbred MA mice injected intraperitoneally with 400 μg of DMBA during the last stages of pregnancy. The authors reported that the life span "in the experimental groups was similar or even longer" than that of the control. Indeed, the experimental females lived statistically significantly longer. To illustrate the large difference, whereas 54% of the control females were alive at 70 weeks of age, in the four groups of F_2 or F_3 experimental females the percentages then alive ranged from 67 to 80%. Because mice were only examined for tumors when they died or were moribund, the longer life span of the experimental mice gave them more time in which to develop tumors. It is thus unclear whether DMBA induced tumor mutations.

Tomatis *et al.* (1975) studied the occurrence of tumors in descendants of inbred BD rats exposed to 20 mg/kg MNU during pregnancy. Rats were killed for histological study if moribund or at 120–130 weeks of age. Curiously, even though mammary gland tumors were found in 43% of the F_2 experimental females but in only 7.1% of the F_2 control females, 84% of experimental females were still alive at 120 weeks of age compared with only 53% in the control, this difference being significant ($P = 0.012$). It thus appears that more experimental females lived long enough to develop mammary gland tumors. Regarding possible induction of tumor mutations, the authors attached most importance to their finding of a few rare types of nerve and kidney tumors in the experimental group. Their argument appeared stronger for nerve tumors, of which they found 1 in 119 F_2 offspring and 2 in 85 F_3 offspring. There were no tumors in 22 control offspring of the same generations. MNU was concluded to induce nerve tumors because no nerve tumors were reported in an analysis of 30,000 BD rats reported in the literature; however, since, in a later paper, Tomatis *et al.* (1982) reported that nerve tumors actually occur in 0.8–1.2% of control BD rats, their claimed effect for MNU no longer appears convincing.

Tomatis *et al.* (1977) administered 40 mg/kg ENU to BDVI rats on d 15 of pregnancy and studied tumor induction in the subsequent three generations. Animals were killed when moribund or at 120 weeks of age, and no difference from controls was seen in the overall percentage of tumor-bearing animals. Frequencies of nerve tumors were 0/18 in F_2s and 3/38 in F_3s in the experimental group and 0/33 and 1/34 in the same two generations in the control. A chi-square test of the frequency of 3/38 versus a control frequency of 1/91 (combined P, F_1, F_2, and F_3 generations), without the necessary correction for continuity, yields a P value of 0.04. It appears that this is the statistical test the authors used to claim that ENU-induced mutations were causing nerve tumors in the F_3 generation because they stated that "$p < 0.05$." Because of possible influences of the environment on such an endpoint, a more appropriate statistical comparison would seem to be between the combined frequency in the F_2 and F_3 generations and that in their matched controls, for which $P = 0.24$ in a 1-tailed Fisher's exact test. Thus, there is little support for the conclusion that ENU induced tumor mutations.

Tomatis *et al.* (1981, 1982) concluded that exposure of male rats to 80 mg/kg ENU induced dominant mutations that caused neurogenic tumors. This conclusion is statistically questionable, however, because it was reached only by selecting two successive subsets of their experimental data (neurogenic tumors out of five tumor categories and week 2 out of 4 weeks of mating) simply because those subgroups had a higher frequency of rats with neurogenic tumors. Unless the initial analysis of data shows evidence of inhomogeneity, which it did not, it is not legitimate to pick out a group for comparison merely because it shows the biggest difference. It could, perhaps, be argued that their earlier unfounded conclusion that ENU induces tumor mutations provided a reason to look at nerve tumors alone, but that still only accounts for one of their two special analyses.

Rao (1982) reported an increase in tumors in the F_2 generation of inbred Swiss albino mice exposed to 500 μg of DMBA on d 17 of pregnancy. Mice in this study were also exposed to the antioxidant butylated hydroxyanisole (BHA). Tumors were classified histopathologically at 50 weeks, even though Tomatis (1965) had reported that tumor mutations induced by DMBA in Swiss mice have a latent period of at least 55 weeks. There is uncertainty about the results of this experiment because the frequency of control mice with tumors seems low based on what is known about cancers in mice in general. Also, if induced tumor mutations are the explanation for the effect reported, it seems odd that there is a strong and statistically significant negative correlation between the fraction of the F_2 offspring with tumors and parity. Thus, there were many fewer tumors in offspring from fifth litters than in those from first litters.

Napalkov *et al.* (1987b) injected outbred SHR mice intraperitoneally with 100 mg/kg DMBA at 17–19 d pc and appear to have induced mutations that caused lung tumors (adenomas or adenocarcinomas of alveologenic origin) in F_2 offspring. The incidences of lung tumors in experimental and control F_2 offspring were 35/79 (44%) and 17/134 (12.7%), respectively. This difference was statistically significant, as was another comparison of similar incidences for two additional groups that received identical treatments but in addition were painted twice weekly for 24 weeks with 12-*O*-tetradecanoylphorbol-13-acetate (TPA), a skin-tumor promoter. All animals were kept until death or killed when moribund; their controls were F_1s and thus were not concurrent.

Napalkov *et al.* (1987a), using the same stock of animals, appear to have shown that DMBA induces skin-tumor mutations, but the mutations only manifested their effects if the F_2 mice were painted with the promoter TPA for many weeks. The frequencies of mice with papillomas or other skin tumors in the untreated control, TPA-only control, DMBA-only F_2s, and DMBA + TPA F_2s were 2/134, 5/76, 4/82, and 15/71, respectively, with the last frequency being statistically significantly higher than the TPA-only control ($P = 0.01$). Interestingly, most papillomas disappeared within 5 weeks after the cessation of skin painting. If tumor mutations were actually induced in these experiments, they would have been induced in gonocytes, pachytene oocytes, or both. No specific-locus data exist for any chemicals in those germ-cell stages.

3.8.2. Results of Nomura

Nomura (1975) first reported the induction of dominant tumor mutations detected in F_2 offspring after he exposed pregnant strain ICR/Jcl mice subcutaneously to 1000 mg/kg urethan on d 16 pc. Almost all tumors were lung tumors. The frequency of F_2 offspring with lung tumors (18/112, 16.1%) in the experimental group was statistically significantly higher than that (10/237, 4.2%) in the control group. There was no suggestion of any increase in tumors in 156 F_3 offspring or in 82 offspring produced when F_2 males were mated to untreated females. In additional urethan experiments reported in the same paper, adult females exposed to 1000 or 1500 mg/kg and adult males exposed to 1500 mg/kg were mated "at 1- to 10-week intervals" to untreated animals, and there was an untreated control. The frequencies of mice with any tumors or with lung tumors (84% of total) were statistically significantly higher in offspring in all three experimental groups. The frequencies of mice with one or more lung tumors were always approximately twice that of the control frequency of 50/809 (6.2%). There was also a statistically significant increase in the frequency of mice with other tumors in two of the three groups; the third was on the borderline of significance.

Nomura (1978) did many additional large experiments in which both sexes were exposed to urethan or X radiation in various doses, and then mated to untreated mice. Offspring were killed at 8 months of age to determine how many had tumors and how many tumors they had. In offspring of males given 1500 mg/kg urethan, there was a statistically significant increase in tumors for matings at 1–7, 8–14, 29–35, 36–42, and 50–56 d after treatment. One of two data points for stem-cell spermatogonia showed no significant induction of tumor mutations by urethan, and that is presumably the one to which Nomura (1984) referred when he argued that there was good agreement between his results and specific-locus results (which were not yet available for postspermatogonial stages). Because one of his two data points for stem cells showed highly significant induction, it is of interest to combine both sets of data in stem cells for the comparison with the control. This comparison shows that there was a statistically significantly higher mutation frequency in stem cells. These results in males conflict with the expectation from specific-locus experiments in which urethan was ineffective both in postspermatogonial stages and in stem cells (L. B. Russell *et al.,* 1987). In offspring of females given 1000 and 1500 mg/kg urethan, there was a statistically significant increase in tumors, at least at one dose or the other, for each weekly interval through d 56, with frequencies of affected mice at the higher dose ranging from 10/97 (10%) to 16/56·(29%).

In the radiation part of this experiment, Nomura (1978, 1982a) exposed adult males and females to 36, 108, 216, 360, or 504 rad of 72-rad/min X radiation administered as a single dose or in 36-rad fractions every 2 h and mated them to untreated mice at weekly intervals as late as 80 d after exposure. For the single exposure to 216 rad, Nomura reported the frequency of tumor-bearing offspring during individual weeks after treatment, and there were statistically significant increases for treated males in the intervals 8–14, 15–21, and 64–80 d and for treated females in the intervals 8–14 and 15–21 d, with no offspring being born after 28 d. Dose–response data, which were sparse at many points, were presented for males in the intervals 1–7, 15–21, and 64–80 d and for females at 1–7 d, and they appeared to be consistent with a linear dose response for both single and fractionated doses in all germ-cell stages in males, with the exception that 504 rad had no effect in spermatogonia when fractionated. The data also suggest that the dose–response curve is concave upward in females for unfractionated exposures, but that the fractionated exposure had no effect. The slope of a simple regression line of dose versus frequency is significantly greater than zero for unfractionated exposures in sperm, stem cells, and mature oocytes. Thus, the radiation data seem to be in good agreement with specific-locus results. Nomura noted that 87% of tumors were papillary adenomas.

A possible weakness in Nomura's experiments is that the same control group, which was collected at an earlier time than most experiments, was used throughout. His control group was killed for study at 32–50 weeks of age in the earlier work (Nomura, 1975), but only those killed at 32 weeks, of which 4.7% had one or more lung tumors, were used in most comparisons. Another feature of that single control, which raises a question about its suitability, is that none of the control mice had pneumonia (Nomura, 1975, 1978) even though 3.2% of experimental F_1s had pneumonia in the experiments done at the same time (Nomura, 1975) and 3.6% had pneumonia in the more recent (Nomura, 1978) experiments on urethan. These differences in the incidences of pneumonia are highly statistically significant and possibly important because they may reveal an important difference in environmental conditions.

Nomura (1978, 1982a,b, 1986) demonstrated transmissibility of lung tumor mutations, although the data from crosses of lung-tumor-bearing offspring mated to non-tumor-bearing offspring were rather limited. He concluded that these tumors are inherited as if they are dominant mutations with about 40% penetrance (Nomura, 1982a). There is no way of knowing, however, whether the mutations shown to be transmitted were induced mutations or were preexisting mutations. Even without this uncertainty, the evidence for transmission of tumor mutations is not nearly as well documented as it is for skeletal, cataract, and visible mutations.

Nomura (1982a) did a breeding test on four females from the untreated control that retrospectively were found to have developed lung tumors, and they gave no indication of transmission since only 2 of their 49 progeny had lung tumors. From this, Nomura (1984) concluded that dominant inheritance was not observed in most spontaneous lung tumors. Extensive data of this type, with adequate sample sizes, could show whether preexisting mutations significantly influenced his conclusions. It would not be surprising, however, if many lung tumors in both the control and experimental groups were nonmutational variants.

A characteristic feature of the tumor mutations is that they have very mild expressivity, with the mice carrying them having only one or two tumor nodules in the lung (Nomura, 1983, 1984) even though the tumor mutation is present in all lung cells. Nomura (1983) found that when F_1 mice in X-ray experiments were postnatally exposed to the tumor promoter urethan, larger clusters of lung tumor nodules formed than in controls with the same postnatal urethan treatment. This strongly suggests that there are tumor mutations present that have much stronger expressivity if the postnatal treatment is given. Nomura attached much importance to a small experiment in which the offspring of males irradiated at 15 d pc had no lung tumors at all without the postnatal urethan treatment (sample of 34) and no large clusters of lung tumors after the postnatal treatment (sample of 19). From this observation, Nomura (1983) concluded that "without tumor mutations, carcinogens rarely produce tumors." However, it appears that his samples were too small to justify such a conclusion. This experiment used the same control as the earlier experiments.

Nomura (1986) reported induction of tumor mutations in germ cells of males of two other mouse strains by exposure to 504 rad of X radiation. For the LT strain, fertility never returned following the onset of sterility, so he reported data for postspermatogonial stages alone. There was a statistically significant increase in the experimental frequencies, compared with those of the untreated control, for tumor-bearing mice (16/75 or 21% versus 37/411 or 9.0%), for mice with lung tumors (12/75 or 16% versus 22/411 or 5.4%), and for mice with leukemia (4/75 or 5.3% versus 4/411 or 1.0%). Only data for irradiated stem cells were reported for the N5 strain, and there was a statistically significant increase in the experimental frequencies, compared with those of the untreated control, for tumor-bearing mice (76/229 or 33% versus 56/244 or 23%) and for mice with leukemia (9/229 or 3.9% versus 1/244 or 0.4%). There were also strong, but not statistically significant, suggestions of induction of mutations causing lung tumors (48/229 or 21% versus 35/244 or 14%, $P = 0.06$) and ovarian tumors (28/123 or 23% versus 18/116 or 16%, $P = 0.16$).

Nomura (1986) found that when offspring in the N5 experiment were randomly mated in each generation to produce F_2 through F_4 generations, the frequencies of mice with leukemia in each generation were as follows: F_1, 9/229 (3.9%); F_2, 2/110 (1.8%); F_3, 3/116 (2.6%); and F_4, 4/152 (2.6%). He concluded that the frequency at generation 4 was still high compared with the control of 1/244. If his conclusion is correct that the effects in the F_1s result from dominant leukemia mutations, those mutations must have very low penetrance in order for the frequency of affected mice to stay similar for four generations of random mating. To illustrate, even with penetrance as high as 0.20, the finding that 3.9% of the F_1s had leukemia implies that $(1 \div 0.2) \times 3.9\%$, or about 20%, of the F_1s were carriers of dominant leukemia mutations. About two-thirds of the time, when breeding pairs were selected at random from such a population, neither parent would have carried a mutation, with the result that the frequency of mice with leukemia would decrease rapidly in each generation. Thus, if Nomura's interpretation is correct, the true penetrance must be much less than 0.20. This analysis does not take into account the further reduction that would occur if the affected mice died from leukemia before producing offspring.

Nomura (1986) found that 23 of 26 X-ray-induced tumors (including 11 lung tumors) in the N5 strain were transplantable. He also concluded that the lung tumors in the N5 strain resulted from dominant mutations with 40% penetrance. Nomura (1984) also stated that, in his experiments on ICR mice, ovarian tumors and lymphocytic leukemias increased in the same ratio as the much more frequent lung tumors.

3.8.3. General Comments on Induction of Tumor Mutations

Although Nomura's experiments suggest that it might be easy to demonstrate induction of tumor mutations by X radiation, Kohn *et al.* (1965) found no effect of irradiation on tumor incidence in offspring from male mice exposed to 525–720 rad of X rays (stem-cell spermatogonia). Both the experimental and control groups developed many tumors. For example, between 58 and 73% of the experimental offspring had tumors when they were autopsied at the time of natural death. Lung tumors were by far the most common tumor type in both control and experimental F_1 males; lymphomas and lung tumors were both common in females.

It seems highly likely that dominant mutations that cause tumors can be induced, but it is unclear whether the few apparently convincing experiments described in this section indicate true induction of tumor mutations or whether, instead, they reflect the presence of preexisting mutations or other difficulties that creep into the interpretation when dealing with endpoints that are not rare. If parents had been randomized and offspring coded in Nomura's urethan experiments, e.g., his results would have been much more convincing. In this regard, however, it is important to note that according to IARC (1986) Scientific Publication No. 83, positive results in the lung tumor bioassay "may be strongly suggestive of carcinogenicity but are not conclusive by themselves." When positive results are found using such an assay, replication of results in another laboratory would greatly increase confidence in the results; and a long-term, full-fledged bioassay might be necessary to remove all doubt.

3.9. Effects on Behavior

3.9.1. Newcombe's Experiment on Maze-Learning Ability

Newcombe and McGregor (1964) attempted to increase their chances of finding an induced genetic effect on maze-learning ability in rats by accumulating mutations from up to 12 generations of exposure of sperm and spermatids to high-dose-rate X radiation. They tested 464 control and 460 experimental progeny following an average accumulated exposure to ancestors of 2587 rad and found a significant decline in ability equivalent to a decline in mean I.Q. in humans from 100 to 94.65. The basis for relating their results to human I.Q. was that 15 points on the human I.Q. scale are approximately equivalent to one standard deviation. This interesting application of their data had limited usefulness, however, because, as they noted, the high level of dominant lethality made the experimental litters much smaller, which led to uncertainty over whether the effect on maze-learning ability resulted from mutations affecting intelligence or, in some way, from the much reduced litter size.

3.9.2. Behavioral Studies of Adams *et al.*

Work done at the University of Texas Medical Branch has led to claims (Fabricant *et al.*, 1983; Adams *et al.*, 1984; Fanini *et al.*, 1984) that a battery of tests for assaying behavioral development provides a sensitive method for detecting mutational damage induced in male germ cells. (These same papers discuss some related literature not reviewed here.) The tests used, and the ages at which they are applied to inbred F344 rats, are as follows: surface righting (3 and 6 d), cliff avoidance (4 and 10 d, at which time eyes are still closed), various indicators of swimming ability (6, 8, 10, 12, and 14 d), negative geotaxis, open-field activity (14 and 21 d), and one-way active avoidance (30 d). In the last-named test, rats are conditioned to avoid an electrical shock by climbing onto a platform. There was no indication that parents were randomized or that offspring were coded in these experiments.

In their first experiment, Adams *et al.* (1981) injected both sexes with 10 mg/kg CP or saline solution 5 d/week for 5 weeks. From rats mated from 3 d through 4 weeks after the final injection, they obtained 35, 9, and 31 pups for behavioral testing, from the control × control, CP × CP, and CP × control crosses, respectively. Significant differences from the control were found in the F_1 offspring for cliff avoidance (70–80% showed a developmental delay in experimental group), swimming ability (poorer in experimental group), and open-field behavior (hyperactive in experimental group). In the active avoidance test, there was no difference in rate of learning to avoid electrical shock, but the experimental group remembered to avoid the shock, after it was no longer applied, significantly longer than the control group (Adams *et al.*, 1984).

Fabricant *et al.* (1983) and Adams *et al.* (1984) also studied the effect of a single i.p. injection of 10 mg/kg CP or saline, with matings made 7–9, 14–16, or 28–30 d after treatment. Even though the total exposure was only 4% of that in the former experiment, statistically significant effects were found, in at least one age group, for cliff avoidance, swimming development, open-field activity (hypoactive instead of hyperactive), and active avoidance (similar to effect seen for the fractionated exposure).

Fanini *et al.* (1984) applied all but the test of active avoidance to Fisher 344 inbred rats exposed to five daily i.p. dosages of ethylene dibromide (EDB) of 1.25, 2.5, 5.0, and 10.0 mg/kg dissolved in corn oil.

For pups conceived 4–5 weeks after the last injection, there were statistically significant effects, compared with the solvent control, for cliff avoidance, swimming development, and open-field activity (hypoactive). For pups conceived 9 weeks after injection, there was statistically significantly less open-field activity than in the control.

Hsu *et al.* (1985, 1987) found changes in the activities of three neurotransmitter enzymes in the brains of rats whose fathers had one or more postspermatogonial stages exposed to CP or EDB. For CP, 6 males and 6 females from each of two dosage groups and their corresponding controls were evaluated in five different brain regions when mice were 90 d old. Of 60 comparisons made with the controls, 17 differed at the 5% significance level, with 11 showing lower activity and 6 showing higher activity. For EDB, examinations of these brain regions were made in mice 7, 14, 21, and 90 d of age, and individual groups consisted of 4–9 mice. Sixteen of fifty-five comparisons showed a significant difference from the control at the 5% significance level; 9 had higher activities and 7 had lower ones. For EDB, enzyme activities seemed to have reverted to normal in adults, with only 1 in 15 comparisons in the adult being significantly different at the 5% significance level.

The authors indicated that they had not yet established any relationship between the many reported differences in enzyme activity and the behavioral effects. The statistical interpretations are also unclear because, e.g., the CP experiment had two separate controls which sometimes differed more from each other than from the experimental values said to be significantly different. Sometimes the changes from the control are in opposite directions for single and multiple CP treatments or, in the EDB experiment, at different ages.

Adams *et al.* (1984) claimed that their behavioral assessments are sensitive enough to detect genotoxic effects transmitted to the second generation. They did find behavioral effects in the F_2 generation; however, they did not demonstrate transmissibility by breeding affected animals to see if their offspring showed the same or similar effects, as has been done for skeletal, cataract, visible, and tumor mutations. Instead, they made sib matings of adult animals, regardless of behavioral phenotype, to produce F_2 progeny, which they found had statistically significantly different responses from the control in swimming development (slower than control), active avoidance (similar response to F_1s), and open-field activity (F_2s were hypoactive but F_1s were hyperactive).

Even though these behavioral tests have yielded many significant differences in spite of the generally small sample sizes, it is unclear whether they will be useful for identifying agents that are mutagenic in mammalian germ cells. One puzzle is that a single exposure of postspermatogonial stages to 10 mg/kg CP (Fabricant *et al.*, 1983) induced more mutations affecting cliff avoidance than 25 such exposures to postspermatogonial stages (Adams *et al.*, 1981). Another puzzle is illustrated by the following example, of which many could be given. Fabricant *et al.* (1983) reported that 70.8% of 8-d-old control offspring could swim in a straight line but that only 18.7% could do so if their fathers were exposed to 10 mg/kg CP 7–16 d before the offspring were conceived. If the reason for this difference is really induced mutations, it follows that the point estimate of the percentage of offspring carrying one or more induced mutations that reduce swimming ability is 52.1% (i.e., 70.8–18.7%). Not only does this mutation frequency seem unbelievably high for a single type of behavior, but it seems odd that so many induced mutations would affect this particular behavior at 8 d of age but none would do so a few days later.

Specific-locus results are in conflict with those of the behavioral tests for EDB and probably for the low dose of CP. No specific-locus mutations were found in 26,242 offspring derived from stem-cell spermatogonia and 2370 offspring derived from postspermatogonial stages, following i.p. injections of 160 or 167 mg/kg EDB (L. B. Russell *et al.*, 1985; W. L. Russell, 1986). There is thus no evidence that EDB induces specific-locus mutations at a much higher dose than was used in the behavioral work. Cumming and Walton (1971) found three specific-locus mutations in 3642 offspring derived from exposed postspermatogonial stages following an i.p. injection of 350 mg/kg CP. This small an effect makes it seem unlikely that the behavioral studies could have found any genetic effect from only 10 mg/kg. Another problem about the meaningfulness of the behavioral studies is that there were so many tests applied to so many treated groups (both different intervals after treatment and different ages) that, by chance alone, some significant differences would be expected. An abstract by Lowery *et al.* (1987) reported effects on motor reflexes and learned behavior in the offspring of male rats exposed to ionizing radiation.

3.10. Miscellaneous Experiments

Charles *et al.* (1960, 1961) looked for several types of induced dominant damage in the first generation after male DBA mice were exposed to X rays and mated to C57 females. A mixture of different germ-cell stages was exposed at dose rates of from 0.1-10 r/d to doses ranging from 0.1–1000 r; there was an unirradiated control. Over 12,000 offspring, mostly from lower doses, were examined in an attempt to detect dominant visibles, death before weaning, chromosomal damage causing small litter size, changes in sex ratio, and damage detected in an autopsy to such things as blood vessels, internal organs, and teeth. A significantly higher proportion of rare abnormalities was found in offspring of exposed males, with "rare" defined as occurring in no more than four of the approximately 12,000 individuals autopsied. The mutation rate for all kinds of damage, including mice with "rare" abnormalities, was 1.16×10^{-4}/r of paternal exposure.

Green's (1968) review and the papers edited by Roderick (1964) described many early attempts to detect genetic damage in populations. Most of that massive amount of research, which gave generally negative results, was done to determine the increase in genetic damage resulting from multiple generations of radiation exposure. Most endpoints studied were measures of fitness such as life span, weight gain, postnatal survival, reproductive productivity, and numbers of offspring present at midgestation, birth, or weaning. Many experiments involved complexities such as somatic exposure of the animals being studied, calculations of ancestral exposures, and exposures of many germ-cell stages. According to Green's review, the only attempt to estimate the frequency of a malformation in multigeneration experiments was Newcombe and McGregor's (1964) work on "dwarfism."

Green (1968) cautioned that "the generally negative results of these studies may be due [1] to the nonexistence of induced mutations having only moderate individual effects in heterozygotes, [2] to the failure to find the right indicator trait, or [3] to relatively small sizes of the experiments so far conducted and their relative lack of power for discriminating small genetic differences in the presence of large amounts of nongenetic variability." Knowledge gained since then has shown that the first alternative is incorrect. Much progress has been made in finding useful indicator traits, and it appears likely that more powerful ways of detecting induction of dominant mutations can be developed.

4. GENETIC RISK ESTIMATION

The two major approaches used for human genetic risk estimation for radiation are the indirect and direct methods (UNSCEAR, 1977; BEIR III, 1980). The indirect method, often called the doubling-dose approach, is a method of relative risk estimation; the direct method is a method of absolute risk estimation.

4.1. Indirect Method

The indirect method rests on the assumption that the likelihood of causing a genetic handicap is approximately the same for induced and spontaneous mutations. This assumption has always been in doubt, and now is increasingly more in question (UNSCEAR, 1986). Animal data are used to estimate the doubling dose, which is the amount of radiation or other mutagen that induces as many mutations as occur spontaneously in the germ cells of one parent. The indirect method also requires (1) estimates of the current incidences of different classes of genetic disorders in the human population, (2) estimates or assumptions about the persistences of different types of mutations in the population, and (3) assumptions about the mutational components of those disorders that have complex patterns of inheritance. The BEIR III report (1980) and the UNSCEAR reports (1977, 1982, 1986) discuss the indirect method in detail. Here, only those cases are discussed in which this method has been applied to dominant mutations included in this review.

Nomura (1983, 1988) calculated radiation doubling doses both for congenital malformations and for tumors. It is important to realize that his results on congenital malformations and tumors, as well as any other results described in this review (with the possible exception of dominant visibles with complete penetrance), are completely inappropriate for the calculation of doubling doses because it is not known what part of the control frequency consists of new spontaneous mutations. For autosomal dominant mutations, the numerator in the doubling-dose calculation is taken as one-half of the frequency of newly occurring spontaneous mutations detected in F_1 offspring (Lüning and Searle, 1971). It would not be surprising if the great majority of the "mutations" found in Nomura's control groups were either preexisting mutations (Haldane, 1936; Section 3.1.2b) or nonmutational variants. Thus, Nomura's use of his control frequencies as the frequencies of newly occurring spontaneous mutations could have led to an extreme overestimation of the doubling dose. The denominator in the doubling-dose calculation is the induced mutation frequency per unit dose, and the induced frequencies used in Nomura's calculations were also too imprecise to permit the calculation of reliable doubling doses. Lüning and Searle (1971) calculated radiation doubling doses for presumed dominant skeletal mutations and dominant visibles, but they expressed reservations about doing so because there was no proof of inheritance for the former and a large subjective element in the detection of the latter.

4.2. Direct Method

The direct method extrapolates from induced phenotypic damage found in F_1 offspring of exposed animals to overall or partial damage in the human population.

4.2.1. Based on Skeleton

The first direct risk estimate made by a committee was based on data on dominant skeletal mutations (UNSCEAR, 1977), and is still used (UNSCEAR, 1982, 1986). UNSCEAR (1977) used both available estimates of the induced dominant skeletal mutation frequency, namely those of 37/2646 (Selby and Selby, 1977a) and 5/754 (Ehling, 1966). Both frequencies were divided by 3 to correct for the dose-rate effect, and the former frequency was divided by 1.9 to correct for the fractionation effect. (These corrections were based on specific-locus results.) Thus, UNSCEAR calculated that the induced mutation frequency of dominant skeletal mutations was 4×10^{-6}/R per gamete.

UNSCEAR (1977), however, wanted to estimate the damage to all body systems. According to McKusick's catalogue (1975), 74 of 328, or roughly 1 out of 5, "proved" and clinically important autosomal dominant disorders in humans involve the skeleton. From this it would appear that multiplication of the skeletal mutation frequency by 5 would yield a rough estimate of the total number of dominant mutations. There is, however, bias of ascertainment for skeletal disorders because it is much simpler to recognize skeletal abnormalities than many others. Thus, the true multiplier must be larger than 5. The finding that many genetic disorders affect several body systems, however, suggests that this multiplier should not be vastly larger than 5. Using this line of reasoning and the opinions of two well-known clinical geneticists, namely McKusick and Carter, UNSCEAR decided to use 10 as the multiplier for expanding from the skeleton to the total body.

UNSCEAR also wanted to restrict its estimate to phenotypic effects of clinical significance. An important feature of the data of Selby and Selby (1977a, 1978a,b) was that they provided details on the effects of a large sample of mutations. A detailed discussion of the 37 mutations by McKusick, Selby, and Russell, at the request of UNSCEAR, led to the conclusion that about 50% of them had effects that would be a serious handicap if the same syndromes occurred in humans. This estimate agreed well with the tentative conclusion reached by Carter based on his study of the Selby and Selby manuscripts.

From multiplication of 4×10^{-6} by 10, by 0.5, and by 1 million (so that risk could be expressed per million live-born), UNSCEAR estimated that paternal exposure to 1 R per generation would lead to 20 serious dominant disorders per million live-born. The BEIR III (1980) Committee used a similar approach except that it preferred to give a range, and assumed that the correct multiplier for extrapolating from the skeleton to the whole body is between 5 and 15, that the correct multiplier for severity is between 0.25 and 0.75, and that, based on the conclusion of W. L. Russell (1977), risk from maternal exposure is between 0 and 0.44 times that from paternal exposure. Accordingly, the BEIR III estimate for induced dominant damage in the first generation, following the same population exposure, was 5–65 serious dominant disorders per million live-born.

4.2.2. Based on Cataracts

Starting with its 1982 report, UNSCEAR (1982) has also made a direct estimate based on the cataract data of Ehling and Kratochvilova. The committee concluded from the two single-dose experiments (Ehling *et al.*, 1982) that the dominant cataract mutation frequency was in the range of $0.45–0.55 \times 10^{-6}$ mutation/R per gamete for high-dose-rate exposures. For low-dose-rate exposures, the mutation rate was divided by 3, for the dose-rate effect (based on specific-locus results), thereby yielding a frequency of $0.15–0.18 \times 10^{-6}$ mutation/R per gamete. All cataracts were considered serious, and the multiplier used to extrapolate from cataracts to all dominant disorders was assumed to be 36.8, based on McKusick's (1978) catalogue with no adjustment for bias of ascertainment. Multiplication of the mutation frequency range by 36.8 and by 1 million yielded a risk estimate of 6–7 serious dominant disorders for 1 R of paternal exposure per generation. These calculations were based on those made by Ehling (1980b; Ehling *et al.*, 1982). Ehling also estimated risk from the 4.55 + 4.55-R experiment (Ehling and Kratochvilova, 1979; Kratochvilova and Ehling, 1979) and, as UNSCEAR noted, concluded that 1 R of paternal irradiation to the population would result in 13 offspring with serious dominant disorders. Averaging the above two estimates, UNSCEAR used 10 as the number.

The UNSCEAR direct-risk estimate for mutations having dominant effects is thus 10–20 serious disorders per million live-born, for 1 R of paternal exposure. The lower estimate is based on cataracts, the upper one on skeletal disorders. Risk from maternal exposure has been estimated to be 0–9 serious disorders by applying the aforementioned conclusion of W. L. Russell. (In addition, there is a separate category for unbalanced products of translocations, which is beyond the scope of this review.)

Ehling (1983, 1984b; Jacobi *et al.*, 1981) argued that direct-risk estimates based on cataracts are better than those based on the skeleton because for the skeleton it was deemed necessary to make corrections for severity and bias of ascertainment. It seems certain, however, that those corrections improved the reliability of risk estimates based on skeletal malformations, and it seems likely that similar corrections would improve risk estimates based on cataracts. Some cataracts in humans are asymptomatic. Some of the cataracts found in mice by Ehling and coworkers are said to have minor effects, and it would be interesting to know how many of the radiation-induced cataracts are similar to cataracts that would be asymptomatic in humans. Regarding the other correction, serious cataracts in people probably have an even higher bias of ascertainment than skeletal malformations. It would be very hard to overlook them.

Compared with the cataract methods, the skeletal methods have the distinct advantages of dealing with a much wider range of abnormalities and with much higher mutation frequencies. In addition, the much greater size and complexity of the skeleton, compared with the lens, offer much greater opportunity for devising effective presumed-mutation criteria and for recognizing preexisting mutations.

4.2.3. Advantages of the Direct Method

Needless to say, the direct and indirect methods of genetic risk estimation both have many uncertainties. Much uncertainty in the direct method could be removed by further experiments. For example, skeletal experiments could be carried out on females, or using protracted irradiation, or in additional species to learn more about the uncertainties involved in extrapolations between species.

Perhaps the biggest advantage of the direct method is that it includes within its scope the irregularly inherited disorders which constitute 85% (i.e., 90,000/105,900 in the 1982 UNSCEAR Report's Table 44) of the serious genetic disorders in humans. Indeed, not only has UNSCEAR (1986) abandoned the use of the indirect method of risk estimation for the irregularly inherited disorders, but it has presented substantial evidence that those disorders with complex patterns of inheritance may affect a much larger portion of the human population than was previously thought.

In contrast with the indirect method, the direct method requires (1) no assumption that spontaneous and induced mutations have approximately the same likelihood of causing harm, (2) no knowledge of the current incidence of serious genetic disorders in the human population, (3) no guess as to how many of the irregularly inherited disorders are multifactorial or dominant disorders with low penetrance, and (4) no knowledge of the persistence of mutations in the population, which is needed to derive a first-generation estimate using the indirect method. Ehling (1980c, 1986) and Selby (1983) have both suggested ways in which the direct method of risk estimation could be applied to chemicals.

4.2.4. Litter-Size Reduction

A major reason for the effort of Selby and Russell (1985) to provide a detailed analysis of the extent of the LSR (Section 3.4) was the realization that some mutations kill offspring before they are old enough for their skeletons or eyes to be examined. This could lead to an underestimation of induced genetic damage. As noted earlier, data of Selby and Russell (1985), Lüning (1972), and Searle and Papworth (1986) were used to make the estimate that for 1 R of paternal irradiation per generation, between five and ten offspring per million live births die from dominant sublethal effects which may kill between birth and early life (UNSCEAR, 1986).

5. SPECIAL CONSIDERATIONS FOR FUTURE STUDY OF INDUCTION OF DOMINANT MUTATIONS

The importance of randomizing parents between experimental and control groups, and of taking other precautions to guard against misinterpretations caused by preexisting mutations, cannot be overstated. Also, for those methods involving some level of subjectivity, offspring should be coded until after classifications have been made. Whether or not such precautions have been taken should be reported.

There are many more uncertainties involved in the study of dominant mutations than there are for specific-locus mutations. A notable example is that preexisting mutations can be a major problem in studies of dominant mutations, but they are not a problem in specific-locus experiments (W. L. Russell *et al.*, 1982). Thus, there is reason to be skeptical of results in tests for dominant mutations that conflict with the expectations from specific-locus results. However, it must be kept in mind that it is not yet known how well the induction of specific-locus mutations correlates with that of dominant mutations. It will be important to pursue this question for a variety of mutagens and for different germ-cell stages.

It seems that the ideal method of collecting data to be used for genetic risk estimation would be to identify all offspring in the first generation that have clinically important effects. Subtraction of the frequency of such offspring in the control group from that in the experimental group would provide an estimate of the induced mutation frequency. By doing this and by making clinically important effects the critical endpoint (instead of proof of transmission or the meeting of particular presumed-mutation criteria), it should be possible to remove the danger that some serious mutations will be overlooked. Selby and Niemann (1984) applied this method for skeletal abnormalities with apparent success. This method could become a much more reliable approach if clinical geneticists helped to classify animals regarding clinical severity and if variability within the stocks was better understood.

For many other research questions, however, it will be important to identify as many of the mutations as possible by breeding tests, NBT methods, or both. Presently, breeding tests and NBT methods both have serious drawbacks if applied by themselves. Extensive experimentation, including breeding tests, could make NBT methods for the skeleton, and perhaps for some other endpoints, much more reliable.

After almost four decades, the mouse visible specific-locus test is still the most efficient method for determining whether a particular chemical can induce gene mutations and small chromosomal deficiencies in mammalian germ cells. In spite of the complexities involved in studying dominant mutations, there is good reason to think that tests for induction of dominant mutations may someday provide an even more useful means of determining whether a particular chemical can induce mutations in mammals. Improved NBT methods for skeletal defects, cataracts, stunted growth, and congenital malformations seem attainable. Some other endpoints discussed in this review, and some others not yet tested, may prove useful. One possible approach might be to conduct a multiple-indicator mutation test using all or most of the above endpoints. No time-consuming breeding tests would be used. Such a test would have the advantage of being based on probably thousands of genetic loci that can mutate to dominant alleles. A negative result could be accepted with a higher degree of confidence than at present, and a positive result could be related to risk estimates made by the direct method.

Selby (1986) and Selby *et al.* (1987) reported that different dominant mutations can interact synergistically to cause abnormalities that neither one can cause alone. Mutations *Dsh* and *Ccd,* which are nonallelic and probably unlinked, were found to have seven synergistic and three antagonistic interactions. Synergistic interactions between dominants could mimic recessive inheritance in human pedigrees if each parent carried a different dominant mutation without a clinically important effect, but an offspring had severe abnormalities because it was heterozygous for both mutations. For theoretical reasons, Selby suspected that synergisms might be likely to occur in double heterozygotes for dominant mutations that have one or a few features in common. So far, only one such combination has been tested, and it has yielded many synergisms. Should this phenomenon prove to be common, it could have important implications for genetic risk estimation beyond the first generation, because it would cause the level of induced damage to increase faster than otherwise expected.

It would be relatively easy to induce large numbers of dominant mutations that would be valuable models of serious human disorders. Dominant visibles, dominant cataract mutations, and dominant skeletal mutations are currently being studied using powerful methods of molecular biology to understand biochemical and regulatory changes that result in the complex phenotypes observed. Results of such studies will have profound effects on human health. For example, for a mutation with low penetrance, the underlying developmental factors are only occasionally shifted in such a way as to cause the crossing of a threshold for expression of the mutant phenotype. An understanding of these factors might well result in the capability of intervening in some way (perhaps even medication during pregnancy) to prevent the expression of certain genetic disorders in people. Methods of detecting carriers of mutations during pregnancy are rapidly improving, but for mutations with low penetrance, abortion would probably be an unacceptable solution if a mutation were identified. Thus, it is especially important to learn much more about the phenomenon of incomplete penetrance. Fortunately, more attention is being given to this phenomenon now that an increased effort is being applied to estimate genetic risk for the irregularly inherited disorders, which account for the great bulk of the genetic load of our species.

ACKNOWLEDGMENT. Research sponsored by the Office of Health and Environmental Research, U.S. Department of Energy, under contract DE-AC05-84OR21400 with the Martin Marietta Energy Systems, Inc.

REFERENCES

Adams, P. M., Fabricant, J. D., and Legator, M. S. 1981. Cyclophosphamide-induced spermatogenic effects detected in the F_1 generation by behavioral testing. *Science* **211**:80–82.

Adams, P. M., Shabrawy, O., and Legator, M. S. 1984. Male transmitted developmental and neurobehavioral deficits. *Teratog. Carcinog. Mutag.* **4**:149–169.

Bailey, D. W. 1978. Sources of subline divergence and their relative importance for sublines of six major inbred strains of mice, in: *Origins of Inbred Mice*, H. C. Morse, III, ed. Academic Press, New York, pp. 197–215.

Baker, T. G. 1973. The effects of ionizing radiation on the mammalian ovary with particular reference to oogenesis, in: *Handbook of Physiology*, Section 7: Endocrinology 2 (Part 1), pp. 349–361.

Bartsch-Sandhoff, M. 1974. Skeletal abnormalities in mouse embryos after irradiation of the sire. *Humangenetik* **25**:93–100.

Batchelor, A. L., Phillips, R. J. S., and Searle, A. G. 1966. A comparison of the mutagenic effectiveness of chronic neutron- and γ-irradiation of mouse spermatogonia. *Mutat. Res.* **3**:218–229.

Batchelor, A. L., Phillips, R. J. S., and Searle, A. G. 1967. The reversed dose-rate effect with fast neutron irradiation of mouse spermatogonia. *Mutat. Res.* **4**:229–231.

BEIR III (Advisory Committee on the Biological Effects of Ionizing Radiation of the United States National Academy of Sciences). 1980. Genetic effects, in: *The Effects on Populations of Exposure to Low Levels of Ionizing Radiation.* National Academy Press, Washington, D.C., pp. 91–180 in typescript ed. and pp. 71–134 in printed ed.

Borum, K. 1961. Oogenesis in the mouse, a study of the meiotic prophase. *Exp. Cell Res.* **24**:495–507.

Brambell, F. W. R. 1927. The development and morphology of the gonads of the mouse. Part I. The morphogenesis of the indifferent gonads and of the ovary. *Proc. R. Soc. London Ser. B* **101**:391–409.

Cacheiro, N. L. A., and Russell, L. B. 1975. Evidence that linkage group IV as well as linkage group X of the mouse are in chromosome 10. *Genet. Res.* **25**:193–195.

Carter, T. C., and Lyon, M. F. 1961. An attempt to estimate the induction by X-rays of recessive lethal and visible mutations in mice. *Genet. Res.* **2**:296–305.

Carter, T. C., Lyon, M. F., and Phillips, R. J. S. 1958. Genetic hazard of ionizing radiations. *Nature* **182**:409.

CCEM (Committee on Chemical Environmental Mutagens). 1983. *Identifying and Estimating the Genetic Impact of Chemical Mutagens.* National Academy Press, Washington, D.C.

Charles, D. R., Tihen, J. A., Otis, E. M., and Grobman, A. B. 1960. Genetic effects of chronic X-irradiation exposure in mice. UR-565, AEC Research and Development Report. University of Rochester Atomic Energy Project, Rochester, N.Y.

Charles, D. R., Tihen, J. A., Otis, E. M., and Grobman, A. B. 1961. Genetic effects of chronic exposure in mice. *Genetics* **46**:5–8.

Clermont, Y., Leblond, C. P., and Messier, B. 1959. Durée du cycle de l'épithélium séminal du rat, *Arch. Anat. Microsc. Morphol. Exp.* **48** bis:37.

Cumming, R. B., and Walton, M. F. 1971. Genetic effects of cyclophosphamide in the germ cells of male mice. *Genetics* **68**:s14.

Deol, M. S., Grüneberg, H., Searle, A. G., and Truslove, G. M. 1957. Genetical differentiation involving morphological characters in an inbred strain of mice. I. A British branch of the C57BL strain. *J. Morphol.* **100**:345–375.

Ehling, U. H. 1965. The frequency of X-ray induced dominant mutations affecting the skeleton of mice. *Genetics* **51**:723–732.

Ehling, U. H. 1966. Dominant mutations affecting the skeleton in offspring of X-irradiated male mice. *Genetics* **54**:1381–1389.

Ehling, U. H. 1967. Transmission of radiation-induced dominant skeletal mutations in mice. Annu. Prog. Rep. Oak Ridge Natl. Lab. Biol. Div. ORNL-4240:155–156.

Ehling, U. H. 1970. Evaluation of presumed dominant skeletal mutations, in: *Chemical Mutagenesis in Mammals and Man*, F. Vogel and G. Röhrborn, eds. Springer, Berlin, pp. 162–166.

Ehling, U. H. 1980a. Comparison of the mutagenic effect of chemicals and ionizing radiation in germ cells of the mouse, in: *Progress in Environmental Mutagenesis*, M. Alacevic, ed. Elsevier/North-Holland, Amsterdam, pp. 47–58.

Ehling, U. H. 1980b. Strahlengenetisches risiko des menschen. *Umsch. Wiss. Tech.* **80:**754–759.

Ehling, U. H. 1980c. Induction of gene mutations in germ cells of the mouse. *Arch. Toxicol.* **46:**123–138.

Ehling, U. H. 1983. Cataracts—indicators for dominant mutations in mice and man, in: *Utilization of Mammalian Specific Locus Studies in Hazard Evaluation and Estimation of Genetic Risk,* F. J. de Serres and W. Sheridan, eds. Plenum Press, New York, pp. 169–190.

Ehling, U. H. 1984a. Variants and mutants. *Mutat. Res.* **127:**189–190.

Ehling, U. H. 1984b. Methods to estimate the genetic risk, in: *Mutations in Man,* G. Obe, ed. Springer, Berlin, pp. 291–318.

Ehling, U. H. 1985. Induction and manifestation of hereditary cataracts, in: *Assessment of Risk from Low-level Exposure to Radiation and Chemicals,* A. D. Woodhead, C. J. Shellabarger, V. Pond, and A. Hollaender, eds. Plenum Press, New York, pp. 345–367.

Ehling, U. H. 1986. The quantification of the frequency of induced dominant and recessive mutations, in: *Risk and Reason: Risk Assessment in Relation to Environmental Mutagens and Carcinogens,* Liss, New York, pp. 95–98.

Ehling, U. H., and Kratochvilova, J. 1979. Direct estimation of genetic risk from radiation in the first generation. VI Int. Congr. Radiat. Res. Abstr., p. 180.

Ehling, U. H., and Randolph, M. L. 1962. Skeletal abnormalities in the F_1 generation of mice exposed to ionizing radiations. *Genetics* **47:**1543–1555.

Ehling, U. H., Favor, J., Kratochvilova, J., and Neuhäuser-Klaus, A. 1982. Dominant cataract mutations and specific-locus mutations in mice induced by radiation or ethylnitrosourea. *Mutat. Res.* **92:**181–192.

Fabricant, J. D., Legator, M. S., and Adams, P. M. 1983. Post-meiotic cell mediation of behavior in progeny of male rats treated with cyclophosphamide. *Mutat. Res.* **119:**185–190.

Fanini, D., Legator, M. S., and Adams, P. M. 1984. Effects of paternal ethylene dibromide exposure on F_1 generation behavior in the rat. *Mutat. Res.* **139:**133–138.

Favor, J. 1983. A comparison of the dominant cataract and recessive specific-locus mutation rates induced by treatment of male mice with ethylnitrosourea. *Mutat. Res.* **110:**367–382.

Favor, J. 1984. Characterization of dominant cataract mutations in mice: penetrance, fertility and homozygous viability of mutations recovered after 250 mg/kg ethylnitrosourea paternal treatment. *Genet. Res.* **44:**183–197.

Favor, J. 1986. A comparison of the mutation rates to dominant and recessive alleles in germ cells of the mouse, in: *Genetic Toxicology of Environmental Chemicals, Part B: Genetic Effects and Applied Mutagenesis,* C. Ramel, B. Lambert, and J. Magnusson, eds. Liss, New York, pp. 519–526.

Favor, J., Neuhäuser-Klaus, A., and Ehling, U. H. 1987. Radiation-induced forward and reverse specific locus mutations and dominant cataract mutations in treated strain BALB/c and DBA/2 male mice. *Mutat. Res.* **177:**161–169.

Favor, J., Neuhäuser-Klaus, A., and Ehling, U. H. 1988. The effect of dose fractionation on the frequency of ethylnitrosourea-induced dominant cataract and recessive specific locus mutations in germ cells of the mouse. *Mutat. Res.* **198:**269–275.

Frölen, H. 1965. The effect on the length of life in the offspring of X-irradiated male mice. *Mutat. Res.* **2:**287–292.

Generoso, W. M., Rutledge, J. C., Cain, K. T., Hughes, L. A., and Braden, P. W. 1987. Exposure of female mice to ethylene oxide within hours after mating leads to fetal malformation and death. *Mutat. Res.* **176:**269–274.

Graw, J., Favor, J., Neuhäuser-Klaus, A., and Ehling, U. H. 1986. Dominant cataract and recessive specific locus mutations in offspring of X-irradiated male mice. *Mutat. Res.* **159:**47–54.

Green, E. L. 1968. Genetic effects of radiation on mammalian populations. *Annu. Rev. Genet.* **2:**87–120.

Grüneberg, H., Bains, G. S., Berry, R. J., Riles, L., Smith, C. A. B., and Weiss, R. A. 1966. A search for genetic effects of high natural radioactivity in South India. *Med. Res. Counc. G. B. Spec. Rep. Ser.* **307.**

Haldane, J. B. S. 1936. The amount of heterozygosis to be expected in an approximately pure line. *J. Genet.* **32:**375–391.

Hitotsumachi, S., Carpenter, D. A., and Russell, W. L. 1985. Dose-repetition increases the mutagenic effectiveness of N-ethyl-N-nitrosourea in mouse spermatogonia. *Proc. Natl. Acad. Sci. USA* **82:**6619–6621.

Hsu, L. L., Adams, P. M., Fanini, D., and Legator, M. S. 1985. Ethylene dibromide effects of paternal exposure on the neurotransmitter enzymes in the developing brain of F-1 progeny. *Mutat. Res.* **147:**197–203.

Hsu, L. L., Adams, P. M., and Legator, M. S. 1987. Cyclophosphamide: effects of paternal exposure on the brain chemistry of the F-1 progeny. *J. Toxicol. Environ. Health* **21:**471–481.

IARC (International Agency for Research on Cancer). 1986. *Long-Term and Short-Term Assays for Carcinogens: A Critical Appraisal,* R. Montesano, H. Bartsch, H. Vainio, J. Wilbourn, and H. Yamasaki, eds. IARC Sci. Publ. No. 83, p. 71.

Jacobi, W., Paretzke, H. G., and Ehling, U. H. 1981. *Strahlenexposition und Strahlenrisiko der Bevölkerung, GSF-Bericht S-710,* Gesellschaft für Strahlen- und Umweltforschung, Neuherberg.

Jenkinson, P. C., Anderson, D., and Gangolli, S. D. 1987. Increased incidence of abnormal foetuses in the offspring of cyclophosphamide-treated male mice. *Mutat. Res.* **188:**57–62.

Johnson, F. M., and Lewis, S. E. 1981. Electrophoretically detected germinal mutations induced in the mouse by ethylnitrosourea. *Proc. Natl. Acad. Sci. USA* **78:**3138–3141.

Johnson, F. M., and Lovell, D. P. 1983. Dominant skeletal mutations are not induced by ethylnitrosourea in mouse spermatogonia. Abstracts of the 14th Annual Meeting of the Environmental Mutagen Society, San Antonio, p. 184.

Kalter, H., and Warkany, J. 1983. Congenital malformations: etiologic factors and their role in prevention. *N. Engl. J. Med.* **308:**424–431.

Kirk, K. M., and Lyon, M. F. 1982. Induction of congenital anomalies in offspring of female mice exposed to varying doses of X-rays. *Mutat. Res.* **106:**73–83.

Kirk, K. M., and Lyon, M. F. 1984. Induction of congenital malformations in the offspring of male mice treated with X-rays at pre-meiotic and post-meiotic stages. *Mutat. Res.* **125:**75–85.

Knudsen, I., Hansen, E. V., Meyer, O. A., and Poulsen, E. 1977. A proposed method for the simultaneous detection of germ-cell mutations leading to fetal death (dominant lethality) and of malformations (male teratogenicity) in mammals. *Mutat. Res.* **48:**267–270.

Kohn, H. I., Epling, M. L., Guttman, P. H., and Bailey, D. W. 1965. Effect of paternal (spermatogonial) X-ray exposure in the mouse: life span, X-ray tolerance, and tumor incidence of the progeny. *Radiat. Res.* **25:**423–434.

Kratochvilova, J. 1981. Dominant cataract mutations detected in offspring of gamma-irradiated male mice. *J. Hered.* **72:**301–307.

Kratochvilova, J., and Ehling, U. H. 1979. Dominant cataract mutations induced by γ-irradiation of male mice. *Mutat. Res.* **63:**221–223.

Kratochvilova, J., Favor, J., and Neuhäuser-Klaus, A. 1988. Dominant cataract and recessive specific-locus mutations detected in offspring of procarbazine-treated male mice. *Mutat. Res.* **198**:295–301.

Lewis, S. E., and Johnson, F. M. 1983. Dominant and recessive effects of electrophoretically detected specific locus mutations, in: *Utilization of Mammalian Specific Locus Studies in Hazard Evaluation and Estimation of Genetic Risk,* F. J. de Serres and W. Sheridan, eds. Plenum Press, New York, pp. 267–278.

Lovell, D. P., Willis, D. B., and Johnson, F. M. 1985. Lack of evidence for skeletal abnormalities in offspring of mice exposed to ethylnitrosourea. *Proc. Natl. Acad. Sci. USA* **82**:2852–2856.

Lowery, M. C., Rithidech, K., Au, W. W., Adams, P. M., and Legator, M. S. 1987. Genetic damage and the expression of behavioral abnormalities in the progeny of male rats exposed to ionizing radiation. *Environ. Mutag.* **9** (Suppl. 8):64.

Lüning, K. G. 1972. Studies of irradiated mouse populations. IV. Effects on productivity in the 7th to 18th generations. *Mutat. Res.* **14**:331–344.

Lüning, K. G., and Eiche, A. 1977. Penetrance and selection. *Mutat. Res.* **44**:451–454.

Lüning, K. G., and Searle, A. G. 1971. Estimates of the genetic risks from ionizing irradiation. *Mutat. Res.* **12**:291–304.

Lyon, M. F., and Morris, T. 1966. Mutation rates at a new set of specific loci in the mouse *Genet. Res.* **7**:12–17.

Lyon, M. F., and Morris, T. 1969. Gene and chromosome mutation after large fractionated or unfractionated radiation doses to mouse spermatogonia. *Mutat. Res.* **8**:191–198.

Lyon, M. F., and Renshaw, R. 1986. Induction of congenital malformations in the offspring of mutagen treated mice, in: *Genetic Toxicology of Environmental Chemicals, Part B: Genetic Effects and Applied Mutagenesis,* C. Ramel, B. Lambert, and J. Magnusson, eds. Liss, New York, pp. 449–458.

Lyon, M. F., and Renshaw, R. 1988. Induction of congenital malformation in mice by parental irradiation: transmission to later generations. *Mutat. Res.* **198**:277–283.

Lyon, M. F., Phillips, R. J. S., and Searle, A. G. 1964. The overall rates of dominant and recessive lethal and visible mutation induced by spermatogonial x-irradiation of mice. *Genet. Res.* **5**:448–467.

Lyon, M. F., Phillips, R. J. S., and Bailey, H. J. 1972. Mutagenic effects of repeated small radiation doses to mouse spermatogonia. *Mutat. Res.* **15**:185–190.

Lyon, M. F., Phillips, R. J. S., and Fisher, G. 1979. Dose–response curves for radiation-induced gene mutations in mouse oocytes and their interpretation. *Mutat. Res.* **63**:161–173.

McGregor, J. F., and Newcombe, H. B. 1961. Dwarfism and eye abnormality in X-irradiated rat populations. *Radiat. Res.* **14**:674–680.

McKusick, V. A. 1975. *Mendelian Inheritance in Man: Catalogs of Autosomal Dominant, Autosomal Recessive and X-linked Phenotypes,* 4th ed. Johns Hopkins University Press, Baltimore.

McKusick, V. A. 1978. *Mendelian Inheritance in Man: Catalogs of Autosomal Dominant, Autosomal Recessive and X-linked Phenotypes,* 5th ed. Johns Hopkins University Press, Baltimore.

McKusick, V. A. 1983. *Mendelian Inheritance in Man: Catalogs of Autosomal Dominant, Autosomal Recessive and X-linked Phenotypes,* 6th ed. Johns Hopkins University Press, Baltimore.

Mintz, B. 1960. Embryological phases of mammalian gametogenesis. *J. Cell. Comp. Physiol.* **56**(Suppl. 1):31–44.

Nagao, T. 1987. Frequency of congenital defects and dominant lethals in the offspring of male mice treated with methylnitrosourea. *Mutat. Res.* **177**:171–178.

Napalkov, N., Likhachev, A., Anisimov, V., Loktionov, A., Zabezhinski, M., Ovsyannikov, A., Wahrendorf, J., Becher, H., and Tomatis, L. 1987a. Promotion of skin tumours by TPA in the progeny of mice exposed pre-natally to DMBA. *Carcinogenesis* **8**:381–385.

Napalkov, N., Loktionov, A., Likhachev, A., Anisimov, V., Zabezhinski, M., and Tomatis, L. 1987b. Persistence of carcinogenic effect in intact progeny of mice treated transplacentally with 7,12-dimethylbenz[*a*]anthracene. *Carcinogenesis* **8**:381–385.

Newcombe, H. B., and McGregor, J. F. 1964. Learning ability and physical well-being in offspring from rat populations irradiated over many generations. *Genetics* **50**:1065–1081.

Nomura, T. 1975. Transmission of tumors and malformations to the next generation of mice subsequent to urethan treatment. *Cancer Res.* **35**:264–266.

Nomura, T. 1978. Changed urethan and radiation response of the mouse germ cell to tumor induction, in: *Tumors of Early Life in Man and Animals*, L. Severi, ed. Perugia University Press, Perugia, pp. 873–891.

Nomura, T. 1982a. Parental exposure to X rays and chemicals induces heritable tumours and anomalies in mice. *Nature* **296**:575–577.

Nomura, T. 1982b. Role of DNA damage and repair in carcinogenesis, in: *Environmental Mutagens and Carcinogens*, T. Sugimura, S. Kondo, and H. Takebe, eds. University of Tokyo Press, Tokyo, pp. 223–230.

Nomura, T. 1983. X-ray-induced germ-line mutation leading to tumors: its manifestation in mice given urethane post-natally. *Mutat. Res.* **121**:59–65.

Nomura, T. 1984. Quantitative studies on mutagenesis, teratogenesis, and carcinogenesis in mice, in: *Problems of Threshold in Chemical Mutagenesis*, Y. Tazima, S. Kondo, and Y. Kuroda, eds. EMS Japan, Shizuoka, pp. 27–34.

Nomura, T. 1986. Further studies on X-ray and chemically induced germ-line alterations causing tumors and malformations in mice, in: *Genetic Toxicology of Environmental Chemicals, Part B: Genetic Effects and Applied Mutagenesis*, C. Ramel, B. Lambert, and J. Magnusson, eds. Liss, New York, pp. 13–20.

Nomura, T. 1988. X-ray- and chemically induced germ-line mutation causing phenotypical anomalies in mice. *Mutat. Res.* **198**:309–320.

Oakberg, E. F. 1956a. A description of spermiogenesis in the mouse and its use in analysis of the cycle of the seminiferous epithelium and germ cell renewal. *Am. J. Anat.* **99**:391–414.

Oakberg, E. F. 1956b. Duration of spermatogenesis in the mouse and timing of stages of the cycle of the seminiferous epithelium. *Am. J. Anat.* **99**:507–516.

Oakberg, E. F. 1969. Radiation response of the testis. Prog. Endocrinol., Proc., 3rd Int. Congr. Endocr. Mexico City, 1968, pp. 1070–1076.

Oakberg, E. F. 1975. Effects of radiation on the testis, in: *Handbook of Physiology*, Section 7: *Endocrinology* **5**:233–243.

Oakberg, E. F. 1979. Timing of oocyte maturation in the mouse and its relevance to radiation-induced cell killing and mutational sensitivity. *Mutat. Res.* **59**:39–48.

Phillips, R. J. S. 1961. A comparison of mutation induced by acute X and chronic gamma irradiation in mice. *Br. J. Radiol.* **34**:261–264.

Rao, A. R. 1982. Inhibitory action of BHA on carcinogenesis in F1 and F2 descendants of mice exposed to DMBA during pregnancy. *Int. J. Cancer* **30**:121–124.

Roderick, T. H., ed. 1964. The effects of radiation on the hereditary fitness of mammalian populations, *Genetics* **50**:1019–1217.

Röhrborn, G., and Vogel, F. 1969. A search for dominant mutations in F_1 progeny of male mice treated with trenimone (triethyleneiminobenzoquinone-1,4). *Humangenetik* **7**:43–50.

Russell, L. B. 1971. Definition of functional units in a small chromosomal segment of the mouse and its use in interpreting the nature of radiation-induced mutations. *Mutat. Res.* **11**:107–123.

Russell, L. B., Bangham, J. W., Carpenter, D. A., Guinn, G. M., Hunsicker, P. R., Maddux, S. C., Phipps, E. L., Sega, G. A., and Stelzner, K. F. 1985. Specific-locus experiments and related studies with six chemicals. Annu. Prog. Rep. Oak Ridge Natl. Lab. Biol. Div. ORNL-6248, pp. 66–69.

Russell, L. B., Hunsicker, P. R., Oakberg, E. F., Cummings, C. C., and Schmoyer, R. L. 1987. Tests for urethane induction of germ-cell mutations and germ-cell killing in the mouse. *Mutat. Res.* **188**:335–342.

Russell, W. L. 1951. X-ray-induced mutations in mice. *Cold Spring Harbor Symp. Quant. Biol.* **16**:327–336.

Russell, W. L. 1954. Genetic effects of radiation in mammals, in: *Radiation Biology,* Volume 1, A. Hollaender, ed. McGraw–Hill, New York, pp. 825–859.

Russell, W. L. 1957. Shortening of life in the offspring of male mice exposed to neutron radiation from an atomic bomb. *Proc. Natl. Acad. Sci. USA* **43**:324–329.

Russell, W. L. 1963. The effect of radiation dose rate and fractionation on mutation in mice, in: *Repair from Genetic Radiation,* F. Sobels, ed. Pergamon Press, Oxford, pp. 205–217, 231–235.

Russell, W. L. 1964. Effect of radiation dose fractionation on mutation frequency in mouse spermatogonia. *Genetics* **50**:282.

Russell, W. L. 1974. Future research in mouse radiation genetics. *Genetics* **78**:135–138.

Russell, W. L. 1977. Mutation frequencies in female mice and the estimation of genetic hazards of radiation in women. *Proc. Natl. Acad. Sci. USA* **74**:3523–3527.

Russell, W. L. 1981. Problems and solutions in the estimation of genetic risks from radiation and chemicals, in: *Measurement of Risks,* G. G. Berg and H. D. Maillie, eds. Plenum Press, New York, pp. 361–380.

Russell, W. L. 1986. Positive genetic hazard predictions from short-term tests have proved false for results in mammalian spermatogonia with all environmental chemicals so far tested, in: *Genetic Toxicology of Environmental Chemicals, Part B: Genetic Effects and Applied Mutagenesis,* C. Ramel, B. Lambert, and J. Magnusson, eds. Liss, New York, pp. 67–74.

Russell, W. L., and Hunsicker, P. R. 1983. Extreme sensitivity of one particular germ-cell stage in male mice to induction of specific-locus mutations by methylnitrosourea. *Environ. Mutag.* **5**:498.

Russell, W. L., and Hunsicker, P. R. 1984. Mutagenic effect of ethylnitrosourea (ENU) on post-stemcell stages in male mice. *Environ. Mutag.* **6**:390.

Russell, W. L., and Hunsicker, P. R. 1988. Dominant mutagenic effect of ENU on first-generation litter size in mice following treatment of spermatogonial stem cells. *Environ. Mol. Mutag.* **11**(Suppl. 11):90.

Russell, W. L., and Russell, L. B. 1959. Radiation-induced genetic damage in mice. *Prog. Nucl. Energy Ser. 6* **2**:179–188.

Russell, W. L., Bangham, J. W., and Gower, J. S. 1958. Comparison between mutations induced in spermatogonial and postspermatogonial stages in the mouse, in: *Proc. 10th Int. Congr. Genet.,* Volume 2. University of Toronto Press, Toronto, pp. 245–246.

Russell, W. L., Kelly, E. M., Hunsicker, P. R., Bangham, J. W., Maddux, S. C., and Phipps, E. L. 1979. Specific-locus test shows ethylnitrosourea to be the most potent mutagen in the mouse. *Proc. Natl. Acad. Sci. USA* **76**:5818–5819.

Russell, W. L., Hunsicker, P. R., Raymer, G. D., Steele, M. H., Stelzner, K. F., and Thompson, H. M. 1982. Dose–response curve for ethylnitrosourea-induced specific-locus mutations in mouse spermatogonia. *Proc. Natl. Acad. Sci. USA* **79**:3589–3591.

Rutledge, J. C., Cain, K. T., Hughes, L. A., Braden, P. W., and Generoso, W. M. 1986. Difference between two hybrid stocks of mice in the incidence of congenital abnormalities following X-ray exposure of stem-cell spermatogonia. *Mutat. Res.* **163:**299–302.

Searle, A. G. 1964. Effects of low-level irradiation on fitness and skeletal variation in an inbred mouse strain. *Genetics* **50:**1159–1178.

Searle, A. G. 1974. Mutation induction in mice. *Adv. Radiat. Biol.* **4:**131–207.

Searle, A. G. 1987. Evidence for induction of early-acting dominants by irradiation of male and female germ-cells in mice. Report submitted to UNSCEAR.

Searle, A. G., and Beechey, C. 1986. The role of dominant visibles in mutagenicity testing, in: *Genetic Toxicology of Environmental Chemicals, Part B: Genetic Effects and Applied Mutagenesis,* C. Ramel, B. Lambert, and J. Magnusson, eds. Liss, New York, pp. 511–518.

Searle, A. G., and Papworth, D. G. 1986. Analysis of pre- and post-natal mortality after spermatogonial irradiation of mice. Report submitted to UNSCEAR.

Selby, P. B. 1979. Radiation-induced dominant skeletal mutations in mice: mutation rate, characteristics, and usefulness in estimating genetic hazard to humans from radiation. Radiat. Res., Proc. 6th Int. Congr. pp. 537–544.

Selby, P. B. 1981. Radiation genetics, in: *The Mouse in Biomedical Research,* Volume I, H. L. Foster, J. D. Small, and J. G. Fox, eds. Academic Press, New York, pp. 263–283.

Selby, P. B. 1982. Dominant skeletal mutations: applications in mutagenicity testing and risk estimation, in: *Mutagenicity—New Horizons in Genetic Toxicology,* J. A. Heddle, ed. Academic Press, New York, pp. 385–406.

Selby, P. B. 1983. Applications in genetic risk estimation of data on the induction of dominant skeletal mutations in mice, in: *Utilization of Mammalian Specific Locus Studies in Hazard Evaluation and Estimation of Genetic Risk,* F. J. de Serres and W. Sheridan, eds. Plenum Press, New York, pp. 191–210.

Selby, P. B. 1986. Synergistic and antagonistic interactions of two unlinked radiation-induced dominant skeletal mutations in mice. *Mouse News Lett.* **75:**43–44.

Selby, P. B. 1987. A rapid method for preparing high quality alizarin stained skeletons of adult mice. *Stain Technol.* **62:**143–146.

Selby, P. B., and Lee, S. S. 1980. Induction and nature of dominant skeletal mutations. Annu. Prog. Rep. Oak Ridge Natl. Lab. Biol. Div. ORNL-5685, pp. 65–66.

Selby, P. B., and Niemann, S. L. 1984. Non-breeding-test methods for dominant skeletal mutations shown by ethylnitrosourea to be easily applicable to offspring examined in specific-locus experiments. *Mutat. Res.* **127:**93–105.

Selby, P. B., and Russell, W. L. 1985. First-generation litter-size reduction following irradiation of spermatogonial stem cells in mice and its use in risk estimation. *Environ. Mutag.* **7:**451–469.

Selby, P. B., and Selby, P. R. 1977a. Gamma-ray-induced dominant mutations that cause skeletal abnormalities in mice. I. Plan, summary of results and discussion. *Mutat. Res.* **43:**357–375.

Selby, P. B., and Selby, P. R. 1977b. Response to K. G. Lüning and A. Eiche, Penetrance and selection. *Mutat. Res.* **44:**453–454.

Selby, P. B., and Selby, P. R. 1978a. Gamma-ray-induced dominant mutations that cause skeletal abnormalities in mice. III. Description of proved mutations. *Mutat. Res.* **50:**341–351.

Selby, P. B., and Selby, P. R. 1978b. Gamma-ray-induced dominant mutations that cause skeletal abnormalities in mice. II. Description of proved mutations. *Mutat. Res.* **51:**199–236.

Selby, P. B., Whitt, B. J. M., Raymer, G. D., and McKinley, T. W., Jr. 1984a. Breeding-test experiment shows transmission of many ENU-induced mutations. Annu. Prog. Rep. Oak Ridge Natl. Lab. Biol. Div. ORNL-6021, pp. 103–104.

Selby, P. B., Whitt, B. J. M., Raymer, G. D., and McKinley, T. W., Jr. 1984b. Breeding-test experiment demonstrates transmissibility of many dominant skeletal mutations induced by ethylnitrosourea. *Environ. Mutag.* **6**:391.

Selby, P. B., Raymer, G. D., McKinley, T. W., Jr., and Niemann, S. L. 1987. Synergistic and antagonistic interactions of two radiation-induced dominant skeletal mutations in mice. Annu. Prog. Rep. Oak Ridge Natl. Lab. Biol. Div. ORNL-6353, pp. 86–87.

Selby, P. B., Raymer, G. D., and Hunsicker, P. R. 1988. High frequency of dominant mutations causing stunted growth is induced in spermatogonial stem cells by ENU. *Environ. Mol. Mutag. 11*(Suppl. 11):93.

Sillence, D. O., Ritchie, H. E., and Selby, P. B. 1987. Animal model: skeletal anomalies in mice with cleidocranial dysplasia. *Am. J. Med. Genet.* **27**:75–85.

Spalding, J. F. 1964. Longevity of first and second generation offspring from male mice exposed to fission neutrons and gamma rays, in: *Proc. Int. Symp. Effects of Ionizing Radiations on the Reproductive System*, W. D. Carlson and F. X. Gassner, eds. Pergamon Press, Oxford, pp. 147–152.

Tomatis, L. 1965. Increased incidence of tumours in F_1 and F_2 generations from pregnant mice injected with a polycyclic hydrocarbon. *Proc. Soc. Exp. Biol. Med.* **119**:743–747.

Tomatis, L. 1979. Prenatal exposure to chemical carcinogens and its effect on subsequent generations. *Natl. Cancer Inst. Monogr.* **51**:159–184.

Tomatis, L., and Goodall, C. M. 1969. The occurrence of tumours in F_1, F_2, and F_3 descendants of pregnant mice injected with 7,12-dimethylbenz[a]anthracene. *Int. J. Cancer* **4**:219–225.

Tomatis, L., Hilfrich, J., and Turusov, V. 1975. The occurrence of tumours in F_1, F_2, and F_3 descendants of BD rats exposed to N-nitrosomethylurea during pregnancy. *Int. J. Cancer* **15**:385–390.

Tomatis, L., Ponomarkov, V., and Turusov, V. 1977. Effects of ethylnitrosourea administration during pregnancy on three subsequent generations of BDVI rats. *Int. J. Cancer* **19**:240–248.

Tomatis, L., Cabral, J. R. P., Likhachev, A. J., and Ponomarkov, V. 1981. Increased cancer incidence in the progeny of male rats exposed to ethylnitrosourea before mating. *Int. J. Cancer* **28**:475–478.

Tomatis, L., Cabral, J. R. P., Likhachev, A. J., and Ponomarkov, V. 1982. Increased cancer incidence in the progeny of male rats exposed to ethylnitrosourea before mating, in: *Environmental Mutagens and Carcinogens*, T. Sugimura, S. Kondo, and H. Takebe, eds. University of Tokyo Press, Tokyo, pp. 231–238.

Trasler, J. M., Hales, B. F., and Robaire, B. 1985. Paternal cyclophosphamide treatment of rats causes fetal loss and malformations without affecting male fertility. *Nature* **316**:144–146.

UNSCEAR. 1977. Report of the United Nations Scientific Committee on the Effects of Atomic Radiation, *Sources and Effects of Ionizing Radiation*. United Nations, New York, Sales No. E.77.IX.1, pp. 425–564.

UNSCEAR. 1982. Report of the United Nations Scientific Committee on the Effects of Atomic Radiation, *Ionizing Radiation: Sources and Biological Effects*. United Nations, New York, Sales No. E.82.IX.8, pp. 425–569.

UNSCEAR. 1986. Report of the United Nations Scientific Committee on the Effects of Atomic Radiation, *Genetic and Somatic Effects of Ionizing Radiation*. United Nations, New York, Sales No. E.86.IX.9, pp. 27–164.

West, J. D., and Fisher, G. 1985. Inherited cataracts in inbred mice. *Genet. Res.* **46**:45–56.

West, J. D., and Fisher, G. 1986. Further experience of the mouse dominant cataract mutation test from an experiment with ethylnitrosourea. *Mutat. Res.* **164:**127–136.

West, J. D., Peters, J., and Lyon, M. F. 1984. Genetic differences between two substrains of the inbred 101 mouse strain. *Genet. Res.* **44:**343–346.

West, J. D., Kirk, K. M., Goyder, Y., and Lyon, M. F. 1985a. Discrimination between the effects of X-ray irradiation of the mouse oocyte and uterus on the induction of dominant lethals and congenital anomalies I. Embryo transfer experiments. *Mutat. Res.* **149:**221–230.

West, J. D., Kirk, K. M., Goyder, T., and Lyon, M. F. 1985b. Discrimination between the effects of X-ray irradiation of the mouse oocyte and uterus on the induction of dominant lethals and congenital anomalies II. Localised irradiation experiments. *Mutat. Res.* **149:**231–238.

West, J. D., Lyon, M. F., Peters, J., and Selby, P. B. 1985c. Genetic differences between substrains of the inbred mouse strain 101 and designation of a new strain 102. *Genet. Res.* **46:**349–352.

Widmaier, R. 1963. Über die postnatale Hodenentwicklung und Keimzellreifung bei der Maus. *Z. Mikrosk. Anat. Forsch.* **70:**215–241.

Wright, S. 1968. *Evolution and the Genetics of Populations*, Volume 1. University of Chicago Press, Chicago.

Issues and Reviews in Teratology **5**:255–282
Plenum Press, New York, 1990, 978-1-4612-7847-4

The Teratology and Developmental Toxicity of Cadmium

6

WILLIAM S. WEBSTER

1. INTRODUCTION

Cadmium is a nutritionally nonessential metal which has been widely studied in biological systems because of its considerable toxicity and its ubiquitous and increasing environmental presence. The literature on Cd-induced reproductive toxicology is dichotomous, consisting on the one hand of studies in which large doses of Cd have been used as a probe to examine teratogenic mechanisms and gonadal and placental function, and on the other of studies concerned with the possible toxicity of environmental levels of Cd. Although there is some overlap in these studies, they are in general separate and the failure to recognize this dichotomy has led to many inappropriate statements regarding the health hazards of exposure to Cd during pregnancy.

The general aim of this review is to examine the developmental toxicity of Cd from the viewpoints of its use as a research tool and of whether excessive exposure to Cd is a reproductive hazard for humans. The period of development to be considered is mainly prenatal but includes early postnatal development because this period encompasses important developmental stages in the central nervous system (CNS) which may be particularly vulnerable to damage by heavy metals such as Cd.

WILLIAM S. WEBSTER • Department of Anatomy, University of Sydney, Sydney, New South Wales 2006, Australia.

2. CADMIUM AS A TERATOGEN

2.1. General Properties

Cadmium was first shown to be a teratogen by Ferm and Carpenter (1967) in golden hamsters. Single i.v. doses of cadmium sulfate administered early on d 8 of gestation induced exencephaly and a high incidence of facial defects including microphthalmia and clefts of the upper lip, palate, and mandible. Similar exposure later on d 8 induced rib and upper limb defects and exposure on d 9, both upper and lower limb defects (Ferm, 1971). The propensity of Cd to induce facial defects and the observation that Cd-induced cleft palate occurred after exposure on d 8 of gestation although closure of the palatine shelves in hamsters takes place on d 12 and 13 led Ferm (1971) to propose that Cd was a site-specific teratogen. A site-specific teratogen was defined as one that induces malformations only in certain organ systems by interfering with a particular enzymatic event of development whereas a nonspecific teratogen primarily affects organogenic events at the time of the insult. The teratogenic effect of Cd has not been related to interference with a particular enzyme system and the concept of site specificity for Cd is in abeyance. Indeed, the distinction of teratogens into site-specific and nonspecific appears inappropriate as the time-dependent sensitivity of organogenic events to teratogens may be due to enzyme-dependent developmental changes.

Cd has been demonstrated to be teratogenic in rats (Barr, 1973; Chernoff, 1973) and mice (Ishizu *et al.*, 1973; Messerle, 1978) as well as birds (Ribas and Schmidt, 1973), amphibia (Keino, 1973), and fish (Eaton, 1974). It can induce a wide range of malformations affecting all organ systems as well as causing intrauterine growth retardation and prenatal death. The response depends on species, strain (e.g., Barr, 1973; Wolkowski, 1974), dose, and time of exposure. The teratogenic effect is dose dependent, covering a relatively small dose range (Ishizu *et al.*, 1973), with the teratogenic period starting, as expected, at the beginning of the organogenic period (Webster and Messerle, 1980). An increase in the incidence of embryonic death, but not the malformation rate, can be induced by Cd exposure at earlier stages (Chiquoine, 1965; Yu *et al.*, 1985; Abraham *et al.*, 1986). As discussed below, the teratogenic period extends to about d 10 in mice at which time it is replaced by a fetolethal period (Messerle and Webster, 1982) during which large doses of Cd destroy the chorioallantoic placenta and cause death of the fetus. This fetotoxic stage lasts for the rest of gestation. Lower doses of Cd administered during this period cause fetal growth retardation with some evidence of a differential organ effect (Chernoff, 1973; Daston and Grabowski, 1979; Daston, 1981; Samarawickrama and Webb, 1981).

It is important to appreciate that Cd is only teratogenic in rodents following acute, high doses (1–4 mg/kg Cd^{2+}) administered by parenteral injection. Doses of 40 mg/kg administered by gavage have also been reported to be teratogenic (Barański *et al.*, 1982; Barański, 1985). All of these amounts are close to maternally lethal doses and result in sudden and very highly elevated maternal blood Cd levels (Record *et al.*, 1982b; Warner *et al.*, 1983). In our unpublished studies, a teratogenic dose (2.4 mg/kg) administered intraperitoneally to pregnant mice produced a maternal blood Cd level of 2.75 μg/ml after 15 min, 1.0 μg/ml after 1 h, and 0.2 μg/ml after 4 h (Fig. 1). The distribution of Cd in the blood also changed with time; initially only about 15% of the Cd separated with the cellular components of the blood but by 8 h this had increased to 67%. As is discussed below, the very high Cd levels that occur in the first hour after parenteral dosing are a feature of this form of dosing and are probably responsible for the teratogenic effect of Cd.

These levels can be compared with the normal blood Cd levels of 0.1–1.2 ng/ml found in pregnant women at the time of delivery (Alessio *et al.*, 1984) and blood Cd levels of 17 ng/ml (range 7–31) in industrially exposed individuals (Roels *et al.*, 1981). It is unlikely that the high blood Cd levels caused by parenteral administration of Cd, and associated with teratogenesis in rodents, could occur in humans except under the most exceptional circumstances. This assertion is supported by the many demonstrations that high doses of Cd administered in food, water, or air throughout pregnancy are not associated with any increase in the incidence of congenital malformation (e.g., Pond and Walker, 1975; Webster, 1978; Prigge 1978; Ahokas *et al.*, 1980; Machemer and Lorke, 1981; Barański, 1985).

Despite the fact that Cd is unlikely to be a human teratogen, it is a useful chemical for inducing congenital malformations in experimental animals and has been used in studies of neural tube defects (Messerle,

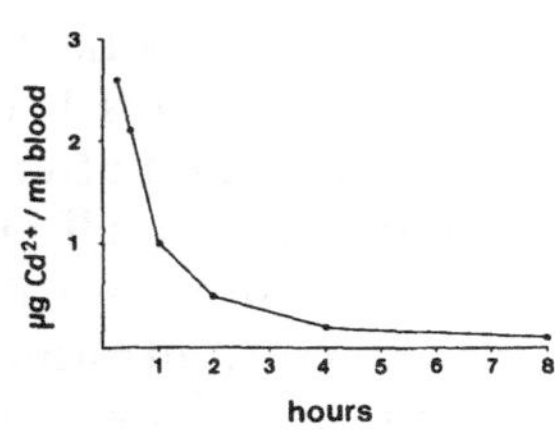

Figure 1. Maternal blood Cd levels in mice (QS) following an i.p. injection of cadmium chloride (2.4 mg Cd^{2+}/kg body) containing 20 μCi^{109}Cd administered on d 9 of gestation. Each point represents the mean of two animals (Webster, unpublished data).

1978; Keino *et al.*, 1978; Webster and Messerle, 1980; Schmid *et al.*, 1985), skeletal dysmorphogenesis (Gale and Ferm, 1973; Padmanabhan and Hameed, 1986), median cleft lip (Tassinari and Long, 1979), cleft palate (Mulvihill *et al.*, 1970), limb malformations (Messerle, 1978; Layton and Layton, 1979; Messerle and Webster, 1982; Christley and Webster, 1983; Kuczuk and Scott, 1984; Milaire, 1985; Feuston and Scott, 1985), lung maturation (Daston and Grobowski, 1979; Daston, 1981), sirenomelia (Ferm, 1969; Hilbelink and Kaplan, 1986), hydrocephalus (Keino and Yamamura, 1974), and gonadal development (Tam and Liu, 1985).

2.2. Mechanisms of Teratogenesis

2.2.1. Direct or Indirect Effect?

Although Cd is probably the most studied elemental teratogen, its mechanism of action is poorly understood. The teratogenic effects may be due to a direct toxic action on the embryo or secondary to a toxic effect on the pregnant female or yolk sac placenta. It is possible that all three mechanisms contribute to the observed embryotoxicity.

A teratogenic dose of Cd causes severe maternal toxicity characterized by anorexia lasting up to 24 h (Record *et al.*, 1982b), hypothermia (Webster, unpublished observations), and a prompt catecholamine-dependent hyperglycemia lasting between 3 and 6 h (Ghafghazi and Mennear, 1973). It has been suggested that maternal toxicity may be the cause of some congenital malformations, such as rib abnormalities and exencephaly (Khera, 1985; Beyer and Chernoff, 1986), but the demonstration that exencephaly in mice can be induced by *in vivo* or *in vitro* Cd exposure (Warner *et al.*, 1984) makes a major role for maternal toxicity in Cd teratogenesis unlikely.

Embryo culture studies provided strong evidence that Cd has a direct effect on the embryo–yolk sac complex, although the first of these studies suggested that serum components other than Cd may contribute to the observed embryotoxicity (Klein *et al.*, 1980). Headfold-stage rat embryos were cultured in sera obtained at various intervals, up to 24 h following treatment, from pregnant rats given 2.13 mg/kg i.p. Sera obtained after 1 or 4 h were lethal to embryos within 12 or 24 h, respectively. Sera taken after 8 h allowed embryonic survival for 48 h, but the embryos were small and the neural tubes remained open. On 16- and 24-h sera, embryos were morphologically normal but smaller than controls. Cd concentrations were not determined in the serum samples but were probably close to control levels by 8 h.

Similar studies were performed by Record *et al.* (1982a,b). They measured the Cd content of sera obtained at 1, 2, 4, and 8 h from rats given an i.p. injection of 4 mg/kgCd (1.6, 1.0, 0.3 μg/ml, and not detectable respectively). They showed that there was a dose-dependent effect on the embryonic growth of rat embryos cultured for 24 h in the sera. The 1-h serum produced severe growth retardation and the yolk sac was described as thickened and aplastic, embryos grown in the 4-h serum were growth retarded but normal in appearance, and those grown in the 8-h serum appeared normal. When embryos cultured in the 4-h serum were compared with those cultured in control serum to which an equivalent amount of Cd had been added, both showed similar growth retardation, indicating that the Cd content of the serum and not some other component was the embryotoxic factor.

In both of these studies embryos were exposed *in vitro* to Cd for 24 h, although a teratogenic dose of Cd administered *in vivo* is cleared from the maternal serum in 4–8 h (Record *et al.*, 1982b; Warner *et al.*, 1983). With this in mind, an *in vitro* study was designed using mouse embryos to examine the parameters determining Cd teratogenicity. Administration of a teratogenic dose of Cd (6 mg/kg) to pregnant mice produced a peak plasma Cd concentration of 5.35 μg/ml with a half life of 30 min (Warner *et al.*, 1983, 1984). Using this information, 9-d mouse embryos in whole embryo culture were exposed for 30 min to Cd levels equal to the maternal peak plasma concentration and were then transferred to Cd-free serum for 48 h. Seventy-two percent of these embryos had neural tube and eye defects similar to those induced by Cd exposure at the same stage *in vivo*. When the *in vitro* dose was reduced to 2.7 μg/ml for 30 min, the malformation rate decreased to 50% (Warner *et al.*, 1983, 1984). In other experiments using rat embryos, Cd was added to the culture medium for the entire duration of the culture period (Record *et al.*, 1982b). A concentration of 320 ng/ml killed the embryos, while 170 and 80 ng/ml caused severe growth retardation but did not cause malformations; 40 ng/ml caused slight growth retardation. These studies provide strong evidence that Cd exerts its teratogenic effect directly on the embryo–yolk sac complex. They also demonstrate that Cd teratogenesis is due to the short-lived peak blood levels associated with administration techniques which cause sudden large elevations in blood Cd. More prolonged exposure to lower concentrations may cause fetal growth retardation, or even death, but not malformations.

These studies have not determined whether Cd damages the embryo directly or exerts its effect by damaging the yolk sac. It has been found *in vivo* that small amounts of Cd (less than 0.02% of the dose/g tissue) enter the embryos following a teratogenic dose during the organogenic period.

This has been shown in hamsters (Ferm *et al.*, 1969), rats (Sonawane *et al.*, 1975; Ahokas and Dilts, 1979), and mice (Wolkowski, 1974; Christley and Webster, 1983). Autoradiographic studies, using radioactive Cd, showed that a teratogenic dose administered on d 9 of gestation in mice rapidly (in less than 1 h) entered all embryonic tissues but was particularly localized in cells of the neural tube, limb buds, and gut (Christley and Webster, 1983). Twelve hours after Cd exposure the cells of the neural tube and limb buds were still labeled and also showed extensive cytolysosomal formation and cell death. If the mice were allowed to continue to term the fetuses had limb and neural tube defects (Messerle and Webster, 1982). This combination of Cd localization, cell damage, and subsequent abnormal development provides evidence that Cd has a direct effect on the embryo and could account for the observed teratogenic effects.

Another possibility is that Cd damages the yolk sac, resulting in a reduced supply of nutrients to the embryo, as was postulated to be the teratogenic mechanism for trypan blue (Beck *et al.*, 1967; Williams *et al.*, 1976). The evidence for this mechanism includes localization of Cd in the yolk sac in concentrations higher than were found in the embryos (Dencker, 1975; Christley and Webster, 1983) and reduced pinocytosis (by 36%) in visceral yolk sacs following their culture in serum containing 0.32 μg/ml Cd (Record *et al.*, 1982a).

2.2.2. Relation of Zn Deficiency to Cd Embryotoxicity

It has been postulated that Cd exerts its teratogenic effect by depriving the embryo of zinc. This suggestion originated from the demonstration that preadministration of Zn (Chiquoine, 1965) or coadministration of Zn (Ferm and Carpenter, 1967; Garcia and Lee, 1981) negated the teratogenic effect of Cd. Protection occurred even if the Zn was administered as late as 6 h after the Cd but not after 12h (Ferm and Carpenter, 1968). These results were not presented in a form that could be analyzed statistically, and it would be reassuring to see them confirmed since they imply that the teratogenic effect of Cd is not irreversibly established 6 h after exposure.

The coadministration of Zn and Cd should result in reduced Cd transfer to the embryo compared with administration of Cd alone because the physicochemical similarities between these two elements result in competitive binding. Reduced Cd transfer was found in 9-d rat embryos (Garcia and Lee, 1981), but in hamsters the same amount of ^{109}Cd entered the embryos whether it was coadministered with Zn or not (Ferm *et al.*, 1969). Also possibly indicating a role for Zn deficiency in Cd embryotoxicity is the finding that maternal Zn deficiency induced malformations similar to those due to Cd (Hurley *et al.*, 1971), with morphologically similar cell death seen in the embryo (Record *et al.*, 1985), and that Zn deficiency exacerbated the teratogenic effects of Cd (Rohrer *et al.*, 1978a; Parzyck *et al.*, 1978; Sato *et al.*, 1985) without increasing the placental transfer of Cd (Rohrer *et al.*, 1978a).

Maternal Zn levels were substantially reduced 8 h after Cd dosing in rats, but in a series of *in vitro* experiments it was clearly shown that it was the Cd content of the serum and not the Zn content that determined the embryotoxic effect. This was demonstrated by culturing rat embryos in serum taken from rats at various times after dosing with 4 mg/kg Cd i.p. (Record *et al.*, 1982a,b). Serum taken after 1, 2, or 4 h was embryotoxic, causing reduced embryonic growth as measured by DNA accumulation. The severity of the retardation was directly related to the Cd content of the serum, which was 1.6, 1.0, and 0.3 μg/ml Cd, respectively. After 8 h the serum Cd was not detectable and the serum Zn was maximally depressed, from a control level of 1.6 μg/ml Zn down to 0.67 μg/ml Zn. When embryos were cultured in the 8-h serum, their development was normal (Record *et al.*, 1982b).

Although these results indicate that maternal Zn deficiency is not the cause of Cd-induced teratogenesis, they do not exclude the possibility that the Cd in the serum prevents the yolk sac from taking up or transporting Zn from the serum to the embryo. As noted above, Record *et al.* (1982a) showed that a Cd concentration of 0.32 μg/ml inhibited pinocytosis by 36% in the rat visceral yolk sac *in vitro*. Since the rate of nutrient supply to the embryo is governed by the rate of pinocytosis rather than the rate of intralysosomal degradation, there must be a nonspecific reduction of all nutrients to the embryo. Just as the teratogenic effect of Cd *in vitro* can be prevented by added Zn (Record *et al.*, 1982b; Warner *et al.*, 1984), so can the effect on pinocytosis (Record *et al.*, 1982a).

There is some evidence that Cd reduces the availability of Zn to the embryo but the effect is not marked. Control rat embryos cultured for 48 h accumulated 133 μg Zn/g protein as determined by ^{65}Zn uptake. When a relatively low level of Cd was added (0.08 μg/ml Cd), Zn accumulation was 122 μg/g protein, and when Cd and an excess of Zn (2 μg/ml) were added, the embryos accumulated 171 μg Zn/g protein (Record *et al.*, 1982b). Zinc deficiency in the embryo would be expected to depress the activity of Zn-dependent enzymes such as thymidine kinase, resulting in reduced DNA synthesis. Although Cd reduced DNA synthesis and the uptake of labeled Zn in 12-d rat embryos (Holt and Webb, 1986), the two events were considered independent because embryos exposed to mercury had the same reduced uptake of Zn but with no effect on DNA synthesis. It was concluded that the inhibition of DNA synthesis was due to a direct effect of the small amount of Cd taken up by the embryo and not to embryonic Zn deficiency.

It is still unproven whether Cd interferes with Zn metabolism by the embryo but there is evidence for a direct action of Cd on the embryo independent of any Zn deficiency. The protective effect of Zn may be due to Zn and Cd competing for binding sites in the embryo and/or yolk sac.

2.2.3. Cd Cytotoxicity

Autoradiographic and gamma counting studies have shown that Cd is present in both the embryo and yolk sac following a parenterally administered teratogenic dose to pregnant animals (Ferm *et al.*, 1969; Wolkowski, 1974; Sonawane *et al.*, 1975; Ahokas and Dilts, 1979; Christley and Webster, 1983). Twelve hours after exposure, embryonic cells showed morphological signs of cytotoxicity, namely, large secondary cytolysosomal formation and extensive cell death (Webster and Messerle, 1980; Messerle and Webster, 1982; Christley and Webster, 1983). The yolk sac did not have morphological signs of cytotoxicity but biochemical studies showed a decrease in pinocytotic activity (Record *et al.*, 1982a). The simplest explanation for the teratogenicity of Cd is that it reaches the embryo and kills or damages cells, thus interrupting developmental processes, the type of malformation induced being dependent upon the critical processes occurring in the embryo at the time of exposure.

From *in vitro* studies (Enger *et al.*, 1986) it is known that cells can protect themselves against the toxic effects of Cd by producing metallothionein, with peak levels being formed within a few hours of exposure. The rapidity of the metallothionein response is a major determinant of the susceptibility of a cell to Cd-induced damage. The fact that BALB/3T3 mouse embryonic cells in culture have very slow inducibility of metallothionein, although they take up Cd as fast as other cells do (Kobayashi and Kimura, 1980), is significant to the teratological situation, because the embryonic unit is exposed to peak Cd levels within minutes of parenteral administration and a 30-min exposure is sufficient to cause abnormal development (Warner *et al.*, 1984).

When the metallothionein response was induced in pregnant animals by prior exposure to Cd (Semba *et al.*, 1974; Ferm and Layton, 1979), selenium (Holmberg and Ferm, 1969), or mercury (Layton and Ferm, 1980), the embryos were protected against a subsequent potentially teratogenic Cd dose. The pretreatment with Cd can be given up to

3 weeks prior to pregnancy and still yield some protection (Layton and Ferm, 1980). Oral pretreatment with α-mercapto-β-(2-furyl)-acrylic acid also protected against Cd-induced malformations but this was probably accomplished by chelation of the Cd ion (Ferm and Hanlon, 1987).

The nutritional status of the cell influences the ease with which Cd enters the cell. Cadmium enters hepatocytes of starved rats much more readily than those of replete rats; pH and competing ions are also important determinants (Müller and Ohnesorge, 1982). These factors may explain the autoradiographic evidence that certain tissues and certain cells in the embryo and yolk sac take up more Cd than others (Christley and Webster, 1983).

The cell membrane must be the initial site of toxic reaction to Cd, resulting in increased lipid peroxidation; but this may not be the main source of toxicity since antioxidants do not prevent Cd-induced cell death (Stacey and Kappus, 1982; Müller and Ohnesorge, 1982). Cd freely enters cells and in culture can accumulate to concentrations 500- to 1000-fold greater than the extracellular level (Enger *et al.*, 1986). Once it is in the cell the known affinity of Cd for sulfhydryl groups and glutathione may determine its distribution. When the cell has large amounts of metallothionein, Cd binds to this molecule and is essentially detoxified, but in the absence of excess metallothionein, Cd is mainly bound to existing proteins and amino acids with a small proportion being bound to nucleic acids (Jacobson and Turner, 1980).

Functionally, Cd is known to enhance the activity of a few enzymes and to inhibit a large number of others (Vallee and Ulmer, 1972), including many involved in nucleic acid metabolism (Jacobson and Turner, 1980). It has been proposed that cells cope with metallothionein-bound Cd by engulfing the molecules in lysosomes and extruding them from the cell by exocytosis (Hayden *et al.*, 1982). If cells can deal with macromolecularly bound Cd in this way, it may account for the excessive number of cytolysosomes seen in otherwise normal-looking embryonic cells after exposure to a teratogenic dose of Cd (Webster and Messerle, 1980; Messerle and Webster, 1982; Christley and Webster, 1983). In greater than a certain threshold concentration, Cd would inhibit too many cellular processes and cause cell death, also a major feature of Cd-exposed embryos. Toxic doses of Cd also interfere with heterolysosome formation and function in adult mouse liver, perhaps by inhibiting the normal rate of primary lysosome formation (Mego and Cain, 1975).

Some teratogens induce a variety of heat shock (stress) proteins in embryonic cells immediately following exposure and give some short-term protection against subsequent exposure to the same or other agents (e.g., German, 1984). Cd induces several heat shock proteins (Levinson *et al.*, 1980; Li *et al.*, 1982), which protect cells against subsequent heat exposure; but heat does not appear to induce protection against Cd (Hahn and Li, 1982).

2.2.4. Acidotic Embryonic Environment

Cd exposure during the period of limb development resulted in a spectrum of postaxial forelimb reduction deformities predominantly involving the left limb in rats (Barr, 1973) and the right limb in mice (Messerle, 1978; Layton and Layton, 1979; Messerle and Webster, 1982). In C57BL/6J mice, this outcome was preceded by large cytolysosomes and cell death in the mesoderm and to a lesser extent the ectoderm of the limb buds (Messerle and Webster, 1982). Other studies showed that alcohol (Webster *et al.*, 1980, 1983), 13-*cis*-retinoic acid (Sulik and Dehart, 1988), and acetazolamide (Layton and Hallesy, 1965; Green *et al.*, 1973; Holmes and Trelstad, 1979) induced similar postaxial limb deformities in C57BL/6J mice. Both alcohol (Webster, 1989) and retinoic acid (Sulik and Dehart, 1988) caused cell death in the limb bud in the hours following exposure, but the acetazolamide-induced limb defects were not preceded by cell death (Holmes and Trelstad, 1979; Datu *et al.*, 1985). It was also found that when Cd and acetazolamide were administered together to induce limb defects, there was no cell death in the limb buds (Milaire, 1985).

This finding indicates that cell death is not a prerequisite for these limb defects, and by extension that its presence in the embryo after many teratogens may be a reflection of the toxicity of the agent rather than an explanation of its teratogenesis. Equally it would be a mistake to assume that because various teratogens induce the same malformation that they all have the same mechanism of teratogenesis. It may be that the limb bud, e.g., has a limited repertoire of responses to any chemical or physical damage.

The observation that both Cd and acetazolamide induced identical forelimb defects with similar mouse strain sensitivity, and that both substances can inhibit the zinc metalloenzyme carbonic anhydrase, led to the hypothesis that Cd-induced limb defects were due to inhibition of carbonic anhydrase (Kuczuk and Scott, 1984). In an experiment specifically designed to test this hypothesis, biochemical and histochemical examination failed to detect any effect of Cd on carbonic anhydrase activity in the embryo (Feuston and Scott, 1985).

In the same experiment, Cd decreased hemoglobin levels in the embryos while increasing the levels in the yolk sac, suggesting that Cd interfered with transport of red blood cells from the yolk sac to the embryo or inhibited vascularization of the yolk sac. This led to the new hypothesis that Cd causes forelimb ectrodactyly by producing an acidotic embryonic environment and that the site of action of Cd is the yolk sac. These considerations might explain the right-sided predominance of the limb defects in mice, which have been associated with differences in the blood supply of the two limb buds (Messerle and Webster, 1982; Milaire, 1985). However, it has recently been demonstrated that the right forelimb bud in rats at this stage has a lower oxygen level than the left (Brown and Coakley, 1987) while Cd induces predominantly left limb defects in rats (Barr, 1973).

Related to the acidotic environment hypothesis is the possibility that Cd may cause vascular damage in the embryo, leading to local hemorrhage and hypoxia. Many of the toxic effects of Cd induced postnatally are related to vascular damage and the vascular endothelium is considered the target tissue in acute Cd toxicity (Nolan and Shaikh, 1986a). Blood vessel damage has been seen in limb buds after Cd exposure but overt hemorrhage was not a feature (Messerle and Webster, 1982). Feather malformations seen in chickens after Cd exposure during development were due to hemorrhagic atrophy of the distal part of the feather (Narbaitz *et al.*, 1983).

2.3. Genetic Considerations

Some of the toxic effects of Cd in the mouse are modulated by the recessive gene cadmium resistance (*cdm*) on chromosome 3. For instance, mouse strains such as the C57BL/6J strain, which are homozygous for the gene, are resistant to Cd-induced testicular necrosis in adults (Taylor *et al.*, 1973), but are more susceptible than strains without the gene to the teratogenic effects of Cd, particularly limb defects (Layton and Layton, 1979), and to the acute toxicity of Cd (Taylor *et al.*, 1973). The primary effect of Cd on the testes is damage to the vasculature leading to ischemia and secondary degeneration of the seminiferous tubules (Aoki and Hoffer, 1978). It is possible that the *cdm* gene determines the extent of vascular binding and hence the severity of the range of vascular-associated Cd toxicities and that resistance to vascular injury is determined by basal metallothionein concentrations in the endothelial cells (Nolan and Shaikh, 1986b). If the teratogenic effect of Cd is due to a direct action on the yolk sac or embryo, then reduced vascular binding may result in higher tissue or embryonic levels. There is further evidence of a major maternal genetic effect on the Cd-induced malformation and resorption rates (Pierro and Haines, 1978) perhaps associated with the rate of metallothionein synthesis and progesterone levels (Wolkowski-Tyl and Preston, 1979).

3. CADMIUM AND THE CHORIOALLANTOIC PLACENTA

In rodents there are two distinct periods of pregnancy during which there is susceptibility to Cd toxicity. The first period is the teratogenic stage, which in mice lasts from gestational d 7 to d 10. The second period, the placental toxic stage, starts late on d 10 and continues until birth. During the latter period a large dose of Cd administered parenterally causes rapid intrauterine death or premature birth of the fetuses. The replacement of a teratogenic by a fetolethal effect coincides in time with the emergence of functional dependence of the embryo on the chorioallantoic placenta (Dencker, 1975). The fetolethal effect of Cd was first demonstrated in rats late in pregnancy and was associated with degeneration of the fetal part of the placenta and a maternal toxemia with renal hemorrhages (Pařízek, 1964, 1965; Pařízek *et al.*, 1968). Autoradiographic studies confirmed that the placental labyrinth is a major site of Cd accumulation (Berlin and Ullberg, 1963; Dencker, 1975), accounting for up to 8% of a dose administered late in pregnancy (Sonawane *et al.*, 1975).

In mice given an i.p. injection of Cd (4 mg/kg) on d 11 of gestation, fetal death usually occurred within 4–8 h. Four hours after exposure, placentas from live and dead fetuses showed many fluid-filled spaces, red blood cells between the trophoblastic layers, and many maternal blood sinuses blocked by thrombi. Similar Cd exposure on d 14 or 17 took longer to kill the fetuses, fetal death or abortion occurring 12–24 h later. Six to eight hours after exposure there was an enormous dilation of the lining cells of the maternal blood sinuses of the chorioallantoic placenta and subsequently an accumulation of platelets and blockage of the sinuses. In all placentas examined, the fetal endothelial layer was the least affected part and damage was restricted to the labyrinth (Webster, unpublished observations).

A similar distribution of damage occurred in rat placentas where the pars fetalis was transformed into an extensive blood clot in 24 h while the pars materna showed slight edema and hyperemia (Pařízek, 1964). Changes were seen in the labyrinth 6–8 h after exposure, with layer II of the trophoblast showing the most severe damage and, as in the mouse, the fetal endothelial layer was remarkable for its lack of damage (Levin *et al.*, 1983; di Sant'Agnese *et al.*, 1983). In contrast, Samarawickrama and Webb (1979) reported that the initial degeneration occurred in the maternal part of the rat placenta 8 h after administration of an LD50 dose of Cd, but this was not confirmed in other studies (Levin *et al.*, 1983).

Although damage to the placenta is the likely cause of the fetal death, it is also possible that damage to the ovaries might be involved.

Injected Cd reaches the ovary in mice (Dencker, 1975, 1976), and in prepubertal rats (Peereboom-Stegeman and Jongstra-Spaamen, 1979) and gerbils (Kaul and Ramaswami, 1970) caused hemorrhage and follicular atresia. Bilateral ovariectomy of pregnant mice at any stage after implantation leads to rapid fetal death and abortion (Harris, 1927), so Cd-induced ovarian damage could be the cause of fetal death. However, in an electron microscopic study in mice, damage to the ovaries was not detected during the period of Cd-induced placental destruction and fetal death (Webster, unpublished observations).

The chorioallantoic placenta does not act as a complete barrier to Cd; small quantities (e.g., 0.015% of the dose/g fetal tissue) were detected in fetuses after treatment of rats with fetotoxic doses on gestation d 12, 15, or 20 (Sonawane *et al.*, 1975). The possibility that the Cd in the fetus is responsible for fetal death was effectively disproved, however. Cadmium chloride injected directly into 18-d rat fetuses to give a fetal Cd concentration eight times greater than that seen after a maternal fetotoxic injection caused only 12% fetal mortality compared with 75% fetal mortality following maternal injection (Levin and Miller, 1980).

Hemodynamic studies have shown a 40% reduction in the uteroplacental blood flow 12–16 h after Cd exposure and 73% reduction after 18–24 h (Levin and Miller, 1981). Blood flow changes were not detected prior to 12 h, suggesting that the placental necrosis precedes the blood flow changes. There were no changes in four placental enzymes after Cd exposure on gestation d 20 in rats (Sonawane *et al.*, 1975) but decreased nutrient transport, particularly of vitamin B_{12}, was found (Danielsson and Dencker, 1984).

Hence, most studies indicate that Cd-induced damage to the placenta leads to decreased uteroplacental blood flow and therefore that the cause of fetal death is probably a combination of anoxia and decreased nutrient supply. These studies on the "peculiar" toxic effects of Cd on the chorioallantoic placenta (Pařízek, 1964, 1965) were initially performed out of scientific curiosity (Pařízek, 1983). The demonstration of Cd accumulation by the human placenta (Roels *et al.*, 1978; Kuhnert *et al.*, 1982) necessitates considering that there are consequences to the fetus other than death.

4. CHRONIC EXPOSURE TO CADMIUM DURING PREGNANCY

4.1. Fetal Growth Retardation

The teratogenic and fetolethal effects of Cd are only produced by acute exposure to high doses. The most common form of human exposure to Cd is chronic low-level intake from cigarette smoke and some foods, with particularly high concentrations in liver, kidney, and shellfish (CEC, 1978; Diehl and Boppel, 1985). Many occupations involve significant additional exposure, particularly primary lead–zinc–cadmium industry (CEC, 1978), and these result in increased Cd assimilation (Lauwerys *et al.,* 1979; Christoffersson *et al.,* 1987); but even this type of exposure does not come close to duplicating the conditions necessary for teratogenesis. Nevertheless, the increasing participation of fertile women in the work force makes it necessary to consider the reproductive consequences of chronic exposure.

"Chronic exposure" is used in this section to refer to repeated daily intake of Cd usually via air, food, or water, but in some instances by gavage or injection. Fetal growth retardation is the major consequence of chronic exposure to Cd throughout pregnancy in experimental animals. Fetal growth retardation occurred following chronic exposure to Cd via food (Pond and Walker, 1975), drinking water (Webster, 1978; Hastings *et al.,* 1978; Ahokas *et al.,* 1980; Laskey *et al.,* 1980; Barański, 1987), air (Prigge, 1978), and gavage (Machemer and Lorke, 1981; Barański *et al.,* 1982; Barański, 1985). Only very high, maternally toxic doses (40 mg/kg), administered by gavage, were associated with the occurrence of congenital malformations (Machemer and Lorke, 1981; Barański *et al.,* 1982; Barański, 1985).

In a multigeneration study of CD mice given 10 ppm Cd in drinking water, the animals died out in two generations (Schroeder and Mitchener, 1971). Much of the toxic effect occurred in the suckling period when Cd can have a severe effect on growth by competitive inhibition of iron uptake and utilization in the mother resulting in severe iron deficiency in the litter. A congenital malformation consisting of a sharp angulation of the distal third of the tail was noted in many of the offspring, but this may have been unrelated to the treatment similar studies in rats using much higher Cd exposure failed to induce this malformation (Pietrzak-Flis *et al.,* 1978).

The degree of fetal growth retardation induced by Cd is dose dependent (Webster, 1978; Barański, 1987) and the effect is demonstrable in mice at intake levels as low as 5 ppm in drinking water throughout gestation (Laskey *et al.,* 1980). The newborn are not only small but have an iron deficiency anemia, which is thought to be the cause of the growth retardation (Webster, 1978). When mice of the same strain (Quackenbush Special) were fed an iron-deficient diet throughout pregnancy, the offspring had the same degree of anemia and the same extent of fetal growth retardation as seen after 40 ppm Cd in the drinking water (Webster, 1979a).

The Cd-induced anemia and growth retardation were completely prevented by parenteral administration of iron (Webster, 1979a), but supplementary ferrous sulfate added to the diet only partly prevented the fetal effects, indicating that Cd interferes with uptake of iron from the maternal duodenum. Cd may also interfere with iron utilization in the pregnant female and its placental transport to the fetus. Zn added to the diet also reduced Cd-induced fetal growth retardation (Ahokas *et al.*, 1980), but not as effectively as parenterally administered iron (Webster, 1979b). The ability of Cd to inhibit iron uptake by the duodenum is well established (Hamilton and Valberg, 1974). It may also inhibit the uptake of other ions since reduced serum levels of iron, Zn, and ceruloplasmin occurred in pregnant rats given 50 ppm Cd in drinking water (Sowa and Steibert, 1985).

There are no reports of human congenital malformations thought to be due to Cd exposure, which is not surprising in view of the acute exposure to very high levels that would be necessary for teratogenesis. A report from Russia noted that a number of women who were industrially exposed to Cd had low-birth-weight babies (Cvetova, 1970, cited by Friberg *et al.*, 1974). An inverse relation was found between the Cd content of babies' hair and their birth weight (Huel *et al.*, 1981), and a similar trend was found in the offspring of women exposed to heavy metals (Huel *et al.*, 1984).

A tenuous link can be postulated between the fetal growth retardation and the increased Cd intake that are both associated with maternal cigarette smoking. In mothers who smoke, the Cd content of the placenta (Roels *et al.*, 1978; Miller and Gardner, 1981; Kuhnert *et al.*, 1982) and amniotic fluid (Siegers *et al.*, 1983) is substantially higher than that of nonsmoking mothers. A similar association between increased Cd intake, hypertension, and reduced birth weight has been postulated (Huel *et al.*, 1981). In neither case, however, has a causal relation been established.

4.2. Cadmium in the Placenta and Fetus

The effect of Cd on fetal growth appears to be secondary to interference with trace element absorption and possibly with utilization and placental transfer. Absorbed Cd may also have a direct toxic effect on the placenta, the conceptus, or both.

Cd absorption from oral exposure is usually low, less than 3% in mice, but may double during pregnancy (Bhattacharyya *et al.*, 1982), probably as an accompaniment of the increased uptake of calcium and iron (Washko and Cousins, 1976; Flanagan *et al.*, 1978). Absorption from inhalation is much higher, about 40% in mice (Moore *et al.*, 1973). Corresponding estimated values for humans are 5 and 13–19% (CEC, 1978). During chronic exposure, most of the absorbed Cd is rapidly bound to metallothionein in the liver, and blood levels are kept low, restricting the amount of Cd available for placental transfer. As metallothionein is inducible, previous exposure to Cd can be an important variable in determining blood levels (Hackett and Kelman, 1983).

The chorioallantoic placenta accumulates Cd and acts as a partial barrier to its transfer to the fetus in rodents (Ferm *et al.*, 1969; Dencker, 1975; Sonawane *et al.*, 1975; Pietrzak-Flis *et al.*, 1978; Ahokas and Dilts, 1979; Christley and Webster, 1983; Webster, 1988) and humans (Roels *et al.*, 1978; Kuhnert *et al.*, 1982; Korpela *et al.*, 1986). In nonindustrially exposed women, the placenta concentrates Cd about tenfold compared with levels in maternal blood (Roels *et al.*, 1978), and levels in umbilical blood are about one-third those in maternal blood (Lauwerys *et al.*, 1978; Roels *et al.*, 1978; Alessio *et al.*, 1984; Korpela *et al.*, 1986).

There is evidence from rodents that Cd is bound to a metallothionein dimer in the placenta (Hanlon *et al.*, 1982), which may effectively detoxify the metal. Acute exposure to Cd interfered with placental transport in rats (Samarawickrama and Webb, 1981) and mice (Danielsson and Dencker, 1984), but this type of insult is not relevant to the usual type of human exposure.

An interesting investigative technique—dual recirculating perfusion of isolated human placental lobules—was recently applied to Cd and the human placenta. The maternal circulation of the isolated lobule was perfused with various concentrations of Cd, and the Cd concentrations in the placenta and fetal circulation was monitored over a 24-h period. At 1.1 μg/ml Cd the placenta accumulated Cd, with little appearing in the fetal circulation, but perfusion with 2.2 μg/ml Cd gave rise to placental concentrations of 11.2 μg/g and placental necrosis in 6–8 h (Miller, 1986). These concentrations are about 1000-fold greater than those seen in pregnant women (e.g., Kuhnert *et al.*, 1982). Further investigations with this experimental system, using blood levels found in human pregnancy, should provide important information on whether environmental levels of Cd exposure can interfere with placental function.

Following chronic or acute exposure during pregnancy in mice or rats, a small amount of Cd crossed the placenta and entered the fetus. The amounts transferred were very low indeed, in the range of 0.0001–0.02% of the dose per gram of fetus for acute doses (Sonawane *et al.*, 1975; Ahokas and Dilts, 1979; Christley and Webster, 1983), and for chronic exposure 0.006% of the ingested dose, or less than 3% of the internally retained dose was passed onto the fetuses (Bhattacharyya *et al.*, 1982). Cd has been detected in human fetuses from nonindustrially exposed mothers (Chaube *et al.*, 1973; Gross *et al.*, 1976); and the estimated amount in newborns was less than 1 μg (Friberg *et al.*, 1974). The toxicological significance of these small amounts of Cd is unclear. Metallothionein is present in rat fetuses in late pregnancy, but fetal metallothionein was not induced by maternal exposure to Cd or by direct injection of Cd into the fetus (Sasser *et al.*, 1985). There is some evidence that metallothionein can be induced by Cd in rabbit fetuses (Waalkes *et al.*, 1982), perhaps because of the longer fetal period in this species.

Cd was not particularly fetotoxic when injected directly into older rat fetuses (Levin and Miller, 1980), but possible subtle deficits were not investigated. Recent autoradiographic studies showed that an i.p. injection of a trace dose of ^{109}Cd on d 17 of gestation in mice could be detected in the fetal brain 24 h later, localized in the choroid plexus and cerebral blood vessels. The same distribution was still evident in the brains of the offspring when examined by autoradiography on postnatal d 15 (Webster and Valois, unpublished observations). The functional significance of the localized Cd is unknown, but cerebral capillary damage was seen in rat fetuses exposed to large doses of Cd late in gestation (Rohrer *et al.*, 1978b).

Prenatal Cd exposure has been shown to lead to a variety of behavioral changes in the offspring (Hastings, 1986), most notably reduced motor activity (Choudhury *et al.*, 1978; Hastings *et al.*, 1978; Barański *et al.*, 1983; Barański, 1984, 1985, 1986; Ali *et al.*, 1986). However, changes in brain structure have not been found and there is no evidence that these behavioral changes are due to a direct effect of Cd on the prenatal CNS. Other possible explanations include the effect of reduced birth weight (e.g., Hastings, 1986), lowered iron, Zn, and copper in the newborn (Sowa and Steibert, 1985), and further Cd exposure during suckling resulting in lowered copper and Zn absorption in the neonates (Barański, 1986). The widespread accumulation of Cd in all maternal tissues during prenatal chronic exposure (Webster, 1988), particularly those with special functions during pregnancy, such as the placenta, uterus, pituitary, and ovary, makes secondary effects on the fetus likely.

5. CADMIUM EXPOSURE IN THE PERINATAL PERIOD

Normally the postnatal period is not included when considering the teratology of a substance. However, birth is an arbitrary mark in the development of the CNS and the developing vasculature of the growing brain has a particular affinity for Cd. Several physiological phenomena associated with the perinatal period make it a susceptible period to heavy metal poisoning (Kostial, 1983). Absorption and retention of an oral dose of Cd is 10–20 times higher in neonatal than in juvenile or adult rats (Sasser and Jarboe, 1977, 1980), and excretion of Cd during the neonatal period is virtually absent (Kostial, 1983). This enhanced absorption of Cd by neonates is not simply a characteristic of the altricial rat with its ability to absorb gammaglobulins from the intestines but is also true of the guinea pig which has minimal capacity for absorption of macromolecules during the neonatal period. Indeed the guinea pig absorbed more Cd from the intestine during this stage than did the rat or pig (Sasser and Jarboe, 1980).

In a recent study using QS mice, the uptake, retention, and distribution of Cd in the brain were compared for two dosages of Cd (a trace dosage and a high dosage) administered intraperitoneally on postnatal d 0, 7, 14, or 42 (Valois and Webster, 1987a,b). Cd absorption by the neonatal brain was much greater than by the adult brain, confirming previous findings (Wong and Klaassen, 1980; Kostial, 1983). Twenty-four-hour brain retention in neonates was 0.3–1.5% for the low and 1.3–2.3% for the high dosage compared with 0.07 and 0.04% for the same dosages in adults. The half-life of the Cd in the brain ranged from 14 to 114 days. Cd exposure on postnatal d 7 resulted in the highest retention in the brain and the longest half-life.

Autoradiographic studies showed that the Cd distribution in the adult-exposed rat or mouse brain was restricted to the choroid plexus and a few regions with a fenestrated endothelium (Arvidson and Tjälve, 1986; Valois and Webster, 1987b), but in the neonatally exposed mouse there was also extensive labeling of the cerebral blood vessels, which did not occur after adult exposure (Valois and Webster, 1987a,b). Cd penetration into the brain parenchyma was greatest following Cd exposure on d 7 or 14 and was minimal after adult exposure. Distribution showed little change with increasing survival time or with the different dosages administered at the same age.

This distribution of Cd in the less developed brain appears to be related to the ontogeny of the brain vascular system. Neonatal brain growth in rats and mice as well as in humans is characterized by extensive capillary growth (Bär, 1980). The immature blood vessels appear to bind Cd in an almost permanent manner unlike their adult counterparts. Maximal parenchymal penetration of the brain coincided with the period of maximum vascular branching. Exposure to acute doses of Cd during the neonatal period caused a characteristic pattern of severe hemorrhage and necrosis in the brain of rabbits and rodents (Gabbiani *et al.,* 1967; Webster and Valois, 1981; Wong and Klaassen, 1982). The distribution of the lesions within the brain varied with the age at exposure and susceptibility ended abruptly with the conclusion of brain growth (Webster and Valois, 1981). The pathology produced by Cd in the neonatal brain is thought to be due to Cd damage to the immature blood vessels, which is in agreement with the autoradiographic studies.

The severe damage to the CNS caused by acute Cd exposure in the neonatal period is, not unexpectedly, associated with a variety of behavioral problems in the surviving animals (Winneke *et al.*, 1979; Webster and Valois, 1981; Smith *et al.*, 1982; Ruppert *et al.*, 1985; Newland *et al.*, 1986), but such dosages are not in the range of human environmental concern. Chronic exposure to Cd in the neonatal period might be expected to result in considerable accumulation of Cd in the blood vessels and choroid plexus of the growing brain and it is noteworthy that such exposure has been associated with changes in behavior and neurotransmitter levels (Rastogi *et al.*, 1977; Smith *et al.*, 1982) although daily exposure levels were very high.

6. SUMMARY

Cd is a teratogen in experimental animals when administered acutely in very high doses. Absorbed Cd has a direct effect on the embryo–yolk sac complex but the precise mechanism underlying its teratogenic effects is unknown. Concern about Cd toxicity in humans is mostly based on its ability to act as a cumulative poison owing to its long biological half-life. On that basis it is unlikely to be a conventional teratogen in humans, however, as the acute parenteral exposure that is necessary for teratogenesis is unlikely to occur. With respect to chronic exposure, the ability of the definitive placenta to sequester Cd means that very little maternally absorbed Cd actually reaches fetuses, which may be fortunate as the fetal cerebral blood vessels and choroid plexus accumulate and retain Cd with a long half-life.

Whether the accumulation of Cd by the placenta results in some compromise of its function awaits further study. Cd may be a contributing factor in the fetal growth retardation associated with cigarette smoking in humans. The neonatal period is characterized by high intestinal absorption of Cd and its extensive transfer to the developing vasculature of the brain and choroid plexus. The long half-life of Cd in the brain and the unknown functional consequences of its presence there may make the neonatal period the most vulnerable stage of mammalian development with respect to Cd toxicity.

ACKNOWLEDGMENTS. I particularly acknowledge the help of my students Karin Messerle, Jeremy Christley, and Angelo Valois and my research assistant Sally Pope, who have all contributed to the many studies of cadmium toxicity carried out in the Department of Anatomy, University of Sydney, Australia.

REFERENCES

Abraham, R., Charles, A. K., Mankes, R., LeFevre, R., Renak, V., and Ashok, L. 1986. In vitro effects of cadmium chloride on preimplantation rat embryos. *Ecotoxicol. Environ. Safety* **12**:213–219.

Ahokas, R. A., and Dilts, P. V. 1979. Cadmium uptake by the rat embryo as a function of gestational age. *Am. J. Obstet. Gynecol.* **135**:219–222.

Ahokas, R. A., Dilts, P. V., and LaHaye, E. B. 1980. Cadmium-induced fetal growth retardation: protective effect of excess dietary zinc. *Am. J. Obstet. Gynecol.* **136**:216–221.

Alessio, L., Dell'Orto, A., Calzaferri, G., Buscaglia, M., Motta, G., and Rizzo, M. 1984. Cadmium concentrations in blood and urine of pregnant women at delivery and their offspring. *Sci. Total Environ.* **34**:261–266.

Ali, M. M., Murthy, R. C., and Chandra, S. V. 1986. Developmental and long term neurobehavioral toxicity of low level in utero cadmium exposure in rats. *Neurobehav. Toxicol. Teratol.* **8**:463–468.

Aoki, A., and Hoffer, A. P. 1978. Reexamination of the lesions in rat testis caused by cadmium. *Biol. Reprod.* **18**:579–591.

Arvidson, B., and Tjälve, H. 1986. Distribution of 109-cadmium in the nervous system of rats after intravenous injection. *Acta Neuropathol.* **69**:111–116.

Bär, T. 1980. The vascular system of the cerebral cortex. *Adv. Anat. Embryol. Cell Biol.* **59**:1–61.

Barański, B. 1984. Behavioral alterations in offspring of female rats repeatedly exposed to cadmium oxide by inhalation. *Toxicol. Lett.* **22**:53–61.

Barański, B. 1985. Effect of exposure of pregnant rats to cadmium on prenatal and postnatal development of the young. *J. Hyg. Epidemiol. Microbiol. Immunol.* **3**:253–262.

Barański, B. 1986. Effect of maternal cadmium exposure on postnatal development and tissue cadmium, copper and zinc concentrations in rats. *Arch. Toxicol.* **58**:255–260.

Barański, B. 1987. Effect of cadmium on prenatal development and on tissue cadmium, copper and zinc concentrations in rats. *Environ. Res.* **42**:54–62.

Barański, B., Stetkiewicz, I., Trzcinka-Ochocka, M., Sitarek, K., and Szymczak, W. 1982. Teratogenicity, fetal toxicity and tissue concentrations of cadmium administered to female rats during organogenesis. *J. Appl. Toxicol.* **2**:255–259.

Barański, B., Stetkiewicz, I., Sitarek, K., and Szymczak, W. 1983. Effects of oral, subchronic cadmium administration on fertility, prenatal and postnatal progeny development in rats. *Arch. Toxicol.* **54**:297–302.

Barr, M. 1973. The teratogenicity of cadmium chloride in two stocks of Wistar rats. *Teratology* **7**:237–242.

Beck, F., Lloyd, J. B., and Griffiths, A. 1967. Lysosomal enzyme inhibition by trypan blue: a theory of teratogenesis. *Science* **157**:1180–1182.

Berlin, M., and Ullberg, S. 1963. The fate of cadmium-109 in the mouse: an autoradiographic study after a single intravenous injection of 109-cadmium chloride. *Arch. Environ. Health* **7**:686–693.

Beyer, P. E., and Chernoff, N. 1986. The induction of supernumerary ribs in rodents: role of maternal stress. *Teratog. Carcinog. Mutag.* **6**:419–429.

Bhattacharyya, M. H., Whelton, B. D., and Peterson, D. P. 1982. Gastrointestinal absorption of cadmium in mice during gestation and lactation. II. Continuous exposure studies. *Toxicol. Appl. Pharmacol.* **66:**368–375.

Brown, N. A., and Coakley, M. E. 1987. Tissue oxygen as a determinant of axially asymmetric teratologic responses: misondazol as a marker for hypoxic cells. *Teratology* **35:**73A.

CEC (Commission of the European Communities). 1978. *Criteria (Dose/Effect Relationships) for Cadmium.* Pergamon Press, Oxford.

Chaube, S., Nishimura, H., and Swinyard, C. A. 1973. Zinc and cadmium in normal human embryos and fetuses. *Arch. Environ. Health* **26:**237–240.

Chernoff, N. 1973. Teratogenic effects of cadmium in rats. *Teratology* **8:**29–32.

Chiquoine, A. D. 1965. Effect of cadmium chloride on the pregnant albino mouse. *J. Reprod. Fertil.* **10:**263–265.

Choudhury, H., Hastings, L., Menden, E., Brockman, D., Cooper, G. P., and Petering, H. G. 1978. Effects of low level prenatal cadmium exposure on trace metal body burden and behavior in Sprague–Dawley rats, in: *Trace Element Metabolism in Man and Animals—*3, M. Kirchgessner. ed. Institut für Ernahrungsphysiologie der Technischen Universität Munchen, Freising-Weihenstephan, pp. 549–552.

Christley, J., and Webster, W. S. 1983. Cadmium uptake and distribution in mouse embryos following maternal exposure during the organogenic period: a scintillation and autoradiographic study. *Teratology* **27:**305–312.

Christoffersson, J. O., Welinder, H., Spang, G., Mattsson, S., and Skerfving, S. 1987. Cadmium concentration in the kidney cortex of occupationally exposed workers measured in vivo using X-ray fluorescence analysis. *Environ. Res.* **42:**489–499.

Danielsson, B. R., and Dencker, L. 1984. Effects of cadmium on the placental uptake and transport to the fetus of nutrients. *Biol. Res. Preg. Perinat.* **5:**93–101.

Daston, G. P. 1981. Effects of cadmium on prenatal ultrastructural maturation of rat alveolar epithelium. *Teratology* **23:**75–84.

Daston, G. P., and Grabowski, C. T. 1979. Toxic effects of cadmium on the developing rat lung. 1. Altered pulmonary surfactant and the induction of respiratory distress syndrome. *J. Toxicol. Environ. Health* **5:**973–983.

Datu, A. R., Nakamura, H., and Yasuda, M. 1985. Pathogenesis of the mouse forelimb deformity induced by acetazolamide: an electron microscopic study. *Teratology* **31:**253–263.

Dencker, L. 1975. Possible mechanisms of cadmium fetotoxicity in golden hamsters and mice: uptake by the embryo, placenta and ovary. *J. Reprod. Fertil.* **44:**461–471.

Dencker, L. 1976. Tissue localization of some teratogens at early and late gestation related to fetal effects. *Acta Pharmacol. Toxicol.* **39**(Suppl. 1):1–131.

Diehl, J. F., and Boppel, B. 1985. Dietary intake of Cd: a re-evaluation. *Trace Elem. Med.* **2:**167–174.

di Sant'Agnese, P. A., Demesey Jensen, K., Levin, A., and Miller, R. K. 1983. Placental toxicity of cadmium in the rat: an ultrastructural study. *Placenta* **4:**149–164.

Eaton, J. G. 1974. Chronic cadmium toxicity to the bluegill (*Lepomis macrochirus Rafinesque*). *Trans. Am. Fish. Soc.* **103:**729–735.

Enger, M. D., Hildebrand, C. E., Seagrave, J., and Tobey, R. A. 1986. Cellular resistance to cadmium. *Handb. Exp. Pharmacol.* **80:**363–396.

Ferm, V. H. 1969. The synteratogenic effect of lead and cadmium. *Experientia* **25:**56–57.

Ferm, V. H. 1971. Developmental malformations induced by cadmium. *Biol. Neonate* **19:**101–107.

Ferm, V. H., and Carpenter, S. J. 1967. Teratogenic effect of cadmium and its inhibition by zinc. *Nature* **216:**1123.

Ferm, V. H., and Carpenter, S. J. 1968. The relationship of cadmium and zinc in experimental mammalian teratogenesis. *Lab. Invest.* **18**:429–432.

Ferm, V. H., and Hanlon, D. P. 1987. Inhibition of cadmium teratogenesis by a mercaptoacrylic acid (MFA). *Experientia* **43**:208–210.

Ferm, V. H., and Layton, W. M., Jr. 1979. Reduction in cadmium teratogenesis by prior cadmium exposure. *Environ. Res.* **18**:347–350.

Ferm, V. H., Hanlon, D. P., and Urban, J. 1969. The permeability of the hamster placenta to radioactive cadmium. *J. Embryol. Exp. Morphol.* **22**:107–113.

Feuston, M. H., and Scott, W. J., Jr. 1985. Cadmium-induced forelimb ectrodactyly: a proposed mechanism of teratogenesis. *Teratology* **32**:407–419.

Flanagan, P. R., McLellan, J. S., Haist, J., Cherian, G., Chamberlain, M. J., and Valberg, L. S. 1978. Increased dietary cadmium absorption in mice and human subjects with iron deficiency. *Gastroenterology* **74**:841–846.

Friberg, L., Piscator, M., Nordberg, G. F., and Kjellstrom, T. 1974. *Cadmium in the Environment*, 2nd ed. CRC, Cleveland.

Gabbiani, G., Baic, D., and Déziel, C. 1967. Toxicity of cadmium for the central nervous system. *Exp. Neurol.* **18**:154–160.

Gale, T. F., and Ferm, V. H. 1973. Skeletal malformations resulting from cadmium treatment in the hamster. *Biol. Neonate* **23**:149–160.

Garcia, M., and Lee, M. 1981. Interaction of cadmium and zinc during prenatal development in the rat. *Biol. Trace Elem. Res.* **3**:149–156.

German, J. 1984. Embryonic stress hypothesis of teratogenesis. *Am. J. Med.* **76**:293–301.

Ghafghazi, T., and Mennear, J. H. 1973. Effects of acute and subacute cadmium administration on carbohydrate metabolism in mice. *Toxicol. Appl. Pharmacol.* **26**:231–240.

Green, M. C., Azar, C. A., and Maren, T. H. 1973. Strain differences in susceptibility to the teratogenic effect of acetazolamide in mice. *Teratology* **8**:143–146.

Gross, S. B., Yeager, D. W., and Middendorf, M. S. 1976. Cadmium in liver, kidney and hair of humans, fetal through old age. *J. Toxicol. Environ. Health* **2**:153–167.

Hackett, P. L., and Kelman, B. J. 1983. Availability of toxic metals to the conceptus. *Sci. Total Environ.* **28**:433–442.

Hahn, G. M., and Li, G. C. 1982. Thermotolerance and heat shock proteins in mammalian cells. *Radiat. Res.* **92**:452–457.

Hamilton, D. L., and Valberg, L. S. 1974. Relationship between cadmium and iron absorption. *Am. J. Physiol.* **227**:1033–1037.

Hanlon, D. P., Specht, C., and Ferm, V. H. 1982. The chemical status of the cadmium ion in the placenta. *Environ. Res.* **27**:89–94.

Harris, R. G. 1927. Effect of bilateral ovariectomy upon the duration of pregnancy in mice. *Anat. Rec.* **37**:83–93.

Hastings, L. 1986. Behavioral teratogenesis resulting from early cadmium exposure, in: *Handbook of Behavioral Teratology*, E. P. Riley and C. H. Vorhees, eds. Plenum Press, New York, pp. 321–333.

Hastings, L., Choudhury, H., Petering, H. G., and Cooper, G. P. 1978. Behavioral and biochemical effects of low level prenatal cadmium exposure in rats. *Bull. Environ Contam. Toxicol.* **20**:96–101.

Hayden, T. L., Turner, J. E., Williams, M. W., Cook, J. S., and Hsie, A. W. 1982. A model for cadmium transport and distribution in CHO cells. *Comput. Biomed. Res.* **15**:97–110.

Hilbelink, D. R., and Kaplan, S. 1986. Sirenomelia: analysis in the cadmium- and lead-treated golden hamster. *Teratog. Carcinog. Mutag.* **6**:431–440.

Holmberg, R. E., and Ferm, V. H. 1969. Interrelationships of selenium, cadmium and arsenic in mammalian teratogenesis. *Arch. Environ. Health* **18**:873–877.

Holmes, L. B., and Trelstad, R. L. 1979. The early limb deformity caused by acetazolamide. *Teratology* **20:**289–296.
Holt, D., and Webb, M. 1986. Comparison of some biochemical effects of teratogenic doses of mercuric mercury and cadmium in the pregnant rat. *Arch. Toxicol.* **58:**249–254.
Huel, G., Boudene, C., and Ibrahim, M. A. 1981. Cadmium and lead content of maternal and newborn hair: relationship to parity, birthweight and hypertension. *Arch. Environ. Health* **36:**221–227.
Huel, G., Everson, R. B., and Menger, I. 1984. Increased hair cadmium in newborns of women occupationally exposed to heavy metals. *Environ. Res.* **35:**115–121.
Hurley, L. S., Gowan, J., and Swenerton, H. 1971. Teratogenic effects of short term and transitory zinc deficiency in rats. *Teratology* **4:**199–204.
Ishizu, S., Minami, M., Suzuki, A., Yamada, M., Sato, M., and Yamura, K. 1973. An experimental study on teratogenic effect of cadmium. *Ind. Health* **11:**127–139.
Jacobson, K. B., and Turner, J. E. 1980. The interaction of cadmium and certain other metal ions with proteins and nucleic acids. *Toxicology* **16:**1–37.
Kaul, D. K., and Ramaswami, L. S. 1970. Effect of cadmium chloride on the ovary of the India desert gerbil *Meriones hurrianae Jerdon. Ind. J. Exp. Biol.* **8:**171–173.
Keino, H. 1973. Effect of cadmium sulfate on closure of neural tubes in frogs. *Teratology* **8:**96–97.
Keino, H., and Yamamura, H. 1974. Effects of a cadmium salt administered to pregnant mice on postnatal development of the offspring. *Teratology* **10:**87A.
Keino, H., Aoki, E., Yamamura, H., and Murakami, U. 1978. Partial inhibition of neural tube formation by cadmium sulphate in ICR-JCL mouse embryos. *Teratology* **18:**149–150A.
Khera, K. S. 1985. Maternal toxicity: a possible etiological factor in embryo–fetal deaths and fetal malformations of rodent–rabbit species. *Teratology* **31:**129–153.
Klein, N. W., Vogler, M. A., Chatot, C. L., and Pierro, L. J. 1980. The use of cultured rat embryos to evaluate the teratogenic activity of serum: cadmium and cyclophosphamide. *Teratology* **21:**199–208.
Kobayashi, S., and Kimura, M. 1980. Different inducibility of metallothionein in various mammalian cells in vitro. *Toxicol. Lett.* **5:**357–362.
Korpela, H., Loueniva, R., Yrjänheikki, E., and Kauppila, A. 1986. Lead and cadmium concentrations in maternal and umbilical cord blood, amniotic fluid, placenta and amniotic membranes. *Am. J. Obstet. Gynecol.* **155:**1086–1089.
Kostial, K. 1983. Specific features of metal absorption in suckling animals, in: *Reproductive and Developmental Toxicity of Metals,* T. W. Clarkson, G. F. Nordberg, and P. R. Sager, eds. Plenum Press, New York, pp. 727–744.
Kuczuk, M. H., and Scott, W. J., Jr. 1984. Potentiation of acetazolamide induced ectrodactyly in SWV and C57BL/6J mice by cadmium sulphate. *Teratology* **29:**427–435.
Kuhnert, P. M., Kuhnert, B. R., Bottoms, S. F., and Erhard, P. 1982. Cadmium levels in maternal blood, fetal cord blood and placental tissues of pregnant women who smoke. *Am. J. Obstet. Gynecol.* **142:**1021–1025.
Laskey. J. W., Rehnberg, G. L., Favor, M. J., Cahill, D. F., and Pietrzak-Flis, Z. 1980. Chronic ingestion of cadmium and/or tritium. II. Effects on growth, development, and reproductive function. *Environ. Res.* **22:**466–475.
Lauwerys, R., Buchet, J. P., Roels, H., and Hubermont, G. 1978. Placental transfer of lead, mercury, cadmium and carbon monoxide in women. I. Comparison of the frequency distributions of the biological indices in maternal and umbilical cord blood. *Environ. Res.* **15:**278–289.
Lauwerys, R., Roels, H., Regniers, M., Buchet, J. P., Bernard, A., and Goret, A. 1979.

Significance of cadmium concentration in blood and in urine in workers exposed to cadmium. *Environ. Res.* **20:**375–391.

Layton, W. M., and Ferm, V. H. 1980. Protection against cadmium-induced limb malformations by pretreatment with cadmium or mercury. *Teratology* **21:**357–360.

Layton, W. M., and Hallesy, D. W. 1965. Deformity of forelimb in rats: association with high doses of acetazolamide. *Science* **149:**306–308.

Layton, W. M., and Layton, M. W. 1979. Cadmium induced limb defects in mice: strain associated differences in sensitivity. *Teratology* **19:**229–236.

Levin, A. A., and Miller, R. K. 1980. Fetal toxicity of cadmium in the rat: maternal vs. fetal injections. *Teratology* **22:**1–5.

Levin, A. A., and Miller, R. K. 1981. Fetal toxicity of cadmium in the rat: decreased uteroplacental blood flow. *Toxicol. Appl. Pharmacol.* **58:**297–306.

Levin, A. A., Miller, R. K., and di Sant'Agnese, P. A. 1983. Heavy metal alterations of placental function: a mechanism for the induction of fetal toxicity by cadmium, in: *Reproductive and Developmental Toxicity of Metals,* T. Clarkson, G. Nordberg, and P. Sager, eds. Plenum Press, New York, pp. 633–654

Levinson, W., Oppermann, H., and Jackson, J. 1980. Transition series metals and sulfhydryl reagents induce the synthesis of four proteins in eukaryotic cells. *Biochim. Biophys. Acta* **606:**170–180.

Li, G. C., Shrieve, D. C., and Werb, Z. 1982. Correlations between synthesis of heat shock proteins and development of tolerance to heat and adriamycin in Chinese hamster fibroblasts: heat shock and other inducers, in: *Heat Shock from Bacteria to Man,* M. J. Schlesinger, M. Ashburner, and M. Tissieres, eds. Cold Spring Harbor Laboratory, Cold Spring Harbor, New York, pp. 395–404.

Machemer, L., and Lorke, D. 1981. Embryotoxic effect of cadmium on rats upon oral administration. *Toxicol. Appl. Pharmacol.* **58:**438–443.

Mego, J. L., and Cain, J. A. 1975. An effect of cadmium on heterolysosome formation and function in mice. *Biochem. Pharmacol.* **24:**1227–1232.

Messerle, K. 1978. Cadmium as a Teratogen. B.Sc.(Med.) thesis, University of Sydney.

Messerle, K., and Webster, W. S. 1982. The classification and development of cadmium-induced limb defects in mice. *Teratology* **25:**61–70.

Milaire, J. 1985. Histological changes induced in developing limb buds of C57BL mouse embryos submitted in utero to the combined influence of acetazolamide and cadmium sulphate. *Teratology* **32:**433–451.

Miller, R. K. 1986. Placental transfer and function: the interface for drugs and chemicals in the conceptus, in: *Drugs and Chemical Action in Pregnancy,* S. E. Fabro and A. R. Scialli, eds. Dekker, New York, pp. 123–169.

Miller, R. K., and Gardner, K. A. 1981. Cadmium in the human placenta: relationship to smoking. *Teratology* **23:**51A.

Moore, W., Stara, J. F., Crocker, W. C., Malanchuk, M., and Itlis, R. 1973. Comparison of 115m cadmium retention in rats following different routes of administration. *Environ. Res.* **6:**473–478.

Müller, L., and Ohnesorge, F. K. 1982. Different responses of liver parenchymal cells from starved and fed rats to cadmium. *Toxicology* **25:**141–150.

Mulvihill, J. E., Gamm, S. H., and Ferm, V. H. 1970. Facial formation in normal and cadmium-treated golden hamsters. *J. Embryol. Exp. Morphol.* **24:**393–403.

Narbaitz, R., Riedel, K. D., and Kacew, S. 1983. Induction of feather malformations in chick embryos by cadmium: protection by zinc. *Teratology* **27:**207–213.

Newland, M. C., Ng, W. W., Baggs, R. B., Gentry, G. D., Weiss, B., and Miller, R. K. 1986. Operant behavior in transition reflects neonatal exposure to cadmium. *Teratology* **34:**231–241.

Nolan, C. V., and Shaikh, Z. A. 1986a. The vascular endothelium as a target tissue in acute cadmium toxicity. *Life Sci.* **39:**1403–1409.

Nolan, C. V., and Shaikh. Z. A. 1986b. An evaluation of tissue metallothionein and genetic resistance to cadmium toxicity in mice. *Toxicol. Appl. Pharmacol.* **85:**135–144.

Padmanabhan, R., and Hameed, M. S. 1986. Exencephaly and axial skeletal dysmorphogenesis induced by maternal exposure to cadmium in the mouse. *J. Craniofac. Genet. Dev. Biol.* **6:**245–258.

Pařízek, J. 1964. Vascular changes at sites of estrogen biosynthesis produced by parenteral injection of cadmium salts: the destruction of the placenta by cadmium salts. *J. Reprod. Fertil.* **7:**263–265.

Pařízek, J. 1965. The peculiar toxicity of cadmium during pregnancy—an experimental 'toxaemia of pregnancy' induced by cadmium salts. *J. Reprod. Fertil.* **9:**111–112.

Pařízek, J. 1983. Cadmium and reproduction: a perspective after 25 years, in: *Reproductive and Developmental Toxicity of Metals,* T. W. Clarkson, G. F. Nordberg, and P. R. Sager, eds. Plenum Press, New York, pp. 301–313.

Pařízek, J., Ošťádalová, I., Beneš, I., and Pitha, J. 1968. The effect of a subcutaneous injection of cadmium salts on the ovaries of adult rats in persistent oestrus. *J. Reprod. Fertil.* **17:**559–562.

Parzyck, D. C., Shaw, S. M., Kessler, W. V., Vetter, R. J., Van Sickle, D. C., and Mayes, R. A. 1978. Fetal effects of cadmium in pregnant rats on normal and zinc deficient diets. *Bull. Environ. Contam. Toxicol.* **19:**206–214.

Peereboom-Stegeman, J. H. J., and Jongstra-Spaapen, E. J. 1979. The effect of a single sublethal administration of cadmium chloride on the microcirculation in the uterus of the rat. *Toxicology* **13:**199–213.

Pierro, L. J., and Haines, J. S. 1978. Cadmium-induced teratogenicity and embryotoxicity in the mouse, in: *Developmental Toxicology of Energy Related Pollutants,* D. D. Mahlum, M. R. Sikov, P. L. Hackett, and F. D. Andrew, eds. United States Department of Energy, Oak Ridge, Tenn., pp. 614–626.

Pietrzak-Flis, Z., Rehnberg, G. L., Favor, M. J., Cahill, D. F., and Laskey, J. W. 1978. Chronic ingestion of cadmium and/or tritium in rats. 1. Accumulation and distribution of cadmium in two generations. *Environ. Res.* **16:**9–17.

Pond, W. G., and Walker, E. F. 1975. Effect of dietary Ca and Cd level of pregnant rats on reproduction and on dam and progeny tissue mineral concentrations. *Proc. Soc. Exp. Biol. Med.* **148:**665–668.

Prigge, E. 1978. Inhalative cadmium effects in pregnant and fetal rats. *Toxicology* **10:**297–309.

Rastogi, R. B., Merali, Z., and Singhal, R. L. 1977. Cadmium alters behavior and the biosynthetic capacity for catecholamines and serotonin in neonatal rat brain. *J. Neurochem.* **28:**789–794.

Record, I. R., Dreosti, I. E., and Manuel, S. J. 1982a. Inhibition of rat yolk sac pinocytosis by cadmium and its reversal by zinc. *J. Nutr.* **112:**1994–1998.

Record, I. R., Dreosti, I. E., Manuel, S. J., and Buckley, R. A. 1982b. Interactions of cadmium and zinc in cultured rat embryos. *Life Sci.* **31:**2735–2743.

Record, I. R., Tulsi, R. S., Dreosti, I. E., and Fraser, F. J. 1985. Cellular necrosis in zinc-deficient rat embryos. *Teratology* **32:**397–405.

Ribas, B., and Schmidt, W. 1973. Der einfluss von cadmium aus die entwicklung von huhnerembryonen. *Gegenbaurs Morphol. Jahrb.* **119:**358–366.

Roels, H. A., Hubermont, G., Buchet, J. P., and Lauwerys, R. R. 1978. Placental transfer of lead, mercury, cadmium and carbon monoxide in women. III. Factors influencing the accumulation of heavy metals in the placenta and the relationship between metal concentrations in the placenta and in maternal and cord blood. *Environ. Res.* **16:**236–247.

Roels, H. A., Lauwerys, R. R., Buchet, J.- P., Bernard, A., Chettle, D. R., Harvey, T. C., and Al-Haddad, I. K. 1981. In vivo measurement of liver and kidney cadmium in workers exposed to this metal: its significance with respect to cadmium in blood and urine. *Environ. Res.* **26**:217–240.

Rohrer, S. R., Shaw, S. M., Born, G. S., and Vetter, R. J. 1978a. The maternal distribution and placental transfer of cadmium in zinc deficient rats. *Bull. Environ. Contam. Toxicol.* **19**:556–563.

Rohrer, S. R., Shaw, S. M., and Lamar, C. H. 1978b. Cadmium induced endothelial cell alterations in the fetal brain from prenatal exposure. *Acta Neuropathol.* **44**:147–149.

Ruppert, P. H., Dean, K. F., and Reiter, L. W. 1985. Development of locomotor activity of rat pups exposed to heavy metals. *Toxicol. Appl. Pharmacol.* **78**:69–77.

Samarawickrama, G. P., and Webb, M. 1979. Acute effects of cadmium on the pregnant rat and embryo–fetal development. *Environ. Health Perspect.* **28**:245–259.

Samarawickrama, G. P., and Webb, M. 1981. The acute toxicity and teratogenicity of cadmium in the pregnant rat. *J. Appl. Toxicol.* **1**:264–269.

Sasser, L. B., and Jarboe, G. E. 1977. Intestinal absorption and retention of cadmium in neonatal rat. *Toxicol. Appl. Pharmacol.* **41**:423–431.

Sasser, L. B., and Jarboe, G. E. 1980. Intestinal absorption and retention of cadmium in neonatal pigs compared to rats and guinea pigs. *J. Nutr.* **110**:1641–1647.

Sasser, L. B., Kelman, B. J., Levin, A. A., and Miller, R. K. 1985. The influence of maternal cadmium exposure or fetal cadmium injection on hepatic metallothionein concentrations in the fetal rat. *Toxicol. Appl. Pharmacol.* **80**:299–307.

Sato, F., Watanabe, T., Hoshi, E., and Endo, A. 1985. Teratogenic effect of maternal zinc deficiency and its co-teratogenic effect with cadmium. *Teratology* **31**:13–18.

Schmid, B. P., Kao, J., and Goulding, E. 1985. Evidence for reopening of the cranial neural tube in mouse embryos treated with cadmium chloride. *Experientia* **41**:271–272.

Schroeder, H. A., and Mitchener, M. 1971. Toxic effects of trace elements on the reproduction of mice and rats. *Arch Environ. Health* **23**:102–106.

Semba, R., Ohta, K., and Yamamura, H. 1974. Low dose preadministration of cadmium prevents cadmium-induced exencephalia. *Teratology* **10**:96–97.

Siegers, C.-P., Jungblut, J. R., Klink, F., and Oberhauser, F. 1983. Effect of smoking on cadmium and lead concentrations in human amniotic fluid. *Toxicol. Lett.* **19**:327–331.

Smith, M. J., Pihl, R. O., and Garber, B. 1982. Postnatal cadmium exposure and longterm behavioral changes in the rat. *Neurobehav. Toxicol. Teratol.* **4**:283–287.

Sonawane, B. R., Nordberg, M., Nordberg, G. F., and Lucier, G. W. 1975. Placental transfer of cadmium in rats: influence of dose and gestational age. *Environ. Health Perspect.* **12**:97–102.

Sowa, B., and Steibert, E. 1985. Effect of oral cadmium administration to female rats during pregnancy on zinc, copper and iron content in placenta, foetal liver, kidney, intestine, and brain. *Arch. Toxicol.* **56**:256–262.

Stacey, N. H., and Kappus, H. 1982. Heavy metal toxicity and lipid peroxidation in isolated rat hepatocytes. *Naunyn Schmiedebergs Arch. Pharmacol. Suppl.* **319**:R27.

Sulik, K. K., and Dehart, D. B. 1988. Retinoic acid-induced limb malformations resulting from apical ectodermal ridge cell death. *Teratology* **37**:527–538.

Tam, P. P. L., and Liu, W. K. 1985. Gonadal development and fertility of mice treated prenatally with cadmium during the early organogenesis stages. *Teratology* **32**:453–462.

Tassinari, M. S., and Long, S. Y. 1979. Cadmium-induced median facial cleft in the hamster. *Teratology* **19**:50A.

Taylor, B. A., Heiniger, H. J., and Meier, H. 1973. Genetic analysis of resistance to cadmium-induced testicular damage in mice. *Proc. Soc. Exp. Biol. Med.* **143**:629–633.

Vallee, B. L., and Ulmer, D. D. 1972. Biochemical effects of mercury, cadmium and lead. *Annu. Rev. Biochem.* **41:**91–128.

Valois, A. A., and Webster, W. S. 1987a. Retention and distribution of cadmium in the mouse brain: an autoradiographic and gamma counting study. *Neurotoxicology* **8:**463–470.

Valois, A. A., and Webster, W. S. 1987b. The choroid plexus and cerebral vasculature as target sites for cadmium following acute exposure in neonatal and adult mice: an autoradiographic and gamma counting study. *Toxicology* **46:**43–55.

Waalkes, M. P., Thomas, J. A., and Bell, J. U. 1982. Induction of hepatic metallothionein in the rabbit fetus following maternal cadmium exposure. *Toxicol. Appl. Pharmacol.* **62:**211–218.

Warner, C. W., Sadler, T. W., Tulis, S. A., and Smith, M. K. 1983. In vivo cadmium teratogenicity reproduced in whole embryo culture. *Teratology* **27:**82A–83A.

Warner, C. W., Sadler, T. W., Tulis, S. A., and Smith, M. K. 1984. Zinc amelioration of cadmium-induced teratogenesis in vitro. *Teratology* **30:**47–53.

Washko, P. W., and Cousins, R. J. 1976. Metabolism of Cd 109 in rats fed normal and low calcium diets. *J. Toxicol. Environ. Health* **1:**1055–1066.

Webster, W. S. 1978. Cadmium-induced fetal growth retardation in the mouse. *Arch. Environ. Health* **33:**36–42.

Webster, W. S. 1979a. Iron deficiency and its role in cadmium-induced fetal growth retardation. *J. Nutr.* **109:**1640–1645.

Webster, W. S. 1979b. Cadmium-induced fetal growth retardation in mice and the effects of dietary supplements of zinc, copper, iron and selenium. *J. Nutr.* **109:**1646–1651.

Webster, W. S. 1988. Chronic cadmium exposure during pregnancy in the mouse: influence of exposure levels on fetal and maternal uptake. *J. Toxicol. Environ. Health* **24:**183–192.

Webster, W. S. 1989. Alcohol as a teratogen: a teratological perspective of the fetal alcohol syndrome, in: *Human Metabolism of Alcohol,* Volume 1, *Pharmacokinetics, Medicolegal Aspects, and General Interest,* K. E. Crow and R. D. Batt, eds. CRC, Cleveland, pp. 133–155.

Webster, W. S., and Messerle, K. 1980. Changes in the mouse neuroepithelium associated with cadmium-induced neural tube defects. *Teratology* **21:**79–88.

Webster, W. S., and Valois, A. A. 1981. The toxic effects of cadmium on the neonatal mouse CNS. *J. Neuropathol. Exp. Neurol.* **40:**247–257.

Webster, W. S., Walsh, D. A., Lipson, A. H., and McEwen, S. E. 1980. Teratogenesis after acute alcohol exposure in inbred and outbred mice. *Neurobehav. Toxicol.* **2:**227–234.

Webster, W. S., Walsh, D. A., McEwen, S. E., and Lipson, A. H. 1983. Some teratogenic properties of ethanol and acetaldehyde in C57BL/6J mice: implications for the study of the fetal alcohol syndrome. *Teratology* **27:**231–243.

Williams, K. E., Roberts, G., Kidston, M. E., Beck, F., and Lloyd, J. B. 1976. Inhibition of pinocytosis in rat yolk sac by trypan blue. *Teratology* **14:**343–354.

Winneke, G., Wurms, F., Krause-Fabricius, G., and Ewers, U. 1979. Neurobehavioral deficit in rats subsequent to either maternal or maternal plus direct postnatal cadmium exposure. *Naunyn Schmiedebergs Arch. Pharmacol. Suppl.* **308:**R46.

Wolkowski, R. M. 1974. Differential cadmium-induced embryotoxicity in two inbred mouse strains. *Teratology* **10:**243–262.

Wolkowski-Tyl, R. M., and Preston, S. F. 1979. The interaction of cadmium-binding proteins (Cd-bp) and progesterone in cadmium induced tissue and embryo toxicity. *Teratology* **20:**341–352.

Wong, K. L., and Klaassen, C. D. 1980. Tissue distribution and retention of cadmium during postnatal development: minimal role of metallothionein. *Toxicol. Appl. Pharmacol.* **53**:343–353.

Wong, K. L., and Klaassen, C. D. 1982. Neurotoxic effects of cadmium in young rats. *Toxicol. Appl. Pharmacol.* **63**:330–337.

Yu, H. S., Tam, P. P. L., and Chan, S. T. H. 1985. Effects of cadmium on preimplantation mouse embryos in vitro with special reference to their implantation capacity and subsequent development. *Teratology* **32**:347–353.

Issues and Reviews in Teratology **5**: 283–316
Plenum Press, New York, 1990, 978-1-4612-7847-4

Epidemiologic Aspects of Down Syndrome

7

Sex Ratio, Incidence, and Recent Impact of Prenatal Diagnosis

CARL A. HUETHER

1. INTRODUCTION

A recent study of more than one million consecutive live births in British Columbia found that 53/1000 had diseases with an important genetic component before approximately 25 years of age (Baird *et al.*, 1988). Down syndrome was only 2.3% of this total (1.22/1000), but studies of severe mental retardation in children over 1 year of age in two Swedish counties found 33% had Down syndrome, out of a prevalence of severe mental retardation at 1 to 16 years of age of 3.5/1000 (Gustavson *et al.*, 1977a,b). Additionally, the prevalence rate of retardation dropped 42% from 1959–1962 to 1967–1970 in these studies, mainly as a result of a 54% decline in births with Down syndrome. More recent reports (Hook, 1983a; Kiely, 1987) supported these results. Baird and Sadovnick (1988a) also showed that the probability of survival of live births with Down syndrome has continued to increase, and that currently 44% survive to age 60. These studies reinforce the substantial contribution births with Down syndrome make to the problem of human health, and the positive result to be obtained from their prevention.

Different views have arisen as to the appropriate terminology for this anomaly, and although minor, they are of some interest. Over the last two decades, there has been essentially universal retreat on racial grounds from use of the term "mongolism" in favor of honoring the anomaly's original describer. Yet as Mikkelsen (1985) and others have

CARL A. HUETHER • Department of Biological Sciences, University of Cincinnati, Cincinnati, Ohio 45221-0006.

ironically pointed out, Langdon Down's racist views were clearly expressed in his original article (Down, 1866). The discovery of its chromosomal basis led some to believe the appropriate terminology is "trisomy 21," but clearly the anomaly is also caused by translocation and mosaicism. As indicated by Hook (1981a), these terms apply "operationally" to its genetic basis, whereas "Down syndrome" refers to the phenotypic diagnosis. Finally, the last decade has seen convergence on "Down syndrome" rather than "Down's syndrome."

This chapter focuses on significant aspects of the epidemiology of Down syndrome over the past two decades. As such, it mainly reviews three topics: (1) the available data on sex ratios, and their relevance to the etiology of Down syndrome, (2) incidence studies published since 1970 which show the dramatic effect of demographic changes, and (3) the effect of prenatal diagnosis in reducing incidence, as well as the effect of the availability of elective abortion, during this time period. The review of sex ratios is presented necessarily in a historical context, and is significant in that this time period has provided a number of studies in which the ascertainment of sex was minimally biased, so that the traditional view of an excess of male births with Down syndrome can be meaningfully evaluated. The use of prenatal chromosome diagnosis beginning about 20 years ago provided an opportunity for "avoiding" births with Down syndrome, and its current success can be examined by reviewing studies whose incidence data may reflect the impact of prenatal diagnosis.

One significant aspect of the epidemiology of Down syndrome not reviewed in depth here is the puzzle of the basis for the maternal-age effect. Two hypotheses have emerged as the main competitors for explaining this well-known association: (1) *prezygotic*, the "older-egg" model, i.e., an increase in nondisjunction with increasing maternal age, and (2) *postzygotic*, the "relaxed selection" model, i.e., a decreasing selection against trisomic conceptuses with increasing maternal age. Modern molecular techniques have not yet determined which (if not both) of these models is correct, but rapid progress is being made as a brief glimpse demonstrates.

Using molecular polymorphisms, DNA haplotypes have been tested for association with meiotic nondisjunction, which would support the prezygotic model, but the few results currently available are conflicting (Antonarakis *et al.*, 1985; Sacchi *et al.*, 1988). Also, several recent studies using molecular polymorphisms with or without cytogenetic markers on chromosome 21 have indicated that reduced recombination rates occur in chromosomes participating in nondisjunction, thus perhaps providing a causal effect, but it is not yet clear whether these are maternal-age related (Antonarakis *et al.*, 1986; Warren *et al.*, 1987; Stewart *et al.*, 1988). While these results could support the prezygotic model if correlated with maternal age, another study of a recurrent trisomy 21 family did not find reduced recombination rates (Hamers *et al.*, 1987), nor have reduced rates so far been found in the few trisomy 13 cases studied (Hassold *et al.*, 1987).

Other recent studies have begun using these same techniques to determine the parental origin of the extra chromosome, which will test the postzygotic hypothesis by determining whether the maternal-age effect remains in cases of paternal origin (Stewart *et al.*, 1985, 1988; Rudd *et al.*, 1988; Bricarelli *et al.*, 1988). Chromosome heteromorphisms have been used over the past 15 years to determine the parental origin of more than 1000 cases of Down syndrome (Stewart *et al.*, 1985), but the technical limitations and lack of knowledge of maternal ages in many of the studies currently provide no clear answers to the relaxed selection model proposed by Stein *et al.* (1975, 1986) and Aymé and Lippman-Hand (1982). The use of modern molecular techniques and chromosome heteromorphisms to test this hypothesis, as suggested by Hassold (1986) and Jacobs and Hassold (1987), should provide the first significant breakthrough by delimiting the etiology of Down syndrome.

For the many other epidemiologic aspects of Down syndrome not covered in this chapter, the reader is referred to several excellent recent reviews (Hassold and Jacobs, 1984; Hassold *et al.*, 1984; Mikkelsen, 1985, 1988; Janerich and Bracken, 1986; Jacobs and Hassold, 1987).

2. SEX RATIO IN DOWN SYNDROME

2.1. Historical Perspective

There is a long-standing belief that the number of males born with Down syndrome exceeds the number of females by a significant amount, both statistically and biologically, as indicated by such statements that there is a "well known excess of males in newborns" (Angell *et al.*, 1984). The idea originated in the early literature, with the great majority of individuals with Down syndrome being ascertained from "institutions for the feebleminded," which harbored a preponderance of males the world over. Church and Peterson (1924) in various editions of their text on *Nervous and Mental Diseases* beginning in 1899 stated that "there are nearly twice as many male as female idiots." Not only may there have been more mentally deficient males born than females, but mentally deficient females were often kept at home and looked after there for economic, social, and personal reasons (Bleyer, 1932). Males were generally viewed as not as easy to care for and not as easy to educate.

In the most extensive (and most cited) treatment of sex ratio to be found in the literature relating to Down syndrome, Hug (1951) emphasized this bias of ascertainment by comparing the sex ratio of individuals with Down syndrome obtained from clinical studies (134.7) with those from institutions (172.8) (his Table III). He argued that clinic studies contain essentially unbiased sex ratios of children with Down syndrome, but that newborns are the best population to study because there is a higher mortality rate of females, particularly during the first year, which is reflected in the clinic results. In four small studies of newborns with Down syndrome made during 1930–1949 he found a sex ratio of 124/68 or 1.82 (his Table X).

Hug also summarized various studies which indicated that the sex ratio in children with Down syndrome is reduced dramatically with maternal age, from 1.74 in women under 30, to 1.57 for women aged 30–39, and 1.22 for women 40 and older (his Table XII). He showed that this is also true for all live births, and argued that its basis is greater mortality of male fetuses in older women. This and other evidence led him to conclude that the primary sex ratio (at conception) is at least 2.00 for Down syndrome, or two such males conceived for every female.

2.2. Etiologic Relevance of Sex Ratio

Perhaps most important, Hug concluded that such a sex ratio would have to have etiologic consequences, and believed that its basis lay in the genetic constitution of the zygote. There are only three other papers of which I am aware that consider the etiology of the increased sex ratio in Down syndrome. Rosanoff and Handy (1934) proposed that the X chromosome in the sperm has "the power of protecting an injured ovum against its tendency to develop into a mongolian child." They felt that not all females were protected because either not all X chromosomes carried the protective genes or damage to the ovum was less severe in some cases and therefore more readily repaired.

Hassold *et al.* (1984) also considered the origin of an increased sex ratio for trisomy 21, and supported the conclusion of Hug that it is likely to occur at conception. They reached this conclusion by associating the increased sex ratio with the increased proportion of cases of paternal origin in trisomy 21, which they believed was not the result of differential intrauterine selection. Data were presented on parental origin and sex, which indicated a significant excess of males associated with nondisjunction at paternal meiosis I (18/7). They concluded that a nondisjunctional mechanism may exist in which the extra chromosome 21 preferentially segregates with the Y.

Other studies were inconsistent in this regard. Mikkelsen (1988) provided data which are very supportive (17/5), but also stated that her own data were in contrast with those of the collaborative European study, which did not show an increased sex ratio for paternal meiosis I failures. Also, Mikkelsen (1982) summarizing data from several studies, found more females than males from paternal meiosis I (14/17). Regardless, the mechanism proposed by Hassold *et al.* (1984) is a different mechanism from that Hug (1951) would have envisioned, but they both accept that some mechanism is acting prezygotically, in contrast with Rosanoff and Handy (1934), whose mechanism would have to be acting postzygotically.

The third paper (Lindsten *et al.*, 1981) proposed two possible explanations for an increased sex ratio in Down syndrome. One was based on the finding of Drew *et al.* (1978) that there was an excess of males when one parent had the hepatitis B antigen. Thus, they suggested "a change in the cellular membrane might make the ovum more likely to become fertilized by a Y sperm, and to undergo an abnormal cell division." No explanation was given as to why the abnormal cell division (i.e., nondisjunction) should also occur. Their second hypothesis was that "delayed fertilization could lead to both nondisjunction and a greater chance for the ovum to become fertilized by a Y sperm." Harlap (1979) provided evidence for an increased sex ratio when conception occurred 2 d after ovulation. The hypothesis requires an increased sex ratio in meiosis II nondisjunctions, regardless of the parental origin of the extra chromosome. Data from Hassold *et al.* (1984) and Mikkelsen (1982) provided some support for this (18/15 and 32/22, respectively, for an overall sex ratio of 1.35), but since meiosis II nondisjunction accounts for only 20–25% of trisomy 21, a much higher ratio would be needed. Both of these hypotheses by Lindsten *et al.* (1981) utilize prezygotic mechanisms, but contrast with those Hassold *et al.* (1984) and Hug (1951) proposed.

2.3. Sex Ratios in Total Live Births

To place the rest of the discussion in perspective, a brief commentary is needed on the sex ratio in live births generally (the secondary sex ratio), and its variations in space and time. The major factor affecting the well-documented secondary sex ratio appears to be race, with a low of 1.02 in American Indians (Khoury *et al.*, 1984), 1.03–1.04 in Caribbean, African, and U.S. blacks (Visaria, 1967), 1.05–1.07 in European and U.S. whites (Hytten, 1982), and higher ratios in Asian races, with a figure of 1.15 reported in Korea (Kang and Cho, 1962). An interesting study of interracial crosses among whites, blacks, and Indians in the United States found the father's race and not the mother's as the significant factor in determining observed racial differences in the sex ratio (Khoury *et al.*, 1984). This could be due to meiotic drive, gametic selection, different coital habits, or differential intrauterine sex survival. Multiple births were also found to have a major effect in six European countries, with significantly lower sex ratios of 0.96 among triplets and other plural births, 1.03 among twins, and 1.06 among single births (Colombo, 1957).

Other factors are also known to be significantly associated with sex ratio in human live births, but their effect is less dramatic, and the underlying basis unknown. Examples include increased sex ratios occurring after many but not all wars (Colombo, 1957), a negative correlation of sex ratio with increasing birth order (Teitelbaum *et al.*, 1971), and an increased sex ratio when coitus occurs before or after ovulation, but a reduced one at the time of ovulation (Guerrero, 1974; Harlap, 1979). Neither paternal nor maternal age appears to be important in affecting sex ratio once birth order is taken into account (Erickson, 1976). The control value used below for comparing the sex ratio in births with Down syndrome was 1.06, obtained by Khoury *et al.* (1984) in analyzing approximately 14 million white U.S. live births.

Several recent studies determining the fetal sex ratio found an increased ratio in normal spontaneous abortions [1.30 by Hassold *et al.* (1983), 1.30 by Byrne and Warburton (1987), and 1.51 by Honoré (1988)], but a much decreased one in malformed or chromosomally abnormal abortuses (1.13 and 0.92 in the first two studies, respectively). Previous studies indicating decreased sex ratios in normal spontaneous abortions were suggested by Hassold *et al.* (1983) to be due to maternal contamination. If a sex ratio of 1.30 is assumed for the 40% of all conceptuses believed to spontaneously abort (including those unrecognized) and 1.06 for the 60% of live births, the primary sex ratio (at conception) would be approximately 1.15.

Data from several studies of induced abortions are consistent with this estimate. For example, Kellokumpu-Lehtinen and Pelliniemi (1984) found a sex ratio of 1.17 in induced abortions (most <15 weeks) in southwest Finland, with the ratio decreasing as the duration of pregnancy increased. Our own (unpublished) data on amniocenteses (16–20 weeks) from Ohio are somewhat lower, with an overall sex ratio for whites of 1.05 (1996/1880), and for nonwhites 0.93 (300/322), while similar data from Atlanta were somewhat higher at 1.13 (2488/2196) for whites and 1.08 (576/534) for nonwhites. Neither of the Atlanta estimates is statistically different from the comparative ratio in Ohio, however, and all are generally consistent with live birth ratios of 1.06 for whites and 1.03 for blacks.

2.4. A Critique of Available Sex Ratio Data for Trisomy 21

Most monographs on the epidemiology of Down syndrome published in the last few decades have been either mute on the question of sex ratio (Lilienfeld, 1969; de la Cruz and Gerald, 1981; Pueschel and Rynders, 1982; Pueschel *et al.*, 1987), or else simply mention a predominance of males (Penrose and Smith, 1966; Smith and Berg, 1976). Two include studies containing data on sex ratios, which will be considered below (Apgar, 1970; Burgio *et al.*, 1981). One of the earlier monographs (Oster, 1953) includes Hug's (1951) data, as well as those from a previous study showing a sex ratio of 1.15 among 880 individuals with Down syndrome in institutions. Oster's own investigations of individuals with Down syndrome found 204 boys and 183 girls admitted to children's hospitals during 1923–1949, 133 males and 102 females among institutionalized patients, and 136 males and 155 females living at home. Oster concluded there was "no significant preponderance of male mongols" (473 boys and 440 girls, a sex ratio of 1.08). These data were not population based, so no prevalence rates were provided.

If the increased sex ratio in Down syndrome is real, rather than apparent, we are no closer to understanding its biologic basis and etiologic significance than was Hug almost 40 years ago. There have been almost no critical evaluations of the available data to judge the validity of this accepted notion, and in the plethora of writings on the etiology of Down syndrome, little to integrate the origin of an increased sex ratio into a biologic model. If the data are such as to justify the acceptance of an increased sex ratio, future work on the etiology of Down syndrome would benefit by considering its basis.

In examining the validity of the sex ratio record, one feels obliged to dismiss essentially all of the earlier data, even those of Hug (1951) and Oster (1953). The obvious bias of ascertainment associated with data from institutions for the mentally retarded, and the unknown representativeness of boys and girls with Down syndrome in clinic populations, the two sources for almost all of the early studies, seem to demand it. Consideration of a potential differential mortality between boys and girls was also not taken into account, although Baird and Sadovnick (1987) did not find significant sex differences in survival when comparing those with Down syndrome who had congenital heart disease and those who did not. However, they did find a significantly higher proportion of females so afflicted, and that those with congenital heart disease had a significantly higher mortality rate.

Additionally, estimates of the prevalence of Down syndrome in these early studies were quite low, even given that they were based on children surviving high infant mortality rates. Bleyer (1932) estimated a prevalence rate of approximately 1/4300, while estimates in Germany were 1/6800 to 1/7200 (Jervis, 1942). Even the small sets of data on newborns that Hug (1951) reported are suspect, in that no information was provided on how they were collected, nor, in two of them, the number of total live births in the population studied. However, it is of interest that for the two populations for which he did provide total live births, the incidence was 1.92/1000, a quite reasonable value for the 1930–1949 time period. Finally, of course, none of these studies included karyotyping as the definitive determination of Down syndrome.

Available data, then, on which the sex-ratio hypothesis can best be evaluated are those from studies conducted in the last 30 years or so, most of which included cytogenetic analysis. All studies containing sex ratios of (mostly) live births with Down syndrome that I could find in the literature are presented in Table I. Of the 30 studies reported in these 27 papers, 3 (listed in Table I footnotes) were subsumed in later papers. Additionally, 2 were based on patients in institutions for the mentally deficient, 6 on children's clinics, and 6 on censuses of mentally handicapped in the population.

Each of the latter 14 studies has either the actual or potential biases associated with the early studies indicated above. That is, they suffer from the method of ascertainment, underascertainment, differential mortality rates of males and females, and/or lack of cytogenetic analysis to exclude false positives and translocations. Additionally, small sample size in several studies limits robustness. The estimated sex ratios ranged from a low of 1.04 to a high of 1.85, with the median between 1.21 and 1.30. Two of the 14 data sets (Huang *et al.*, 1967; Bernheim, *et al.*, 1979) gave ratios statistically higher than the 1.06 sex ratio of live birth controls. Huang *et al.* (1967) contacted "physicians, hospitals, clinics, orphanages, and retarded children's classes" to solicit cases, with no attempt at complete ascertainment. In the Bernheim *et al.* (1979) study, all children included were karyotyped, approximately half of whom were less than 6 months old, and half over 6 months, with increasing sex ratios up to the age of 10. Essentially no information was provided on how the sample was obtained, and no incidence rates were given, as it was not a population-based study. Thus, none of these 14 studies provided adequate data to support a significantly increased sex ratio in live births with Down syndrome.

The 13 remaining studies were all based on live birth data, on examination of consecutive live births in specific hospitals (Wahrman and Fried, 1970; Hafez *et al.*, 1984), on multiple sources in the postnatal period (Kashgarian and Rendtorff, 1969; Gallagher and Lowry, 1975), or on various live birth registries. However, in four of these studies the data appear suspect. Although Kashgarian and Rendtorff (1969) ascertained births with Down syndrome from several sources, their observed incidence of 0.95/1000 for Caucasoids during the 1955–1966 time period is well below expectations. Also, an unknown (but probably small) number were cytogenetically analyzed, so that potential false positives were included. Mikkelsen *et al.* (1980) did not provide incidence data for the 78 patients with Down syndrome registered with the Danish government's service for the mentally retarded on the island of Funen. Thus, the level of ascertainment cannot be determined, nor whether sex biases in registration with the service, or with the 20% not chromosomally analyzed, occurred.

Lejeune and Prieur (1979) used only trisomy 21 cases in a retrospective study aimed at determining the frequency of oral contraceptive use in their mothers. No information was provided on how the cases, all less than 2 years old, were obtained by the cytogenetic laboratory. Lastly, Nielsen *et al.* (1981) used data only from the Danish Cytogenetic Central Register, which contained "the great majority" of chromosome abnormalities from Denmark's six cytogenetic laboratories, but no incidence data were provided, and potential sex biases in ascertainment cannot be determined. One of these four studies showed a statistically significantly increased sex ratio (1.27, Nielsen *et al.*, 1981), and two others were high (1.38 and 1.50) but not significantly so. Lejeune and Prieur (1979) found significant heterogeneity between the increased sex ratio (1.16) in the children of mothers not taking oral contraceptives and the reduced sex ratio (0.81) in those of the user group, with an overall sex ratio of 1.07.

The remaining nine studies, or 30% of those found in the literature of the last 30 years to have data on sex ratio in Down syndrome, appear to have high or complete ascertainment of karyotyped live births with trisomy 21, and thus contain few if any biases. All or most of the cases were chromosomally analyzed in all but one study (Gallagher and Lowry, 1975). Statistically, two of the nine showed a significantly increased sex ratio (1.27, Iselius and Lindsten, 1986; 1.35, Mikkelsen, 1985). However, Mikkelsen stated that when the population was divided into areas of high and low incidence rates, the sex ratio was significantly increased (1.7) only in the high-incidence area, suggesting that the birth of a surplus of males may be associated with high incidences. Two additional studies had increased ratios (1.24, Evans *et al.*, 1978; 1.28, Hafez *et al.*,

Table I. Studies Published since 1960 That Include Data on the Sex Ratio of Live Births with Down Syndrome

Time period	References	Location	Source	Males	Females	Sex ratio	Statistical significance[a]
1901–1979	Nielsen *et al.* (1981)	Denmark	Survivors in cytogenetic register	1056	875	1.21	**
1960–1979	Nielsen *et al.* (1981)	Denmark	Live births in cytogenetic register	444	350	1.27	**
1955–1959	Forssman and Akesson (1965)	Sweden	Patients in MR[b] institutes	681	582	1.17	ns
1955–1966	Kashgarian and Rendtorff (1969)	Memphis-Shelby Co., Tenn.	Multiple sources of live births	58(wh) 43(bl)	42 49	1.38 0.88	ns ns
1952–1971	Gallagher and Lowry (1975)	British Columbia	Live births registry	479	448	1.07	ns
1959–1970	Gustavson *et al.* (1977a)	Uppsala, Sweden	Census of MR persons born 1959–1970	24	15	1.60	ns
1959–1970	Gustavson *et al.* (1977b)	Vasterbotten, Sweden	Census of MR persons born 1959–1970	32	21	1.52	ns
Early 1960s	Chitham and MacIver (1965)	Carshalton, Great Britain	Adults in MR institutes	57	48	1.19	ns
Early 1960s	Huang *et al.* (1967)	Taipei, Taiwan	Population of Taipei	48	26	1.85	*
1961–1969	Uchida (1970)	Manitoba, Canada	Total population of Manitoba	253	237	1.07	ns
1960–1971	Mikkelsen *et al.* (1976)	Copenhagen, Denmark	Live births registry	77	85	0.90	ns
1952–1981	Baird and Sadovnick (1987)	British Columbia	Health surveillance registry (live births)	703	638	1.10	ns

1965–1969	Wahrman and Fried (1970)	Jerusalem, Israel	Hospital live births	28	25	1.12	ns
1960–1975	Mikkelsen *et al.* (1980)	Danish island of Funen	Live births registry	36	24	1.50	ns
1976–1979	Mikkelsen *et al.* (1980)	Danish island of Zealand	Clinic	27	18	1.50	ns
1960–1976	Bernheim *et al.* (1979)	Paris, France	Clinic and cytogenetic laboratory	384	283	1.36	**
1965–1974	Evans *et al.* (1978)	Manitoba, Canada	Live births registry for congenital anomalies	130	105	1.24	ns
1968–1977	Lejeune and Prieur (1979)	Paris, France	Clinic and cytogenetic laboratory	377	351	1.07	ns
1968–1982	Sharov (1985)	Jerusalem, Israel	Clinic	29	26	1.12	ns
1968–1982	Iselius and Lindsten, (1986)[c]	Sweden	Multiple national registers of live births	1110	873	1.27	**
1970–1980	Verma and Huq (1987)	Brooklyn, N.Y.	Clinic	42	33	1.30	ns
1975–1985	Murthy *et al.* (1987)	Gufarat, India	Clinic	34	26	1.31	ns
1978–1983	Hafez *et al.* (1984)	Mansoura, Egypt	Hospital live births	18	14	1.28	ns
1978–1983	Hafez *et al.* (1984)	Mansoura, Egypt	Clinic	126	110	1.14	ns
1978–1983	Mikkelsen (1985)[d]	Denmark	Live births registry	167	123	1.35	
1981	Mulcahy and Reynolds (1985)	Ireland	Census of MR persons in population	1812	1747	1.04	ns
1984–1985	Mikkelsen (1988)[e]	Denmark	Live births registry	56	49	1.14	ns

[a]ns, not significant at $P > 0.05$; *, significant at $P < 0.05$; **, highly significant at $P < 0.01$.
[b]MR, mental retardation.
[c]Includes data for 1968–1977 from Lindsten *et al.* (1981), which includes data for 1968–1970 from Lindsjo (1974).
[d]Includes data for 1980–1982 from Pilgaard and Mikkelsen (1985).
[e]Excludes data for 1980–1983 presented in Mikkelsen (1985).

1984), but they were not statistically significant, while five studies were similar to the control expectation of 1.06 (1.12, Wahrman and Fried, 1970; 1.07, Gallagher and Lowry, 1975; 0.90, Mikkelsen *et al.*, 1976; 1.10, Baird and Sadovnick, 1987; 1.14, Mikkelsen, 1988).

Thus, the whole of our long-standing acceptance of the higher sex ratio in live births with Down syndrome is based on approximately nine epidemiologically sound studies, of which two showed statistical significance, two were suggestive, and five provided no support. The average sex ratio for these nine studies was 1.17, but this cannot appropriately be tested against the control value of 1.06 since there was heterogeneity among the data sets. There are 15 additional studies with suspect data, of which only two showed statistically significant higher ratios, and half provided no support. Although these studies certainly should not be dismissed, neither does it appear appropriate to conclude that an increased sex ratio in Down syndrome has been conclusively demonstrated.

2.5. Current Live-Birth Data from Atlanta and Southwest Ohio

Two highly ascertained, but as yet not fully published, data sets contain additional information on the question of sex ratio in live births with Down syndrome. The Atlanta data are from the Metropolitan Atlanta Congenital Defects Program (Edmonds *et al.*, 1981), and the SW Ohio data were obtained through multiple sources of ascertainment (Krivchenia, 1988). Both are for the period 1970–1985, and gave incidence rates for white live births with Down syndrome of 1.08/1000 and 1.21/1000, respectively. Sex ratios in these two data sets for the 16-year period were 1.21 (151/125) and 0.97 (255/264), respectively, neither statistically different from the control ratio of 1.06. Thus, these data support the above conclusion that there is no statistical evidence for an increased sex ratio, although the Atlanta data are suggestive. Other data from Ohio during 1970–1987 are useful for showing the type of biases that may exist within data sets. The sex ratio based on individuals ascertained only through cytogenetic laboratories was 1.42 (701/493), while the ratio for those ascertained only through birth certificates was 0.83 (224/269). The two ratios are highly significantly different ($\chi^2 = 24.8$, $P < 0.01$). These data showed a temporal effect as well, in that this difference was more pronounced in the early 1970s and diminished into the 1980s. They suggest, for the early to mid 1970s particularly, that physicians and parents may have been more interested in a definitive diagnosis for males thought to have Down syndrome, while being less willing to publicize it, whereas the opposite was true for females.

2.6. Sex Ratio in Translocation Down Syndrome

Only eight publications (shown in Table II) were found in which the sex ratio of at least ten individuals with translocation Down syndrome was given. Three other studies had seven males and six females (Tonomura *et al.*, 1966; Huang *et al.*, 1967; Slavin *et al.*, 1967). Of the former eight, three have been suggested as having potentially biased data (Nielsen *et al.*, 1981; Uchida, 1970; Bernheim *et al.*, 1979). Three additional studies also had potential ascertainment biases, because of a deliberately selected group (Mikkelsen, 1970) or an unknown population base and method of ascertaining cases (Hook, 1981b; Pulliam and Huether, 1986). Three of the four studies providing data for both translocation and trisomy Down syndrome showed a lower sex ratio for translocations than for trisomies (0.82 versus 1.21; 1.20 versus 1.36; 1.10 versus 1.29).

The two remaining studies appeared to have essentially complete ascertainment and high percentages of chromosome analysis (84%, Iselius and Lindsten, 1986; > 90%, Mikkelsen, 1988). Ironically, they represent the extremes of sex ratios in translocation Down syndrome (0.87, 46/53; and 5.67, 17/3, respectively), although small numbers are an obvious limitation in the latter study. The sex ratio for trisomy Down syndrome in these two studies was 1.27 and 1.33, respectively. In sum, the data in these eight studies appear to be no better than the trisomy data in allowing definitive conclusions regarding the sex ratio in translocation Down syndrome. Thus, the null hypothesis that the translocation sex ratio is no different than that of all live birth controls should not be considered either accepted or refuted at this time.

Nielsen *et al.* (1981) pointed out the relevance of these data to the question of etiology by indicating that factors influencing sex ratio might be different in trisomy and translocation Down syndrome. Hassold *et al.* (1983) went further and suggested that the presumed sex ratio differences mean that no differential selection is occurring *in utero* against the male fetus with Down syndrome, and thus that the excess of males is present at conception. This, coupled with the view that more males are associated with nondisjunction at paternal meiosis I, as suggested above, is what led Hassold *et al.* (1984) to the suggestion of a nondisjunctional mechanism in which the extra chromosome 21 preferentially segregates with the Y chromosome. Further, the greater frequency of trisomy 21 associated with paternal origin, and the differences in maternal-age effect among all trisomies, led Hassold (1986) to propose that different mechanisms of nondisjunction exist, particularly among the different autosomal chromosomes.

Table II. Studies That Include Data on the Sex Ratio of Live Births with Translocation Down Syndrome

Time period	References	Location	Source	Translocations			Trisomics		
				Male	Female	Sex ratio	Male	Female	Sex ratio
1901–1979	Nielsen *et al.* (1981)	Denmark	Survivors in cytogenetic register	65[a]	79[a]	0.82	1056	875	1.21
1960s	Mikkelsen (1970)	Denmark	Referrals and screening of young mothers	12	14	0.86	—	—	—
1961–1969	Uchida (1970)	Manitoba, Canada	Total population of Manitoba	8	7	1.14	253	237	1.07
1960–1976	Bernheim *et al.* (1979)	Paris, France	Clinic and cytogenetic laboratory	18	15	1.20	384	283	1.36
1969–1980	Hook (1981b)	New York and New England	New York State chromosome registry	182	117	1.56	—	—	—
1968–1982	Iselius and Lindsten (1986)	Sweden	Live births registry	46	53	0.87	1110	873	1.27
1970–1981	Pulliam and Huether (1986)	Ohio	Cytogenetic laboratory	34	31	1.10	664	513	1.29
1980–1985	Mikkelsen (1988)	Denmark	Live births registry	17	3	5.67	156	117	1.33

[a]Includes mosaics.

2.7. Effect of Maternal Age on Sex Ratio in Down Syndrome

Based on both the lack of quality data and the inconsistencies in the quality data that are available, it is not surprising that the literature is unclear on the relationship between maternal age and sex ratio. Several studies prior to 1960 calculated a separate average maternal age for female and male births with Down syndrome (see Perry, 1971, for a tabular summary), but the results were inconclusive and the data sources of questionable validity. Brief accounts of five data sets were published in 1971 (Largey and Largey, 1971; Spencer, 1971; Perry, 1971; Qazi and Lanman, 1971; Ros *et al.*, 1971), one showing a significantly lower maternal age for females, one a significantly higher, two no difference, and one untested. Since these data sets were simply indicated to originate from a hospital, clinic, institution, and/or cytogenetic laboratory, their validity is suspect.

Only two studies published during the past 20 years included sex ratio data on live births with Down syndrome by maternal age. Verma and Huq (1987) presented data that unfortunately were referrals to a medical center in Brooklyn, i.e., were not population based, and thus potentially contain unknown biases. Therefore, their data [showing a sex ratio of 2.06 (33/16) for mothers <35 years old and 0.36 (5/14) for mothers ≥35] are of little value. By contrast, Iselius and Lindsten (1986) presented data that appear to be of excellent quality, with essentially complete ascertainment of live births with Down syndrome in Sweden during 1968–1982. Their data show a sex ratio of 1.25 (769/613) to mothers <35 years of age, and 1.31 (341/260) to mothers ≥35, indicating good agreement between these two maternal-age categories. However, mothers <25 years old had a sex ratio of 1.09 (224/205), with maternal age quinquennia between 25–29 and40–44 varying from 1.28 to 1.34, and those mothers ≥45 having a ratio of 1.71 (24/14), rather clearly indicating an excess of males to older women, and no excess to women <25 years of age.

Two other studies are supportive of these results. Mikkelsen (1985) did not present sex ratios by maternal age, but divided the sex ratio data of 1.35 (167/123) for Denmark during 1978–1983 into areas of high and low incidence. In the high-incidence areas the sex ratio was 1.70, while in the low-incidence area 1.16, not statistically different from live birth controls of 1.06. This might best be viewed as indirect supporting evidence, in that other explanations besides increased maternal age could be the basis of the high incidence. More directly, the (unpublished) data from SW Ohio show a significantly different sex ratio of 0.87 (177/204) for women <35 years of age compared with a ratio of 1.30 (78/60) for those ≥35 (heterogeneity $\chi^2 = 4.1$, $P < 0.05$).

Should this turn out to be a general phenomenon, it may help explain the broad range of sex ratios found in even the epidemiologically sound studies as due to substantial differences in maternal-age structure of the populations, both geographically and temporally. It would also give at least the next level of understanding as to why there is an increase in the sex ratio of live births with Down syndrome, assuming this is eventually shown to be true. Clearly, researchers who have access to high-quality data sets on Down syndrome should look at this possibility.

2.8. Sex Ratio in Trisomy 21 Fetuses Diagnosed Prenatally

Publications of five studies were found that have reported the sex ratio in trisomy 21 fetuses (summarized in Table III), four based on amniocentesis, and one on chorionic villus sampling (CVS). Concerns regarding ascertainment bias associated with live births are not applicable here, although there remain the obvious biases concerning the reasons women choose to have prenatal diagnosis. Most notably these include advanced maternal age and family or personal history of genetic or chromosomal anoalies, but also knowledge about the procedure and accessibility to it. However, as indicated in Section 2.3, overall amniocentesis data available show a sex ratio only slightly higher (approx. 1.10 for whites) than that for live births, and thus are consistent with them. Differences in the maternal-age distribution of women utilizing prenatal diagnosis in these five studies could perhaps account for a small portion of their variation in sex ratios, assuming at least a partial correlation between the two. The combined Ohio and Atlanta data suggested such a correlation, in that all fetuses carried by women <35 had a sex ratio of 1.04 (1646/1579), while women >35 had fetuses with a ratio of 1.10 (4386/3981). Only the CVS study reported the percentage of women >35 years of age utilizing the procedure (78%), but it is safe to assume that amniocentesis was also used by a large majority of women >35 in the other studies.

One of the data sets (Mikkelsen, 1981) gave a sex ratio of 1.95 for fetuses with trisomy 21; this is statistically significantly different from the live birth control of 1.06. Unfortunately, little more can be said regarding these data, as the original German Chromosome Registry report of 1977 is unavailable to me. The results of the other four studies were reasonably consistent, and quite similar to those obtained with live births. It is of interest to note that the sex ratio of fetuses with Down syndrome found through CVS was 1.03 (Mikkelsen and Aymé, 1987), not different from that found in normal live births. Additional data from

Table III. Studies That Include Data on the Sex Ratio of Trisomy 21 Fetuses Detected through Prenatal Diagnosis

Time period	References	Location	Source	Males	Females	Sex ratio	Statistical significance[a]
1970s	Mikkelsen (1981)	Germany	German registry	41	21	1.95	*
1970–1979	Nielsen *et al.* (1981)	Denmark	Cytogenetic registry	27	25	1.08	ns
1968–1982	Iselius and Lindsten (1986)[b]	Sweden	Cytogenetic registry	86	68	1.26	ns
1980–1982	Pilgaard and Mikkelsen (1985)	Denmark	Cytogenetic registry	32	31	1.03	ns
1980s	Mikkelsen and Aymé (1987)	Worldwide	CVS[c] collaborative study	31	30	1.03	ns
			Total	217	175	1.24	

[a]ns, not significant at $P > 0.05$; *, significant at $P < 0.05$.
[b]Includes data from Lindsten *et al.* (1981).
[c]CVS, chorionic villus sampling.

our work in Ohio on 131 fetuses with Down syndrome ascertained through amniocenteses performed during 1972–1987 showed a sex ratio of 0.90 (62/69), somewhat lower than the published ratios. Data from metropolitan Atlanta for the same time period showed an even lower sex ratio, 0.64 (23/35). The mean sex ratio for the seven studies (including Ohio and Atlanta) is 1.08, which is not significantly different from the 1.17 ratio obtained by averaging the nine studies of live births with Down syndrome discussed earlier. Clearly, many more data have been and are being collected through prenatal diagnosis, and it is hoped that future reports will include the sexes of fetuses diagnosed with Down syndrome, as well as their mothers' ages.

Of relevance here also are the data on sex ratios observed in studies of spontaneous abortions. Hassold *et al.* (1983) estimated the sex ratio among chromosomally normal spontaneous abortuses to be 1.32 (238/180), and among all autosomal trisomies to be 1.15 (447/390). They also summarized five studies having data on the sex ratio of aborted fetuses with trisomy 21, finding a value of 1.67 (50/30). Assuming these rather small numbers accurately reflect the true ratio, this would be good supporting evidence for the position of Hug (1951) and Hassold *et al.* (1984) presented earlier for an increased sex ratio in conceptions with Down syndrome, as these data clearly suggest a differential loss of male fetuses during gestation. However, it is in direct contrast with the sex ratio of 1.03 for fetuses with Down syndrome found through CVS, particularly as Hassold *et al.* (1980) found that 24 of the 26 fetuses with Down syndrome in their study spontaneously were aborted after 11 weeks of gestation (with a sex ratio of at least 1.40). Only 6.6% of the CVS occurred after 11 weeks (Mikkelsen and Aymé, 1987). Thus, to have the substantial differential loss of male fetuses with Down syndrome suggested by the spontaneous abortion data, and still obtain a live birth ratio approximating 1.17, would require a sex ratio at CVS well above what this admittedly small set of data has indicated. The same would be true for the five amniocentesis studies with a mean sex ratio of 1.16, since over a quarter (7/26) of the spontaneous abortions in the Hassold *et al.* (1980) study were beyond 20 weeks. Clearly, this will be an exciting area for research activities in the immediate future.

2.9. Summary and Conclusions for Sex Ratio in Down Syndrome

Historically, there has been widespread acceptance in the literature that there is an excess of male births with Down syndrome. Few of the studies considered the question of the etiology of this putative increased sex ratio, but authors who did so mostly proposed a prezygotic rather than postzygotic mechanism. A critique of some 30 studies published since 1960 and two unpublished data sets revealed that only 11 of these studies may have epidemiologically sound data sets containing information on the sex ratio in Down syndrome live births. Of these, 2 showed a statistically significant increased sex ratio, 3 were suggestive of an increased ratio but not statistically different from controls, and 6 provided no support for an increased ratio.

There are even fewer valid studies providing data on the sex ratio for translocation Down syndrome, sex ratio by maternal age, trisomy 21 fetuses diagnosed prenatally, and the sex ratio of fetuses with Down syndrome spontaneously aborted. The null hypothesis that none of these sex ratios is different from controls has yet to be definitively rejected. A suggestion to emerge from three recent studies is that there may be a maternal age effect, with older women having a higher sex ratio in births with Down syndrome than younger women. Final determination of the sex ratio in Down syndrome and its importance in understanding the etiology of this condition will come only through publication of appropriate data from the relatively few completely ascertained data sets for Down syndrome that exist throughout the world.

3. INCIDENCE OF DOWN SYNDROME AND EFFECTS OF DEMOGRAPHIC CHANGES

3.1. Historical Perspective

The considerable literature currently available on the incidence of Down syndrome was initiated by Jenkins (1933). His estimate of 1.57/1000 births was surprisingly close to the incidence expected according to the demographics of that time, given that his estimate was based on a total of *six* births with Down syndrome! Since then, well over 100 studies have been reported, almost all of them ascertaining many more births with Down syndrome than did Jenkins, but many also with estimates widely missing the mark. The lowest estimate of which I am aware was 0.32 (Gentry *et al.*, 1959), based on birth certificate ascertainment of more than one million live births in New York State. The highest is more than an order of magnitude above this, being 3.64 in a recent study in Kuwait based on 25 births with Down syndrome (Farag *et al.*, 1988), which undoubtedly represents little more than a random cluster.

Even though almost all studies reported estimates of 1.00 to 2.00 during the 1920s–1980s, little appreciation is gained for the enormous change in incidence of Down syndrome that occurred throughout most of the industrialized world during this 60-year period. The reason for this failure can be found in one or more of the numerous pitfalls encountered in all but a few of these studies. As pointed out most recently by Baird and Sadovnick (1988b), the major confounding factors in incidence studies such as these are underascertainment (with a potential problem of false positives as well as false negatives), small number of births resulting in considerable random error, lack of consideration of population age structure and age-specific fertility rate changes, and, most recently, inability to account for the impact of prenatal chromosome diagnosis. In this light, it is surprising there is the high level of uniformity generally found among these estimates. The evidence indicates that one or more of these four factors can account for all of the differences occurring among these studies.

3.2. Effects of Demographic Changes

The existence of problems with ascertainment and small sample size has of course been widely recognized from the beginning. A major advantage of studies reported during the past 20 years lies not only in better ascertainment and larger samples, but also in the increasing awareness of the importance of demographic changes as a critical determinant of the incidence of Down syndrome. Evidence for their importance is shown by 11 studies published since 1970, which allow comparisons between the reduction in incidence rates over time with reductions in the percentage of births with Down syndrome to women ≥ 35 years of age (see Table IV). It is a striking indication that as fewer live births have occurred to older women, their percentage of births with Down syndrome has dropped precipitously, which in turn has resulted in substantial reductions in incidence rates. These studies, and particularly the study of the changes in the United States by Huether and Gummere (1982), reinforce that demographic events, modified by increasing use of amniocentesis, adequately account for the incidence rate changes observed.

Two factors have been behind the fewer number of live births to older women, particularly in the United States, from the mid-1940s through the 1970s. One has been the sizable shift in the population age structure, caused by the post-World War II baby boom, which has led to

Table IV. Comparison of Changes in Birth Incidence of Down Syndrome and Percentage of Such Births to Women ≥ 35 Years Old among Studies in Different Populations and Time Periods

Region	Reference	Period	Changes in incidence of Down syndrome (per 1000 live births)	Percentage of births with Down syndrome to women ≥ 35 years old
United States	Huether and Gummere (1982)	1920–1979	2.42 → 1.14	64% → 25%
British Columbia	Lowry *et al.* (1976)	1952–1973	1.38 → 1.17	54% → 22%
South Australia	Sutherland *et al.* (1979)	1955–1977	1.25 → 0.90	66% → 30%
Copenhagen	Mikkelsen *et al.* (1976)	1960–1971	constant (~1.10)	34% → 20%
Ontario	Zarfas and Wolf (1979)	1960–1974	1.34 → 0.83	46% → 8%
Japan	Tanaka (1969), Kuroki *et al.* (1977)	1962–1975	1.11 → 0.98	32% → 20%
Manitoba	Evans *et al.* (1978)	1965–1974	constant (~1.14)	constant (~35%)
England	Owens *et al.* (1983)	1961–1979	1.69 → 1.07	55% → 28%
British Columbia	Baird and Sadovnick (1988b)	1964–1983	1.53 → 1.28	~50% → 28%
South Belgium	Koulischer and Gillerot (1980)	1971–1978	1.27 → 1.19	52% → 33%
Sweden	Iselius and Lindsten (1986)	1968–1982	constant (~1.46)	constant (~38%)

there being many more young women of childbearing age. The second factor has been a long-term decline in the age-specific birth rates of older women, which actually began in the early part of this century, and has only in the last decade or so begun to be reversed, as a result of sociological changes. Entries in Table IV reflect all of this, not only for the United States, but for the other countries indicated as well. In the two populations (Manitoba and Sweden) where the percentage of births with Down syndrome were constant, so too were the incidence rates. In eight of the nine populations where the percentage dropped, the incidence was also lowered. In the one population (Copenhagen) where the incidence did not change, the authors themselves suggested that ascertainment in 1960–1962 was not likely to have been complete (Mikkelsen *et al.*, 1976).

3.3. Incidence Rate Projections in the United States

Because of these important demographic effects on the past and current incidence of Down syndrome, it is instructive to consider briefly what may be in store for the next decade. Two studies have projected the number and incidence of births with Down syndrome in the United States for this time period (Huether, 1983; Goodwin and Huether, 1987). They project "baseline" births, on which there is assumed to be no effect of prenatal diagnosis. Figure 1 is from the earlier paper, which graphically indicates the historical perspective. Both this and the more recent paper projected an increase in incidence from approximately 1.18/1000 in 1975 to 1.32/1000 in 1990 and 1.42/1000 in the year 2000, or a 20% increase in 25 years. In the United States, the number of births with Down syndrome changes more dramatically from 1975 to 1990, increasing from approximately 3700 to 5200, a 40% increase, but actually drops slightly during the next 10 years. These are based on Census Bureau projected changes in population age structure and age-specific birth rates only.

The major effect of these demographic changes will be that births with Down syndrome to women ≥ 35 years of age will increase from approximately 29% of all Down syndrome births in 1975 to 32% in 1990 and to 39% by the year 2000, an increase of one third. Past and present utilization of amniocentesis being primarily by these women suggests an increased impact of this procedure in avoiding births with Down syndrome. Projected reductions made by these studies will be presented below, along with results observed to date.

Figure 1. Baseline incidence (*—*) and number (▷—▷) of predicted births with Down syndrome 1920–1978, and projected births with Down syndrome 1979–2000 in the United States, assuming no prenatal diagnosis utilization. Reprinted with permission of the *American Journal of Public Health.*

3.4. Environmental Effects on Incidence Rates

Consistent with demographic changes accounting for changes in incidence and number of births with Down syndrome are the numerous studies that have considered but not found environmental agents as causative factors in Down syndrome. Radiation exposure and viral infections are the most favored of such influences, but studies have also considered high fluoride content of water, atmospheric pollution, cigarette smoking and alcohol use by mothers, eating and sexual habits, socioeconomic level, population size, presence of wild animals in the household at the time of conception, and the number of people arrested annually for alcohol intoxication, to name a few! The International Commission for Protection against Environmental Mutagens and Carcinogens recently concluded that there is no clear evidence for any single etiologic factor or combination of factors being responsible for affecting the incidence of Down syndrome (ICPEMC, 1986).

3.5. Effects of Elective Abortion on Incidence Rates

A few studies have been published that attempt to assess whether the availability of elective abortions has an effect on the incidence of Down syndrome. For the purposes here, this procedure may be viewed simply as a means of fertility control. *If* fetuses with Down syndrome are equally likely to be spontaneously aborted at all maternal ages after the usual time of elective abortion (approx. 8–12 weeks), then the question is simply whether the procedure is differentially used by women of different ages. This of course could be asked of any device used by women to control their fertility, and is in some ways a pointless question. Most of the studies were carried out in the United States to determine the effect of abortion reform during the 1970s. Since it is quite clear that elective abortions prevent the birth of many individuals with Down syndrome (as well as of normal infants), the only question of substance is whether they affect the *incidence* of Down syndrome through differential use at different maternal ages. This point is not always clear in the studies available.

The first study was carried out during 1971–1975 in New York to evaluate abortion reform in 1970 (Hansen, 1978). Although incidence rates for Down syndrome were not presented, more pregnancies were aborted in older women, suggesting the incidence would have been higher had these pregnancies been carried to term. However, the conclusion that ". . . abortion reform may have made a significant contribution to the reduction of severe mental retardation" obscures the distinction between causing fewer births overall and reducing the incidence of Down syndrome.

Two other studies (Luthy *et al.*, 1980; Smith *et al.*, 1980) in the states of Washington and Hawaii similarly assessed the impact of elective abortions, and similarly concluded that a positive effect occurred by finding a decline in the incidence of Down syndrome. The relative importance of demographic changes and elective abortions as the basis for this decline is difficult to determine, however. Also, the conclusion by Luthy *et al.* (1980) that ". . . elective abortion has a greater impact on averting DS births than does the antenatal detection program" says little more than reinforce the fact that abortion is an effective method of fertility control. The same conclusion was reached by Mikkelsen *et al.* (1983) when they stated that unrestricted abortion averted 61 births with Down syndrome, in contrast with 31 prevented by amniocentesis.

4. EFFECTS OF PRENATAL DIAGNOSIS ON INCIDENCE RATES OF DOWN SYNDROME

The development of prenatal chromosome diagnosis through amniocentesis around 1970 signaled a new era in preventing the birth of many anomalies, but particularly Down syndrome. As early as 1973, estimates of its potential impact were being made for different levels of usage by women of various ages (Stein *et al.*, 1973). More recent studies proposed similar models of prenatal diagnosis utilization rates among different maternal age groups, and combined these with projections of births with Down syndrome in the United States to determine potential future reductions (Huether, 1983; Goodwin and Huether, 1987). The most recent data showed that an overall reduction of 35–39% in births with Down syndrome can be achieved during the 1990s with utilization rates of 50% for women aged 30–34, and 70% for women ≥ 35 years of age (Goodwin and Huether, 1987). Only the highest rates reported in the world for women 35 and older have currently reached this level. Mikkelsen (1988) reported that a 70% utilization by these women has been achieved for Denmark as a whole, with 75–85% utilization in certain areas.

Estimates of current prenatal diagnosis utilization anywhere in the United States do not approach these values. I could find reports of ten studies in the literature through 1988 that estimate the percentage reduction in births with Down syndrome occurring as a result of prenatal diagnosis (Table V). Such studies ideally require essentially complete ascertainment of both live births as well as fetuses with Down syndrome. Relatively few such population data bases are available in the world that meet these requirements. Although not all published studies on this topic have such data bases, they are nevertheless included in Table V. Some studies (e.g., Luthy *et al.*, 1980) determined the number of births with Down syndrome averted through amniocentesis, but did not include the effect on incidence reduction, and thus were excluded from Table V.

As expected, for most studies reporting data collected since the 1970s, the impact of prenatal diagnosis was minimal, being less than 10% in all studies but one. This exception was Denmark, where even during 1979–1980 the overall reduction in births with Down syndrome through prenatal diagnosis was 20.6% (Mikkelsen *et al.*, 1983). These authors estimated the amniocentesis utilization rate in this time period to be 61% for women ≥ 35 years old. Given this early "lead," it is not surprising that later data from Denmark, for 1980–1982, showed a 26.1% reduction, and for 1983–1985, 27.8%, presumably among the

Table V. Studies That Estimated the Effect of Prenatal Diagnosis (PND) on the Incidence Reduction of Down Syndrome (DS)[a]

Time period	References	Location	Fetuses with DS terminated	% reduction in DS for women: ≥ Age 35	% reduction in DS for women: All ages	Incidence: Without PND	Incidence: With PND
1969–1979	Owens *et al.* (1983)	United Kingdom	**4**	—	0.9	1.40	1.39
1975–1980	Kučera (1982)	Czechoslovakia	**12**	—	**1.0**	—	—
1976–1981	Ferguson-Smith (1983)	W. Scotland	**23**	—	6.3	**1.20**	1.12
1977–1981	Mulcahy (1983)	W. Australia	**14**	27.4	10.8	0.87	**0.78**
1979–1980	Mikkelsen *et al.* (1983)	Denmark	**37**	57.6	20.6	1.08	0.86
1981–1983	Bell *et al.* (1986)	Queensland, Australia	**10**	11.8	4.4	1.31	**1.25**
1974–1978	Baird and Sadovnick (1988b)	British Columbia	**14**	**16.3**	**5.0**	**1.07**	**1.02**
1979–1983			**23**	**23.9**	**7.0**	**1.12**	**1.05**
1978–1981	Valker and Howard (1986)	United Kingdom	**29**	—	**11.3**	**1.13**	**1.00**
1982–1984			**42**	**44.1**	**19.3**	**1.29**	**1.04**
1980–1982	Mikkelsen (1988)	Denmark	**63**	73.3	26.1	1.04	**0.77**
1983–1985			**84**	84.1	27.8	1.35	**0.98**
1978–1981	Priest *et al.* (1988)	Atlanta, Ga.	**10**	**31.8**	**6.0**	1.04	0.98
1982–1985			**21**	**49.5**	**12.9**	1.19	1.04
1986			**12**	**62.7**	**25.9**	1.34	0.99

[a]Numbers in **boldface** are actual values given in each paper. These were derived by multiplying the number of terminated fetuses with Down syndrome by 0.7, which is the probability a fetus with Down syndrome will survive to birth following amniocentesis (Hook, 1983b). Other values shown have been similarly calculated for this table, but were not in the original paper.

highest in the world (Mikkelsen, 1988). The only other published percentage reduction through prenatal diagnosis close to this is for Atlanta during 1986, where the numbers are small but impressive (Priest *et al.*, 1988). They found a greater than 60% reduction for women ≥ 35 years of age and 25.9% for all women.

These data suggest that, for at least some populations and metropolitan areas, we are within striking distance of the maximum reduction in births with Down syndrome currently envisioned through prenatal diagnosis. The Denmark data for the 1980s are consistent with this view, in that the change in percentage reduction is clearly decreasing, so the country might be expected to stabilize by the end of the decade in preventing around a third of the births with Down syndrome. The other studies listed in Table V indicate that most of the rest of the world is well behind this level of prevention. However, from a positive perspective, these higher levels of reduction represent useful paradigms of what can be achieved.

Additionally, none of these studies considered the rapid expansion of prenatal diagnosis currently taking place through chorionic villus sampling, ultrasonography, and the more sensitive screening tests recently available. The work during the 1980s to utilize maternal serum alphafetoprotein levels at younger maternal ages as a means of improving predictive risk could make a substantial contribution to avoidance if widely implemented. Also, the recent work by Wald *et al.* (1988) proposing the use of chorionic gonadotropin and unconjugated estriol concentrations in maternal serum as additional screening devices should be viewed as further opportunities technology is providing. Our challenge is to incorporate these advances into improved genetic services, and to encourage societies to use them for more informed decision making.

5. CONCLUDING REMARKS

Edwards (1988) reminds us that genetic disease is by definition the consequence of a mutational event, and that prevention can only be achieved through preventing mutational events. Thus, one obvious goal is to minimize mutational exposure. Yet the more important reality is that biologic evolution over the millennia has been built on creating and maintaining genetic variation. Without a biologic system capable of mutation, there is no continued capacity to evolve. Therefore, our existence is owed to the ability of our genome to mutate and to maintain a store of genetic variability to deal with past and future environmental changes. In this light, genetic diseases, and aneuploidy and Down syndrome particularly, are more a reality of our biologic than our cultural evolution. Even with the best of intents to control our environmental hazards (hardly where we currently stand), complete prevention is not possible, or at least a long time off.

The significant point is apparently to distinguish between prevention and avoidance. Ferguson-Smith *et al.* (1978) do so by defining prevention as prezygotic, in contrast with allowing postzygotic death to precede birth, which is avoidance. In the first volume of this series, Warkany (1983) made this same distinction more poetically by arguing that prevention of disorders by removal of the patient is hardly a generally accepted principle of medicine. Yet, he went on to emphasize that spontaneous abortion of deformed embryos has long been recognized to be an effective means of prevention (avoidance). Indeed, much of his work and that of others have been devoted to discovering appropriate manipulations to aid in this process (Warkany, 1978). Is there any less subtle a difference between this approach of "spontaneous" versus "therapeutic" abortions, and that of "prevention" versus "avoidance"?

On the more general level of family planning, there is widespread agreement on both sides of the abortion firestorm that prevention of unwanted births through either abstinence or contraception is preferable to elective abortion. But each year recently in the United States, 1.5 million pregnancies out of 5.2 million recognized are electively terminated. This suggests that the prevention–avoidance argument of academics may not be a significant issue in daily life, and that pregnancy termination may provide an effective method for widespread real reductions in the incidence of Down syndrome, given increased service support. The data presented above from Denmark and Atlanta indicate that, at least in these two locations, avoidance of a significant magnitude is already occurring.

REFERENCES

Angell, R. R., Sandison, A., and Bain, A. D. 1984. Chromosome variation in perinatal mortality: a survey of 500 cases. *J. Med. Genet.* **21**:39–44.

Antonarakis, S. E., Kittur, S. D., Metaxotou, C., Watkins, P. C., and Patel, A. S. 1985. Analysis of DNA haplotypes suggests a genetic predisposition to trisomy 21 associated with DNA sequences of chromosome 21. *Proc. Natl. Acad. Sci. USA* **82**:3360–3364.

Antonarakis, S. E., Chakravarti, A., Warren, A. C., Slaugenhaupt, S. A., Wong, C., Halloran, S. L., and Metaxotou, C. 1986. Reduced recombination rate on chromosome 21 that have undergone nondisjunction. *Cold Spring Harbor Symp. Quant. Biol.* **51**:185–190.

Apgar, V., ed. 1970. Down's Syndrome (Mongolism). *Ann. N.Y. Acad. Sci.* **171**:303–688.

Aymé, S., and Lippman-Hand, A. 1982. Maternal-age effect in aneuploidy: does altered embryonic selection play a role? *Am. J. Hum. Genet.* **34**:558–565.

Baird, P. A., and Sadovnick, A. D. 1987. Life expectancy in Down syndrome. *J. Pediatr.* **110**:849–854.

Baird, P. A., and Sadovnick, A. D. 1988a. Life expectancy in Down syndrome adults. *Lancet* **2**:1354–1356.

Baird, P. A., and Sadovnick, A. D. 1988b. Maternal age-specific rates for Down syndrome: changes over time. *Am. J. Med. Genet.* **29**:917–927.

Baird, P. A., Anderson, T. W., Newcombe, H. B., and Lowry, R. B. 1988. Genetic disorders in children and young adults: a population study. *Am. J. Hum. Genet.* **42**:677–693.

Bell, J., Hilden, J., Bowling, F., Pearn, J., Brownlea, A., and Martin, N. 1986. The impact of prenatal diagnosis on the occurrence of chromosome abnormalities. *Prenat. Diagn.* **6**:1–11.

Bernheim, A., Chastang, C., de Heaulme, M., and de Grouchy, J. 1979. Excès de garcons dans la trisomie 21. *Ann. Genet.* **22**:112–114.

Bleyer, A. 1932. The frequency of mongoloid imbecility: the question of race and the apparent influence of sex. *Am. J. Dis. Child.* **44**:503–508.

Bricarelli, F. D., Pierluigi, M., Perroni, L., Grasso, M., Arslanian, A., and Sacchi, N. 1988. High efficiency in the attribution of parental origin of non-disjunction in trisomy 21 by both cytogenetic and molecular polymorphisms. *Hum. Genet.* **79**:124–127.

Burgio, G. R., Fraccaro, M., Tiepolo, L., and Wolf, U., eds. 1981. *Trisomy 21.* Springer, Berlin.

Byrne, J., and Warburton, D. 1987. Male excess among anatomically normal fetuses in spontaneous abortions. *Am. J. Med. Genet.* **26**:605–611.

Chitham, R. G., and MacIver, E. 1965. A cytogenetic and statistical survey of 105 cases of mongolism. *Ann. Hum. Genet.* **28**:309–315.

Church, A., and Peterson, F. 1924. *Nervous and Mental Diseases,* 9th ed. Saunders, Philadelphia.

Colombo, B. 1957. On the sex ratio in man. *Cold Spring Harbor Symp. Quant. Biol.* **22**:193–202.

de la Cruz, F., and Gerald, P. S., eds. 1981. *Trisomy 21 (Down Syndrome), Research Perspectives.* University Park Press, Baltimore.

Down, J. L. H. 1866. Observations on an ethnic classification of idiots. *London Hosp. Rep.* **3**:259–262.

Drew, J. S., London, T. W., Lustbader, E. D., Hesser, J. E., and Blumberg, B. S. 1978. Hepatitis B virus and sex ratio of offspring. *Science* **201**:687–692.

Edmonds, L. D., Layde, P. M., James, L. M., Flynt, J. W., Erickson, J. D., and Oakley, G. P. 1981. Congenital malformations surveillance: two American systems. *Int. J. Epidemiol.* **10**:247–252.

Edwards, J. H. 1988. The importance of genetic disease and the need for prevention. *Philos. Trans. R. Soc. London Ser. B* **319**:211–227.

Erickson, J. D. 1976. The secondary sex ratio in the United States 1969–71: association with race, parental ages, birth order, paternal education and legitimacy. *Ann. Hum. Genet.* **40**:205–212.

Evans, J. A., Hunter, G. W., and Hamerton, J. L. 1978. Down syndrome and recent demographic trends in Manitoba. *J. Med. Genet.* **15**:43–47.

Farag, T. I., Al-Awadi, A. A., Al-Othman, S. A., Sundareshan, T. S., Krishna Murthy, D. S., Usha, R., Mady, S. A., and Uma, R. 1988. Down syndrome and trisomy 18 in the Bedouins. *Am. J. Med. Genet.* **29**:943–944.

Ferguson-Smith, M. A. 1983. Prenatal chromosome analysis and its impact on the birth incidence of chromosomal disorders. *Br. Med. Bull.* **39:**355–364.

Ferguson-Smith, M. A., May, H. M., Vince, J. D., Robinson, H. P., Rawlinson, H. A., Tait, H. A., Gibson, A. A. M., and Ratcliffe, J. G. 1978. Avoidance of anencephalic and spina bifida births by maternal serum alpha-fetoprotein screening. *Lancet* **1:**1330–1333.

Forssman, H., and Akesson, H. O. 1965. Mortality in patients with Down's syndrome. *J. Ment. Defic. Res.* **9:**146–149.

Gallagher, R. P., and Lowry, R. B. 1975. Longevity in Down's syndrome in British Columbia. *J. Ment. Defic. Res.* **19:**157–163.

Gentry, J. T., Parkhurst, E., and Bulin, G. V., Jr. 1959. An epidemiological study of congenital malformations in New York State. *Am. J. Public Health* **49:**497–513.

Goodwin, B. A., and Huether, C. A. 1987. Revised estimates and projections of Down syndrome births in the United States, and the effects of prenatal diagnosis utilization, 1970–2002. *Prenat. Diagn.* **7:**261–271.

Guerrero, R. 1974. Association of the type and time of insemination within the menstrual cycle with the sex ratio at birth. *N. Engl. J. Med.* **291:**1056–1059.

Gustavson, K. H., Hagberg, B., Hagberg, G., and Sars, K. 1977a. Severe mental retardation in a Swedish county: I. Epidemiology, gestational age, birth weight and associated CNS handicaps in children born, 1959–70. *Acta Paediatr. Scand.* **66:**373–379.

Gustavson, K. H., Holmgren, G., Jonsell, R., and Son Blomquist, H. K. 1977b. Severe mental retardation in children in a northern Swedish county. *J. Ment. Defic. Res.* **21:**161–180.

Hafez, J., El-Tahan, M., Zedan, M., and Eisa, M. 1984. Demographic trends of Down's syndrome in Egypt. *Hum. Biol.* **56:**703–712.

Hamers, A., Vaes-Peeters, G., Jongbloed, R., Millington-Ward, A., Meijer, H., de Die-Smulders, C., and Geraedts, J. 1987. On the origin of recurrent trisomy 21: determination using chromosomal and DNA polymorphisms. *Clin. Genet.* **32:**409–413.

Hansen, H. 1978. Decline of Down's syndrome after abortion reform in New York State. *Am. J. Ment. Defic.* **83:**185–188.

Harlap, S. 1979. Gender of infants conceived in different days of menstrual cycle. *N. Engl. J. Med.* **300:**1445–1448.

Hassold, T. J. 1986. Chromosome abnormalities in human reproductive wastage. *Trends Genet.* **2:**105–110.

Hassold, T., and Jacobs, P. A. 1984. Trisomy in man. *Annu. Rev. Genet.* **18:**69–97.

Hassold, T., Chen, N., Funkhouser, J., Jooss, T., Manuel, B., Matsuura, J., Matsuyama, A., Wilson, C., Yamane, J. A., and Jacobs, P. A. 1980. A cytogenetic study of 1000 spontaneous abortions. *Ann. Hum. Genet.* **44:**151–178.

Hassold, T., Quillen, D. S., and Yamane, J. A. 1983. Sex ratio in spontaneous abortions. *Ann. Hum. Genet.* **47:**39–47.

Hassold, T., Chiu, D., and Yamane, J. A. 1984. Parental origin of autosomal trisomies. *Ann. Hum. Genet.* **48:**129–144.

Hassold, T., Jacobs, P. A., Leppert, M., and Sheldon, M. 1987. Cytogenetic and molecular studies of trisomy 13. *J. Med. Genet.* **24:**725–732.

Honoré, L. 1988. Male excess among anatomically normal fetuses in spontaneous abortions. *Am. J. Med. Genet.* **30:**843–844.

Hook, E. B. 1981a. Down syndrome: frequency in human population and factors pertinent to variation in rates, in: *Trisomy 21 (Down Syndrome), Research Perspectives,* F. de la Cruz and P. S. Gerald, eds. University Park Press, Baltimore, pp. 3–67.

Hook, E. B. 1981b. Unbalanced robertsonian translocations associated with Down's syndrome or Patau's syndrome: chromosome subtype, proportion inherited, mutation rates, and sex ratio. *Hum. Genet.* **59:**235–239.

Hook, E. B. 1983a. Perspectives in mutation epidemiology: 3. Contribution of chromosome abnormalities to human morbidity and mortality and some comments upon surveillance of chromosome mutation rates. *Mutat. Res.* **114:**389–423.

Hook, E. B. 1983b. Chromosome abnormalities and spontaneous fetal death following amniocentesis: further data and associations with maternal age. *Am. J. Hum. Genet.* **35:**110–116.

Huang, S. W., Emanuel, I., Lo, J., Liao, S. K., and Hsu, C. C. 1967. A cytogenetic study of 77 Chinese children with Down's syndrome. *J. Ment. Defic. Res.* **11:**147–152.

Huether, C. A. 1983. Projection of Down's syndrome in the United States 1979–2000, and the potential effects of prenatal diagnosis. *Am. J. Public Health* **73:**1186–1189.

Huether, C. A., and Gummere, G. R. 1982. Influence of demographic factors on annual Down's syndrome births in Ohio, 1970–1979, and the United States, 1920–1979. *Am. J. Epidemiol.* **115:**846–860.

Hug, E. 1951. Das Geschlechtsverhaltnis beim mongolismus. *Ann. Paediatr.* **177:**31–54.

Hytten, F. E. 1982. Commentary: boys and girls. *Br. J. Obstet. Gynaecol.* **89:**97–99.

ICPEMC, 1986. International commission for protection against environmental mutagens and carcinogens. ICPEMC meeting report no. 3. Is the incidence of Down syndrome increasing? *Mutat. Res.* **175:**263–266.

Iselius, L., and Lindsten, J. 1986. Changes in the incidence of Down syndrome in Sweden during 1968–1982. *Hum. Genet.* **72:**133–139.

Jacobs, P. A., and Hassold, T. J. 1987. Chromosome abnormalities: origin and etiology in abortions and livebirths, in: *Human Genetics, Proceedings of the 7th International Congress, Berlin, 1986,* F. Vogel and K. Sperling, eds. Springer, Berlin, pp. 233–244.

Janerich, D. T., and Bracken, M. B. 1986. Epidemiology of trisomy 21: a review and theoretical analysis. *J. Chron. Dis.* **39:**1079–1093.

Jenkins, R. L. 1933. Etiology of mongolism. *Am. J. Dis. Child.* **45:**506–519.

Jervis, G. A. 1942. Recent progress in the study of mental deficiency mongolism: a review of the literature of the last decade. *Am. J. Ment. Defic.* **46:**467–481.

Kang, Y. S., and Cho, W. K. 1962. The sex ratio at birth and other attributes of the newborn from maternity hospitals in Korea. *Hum. Biol.* **34:**38–48.

Kashgarian, M., and Rendtorff, R. C. 1969. Incidence of Down's syndrome in American Negroes. *J. Pediatr.* **74:**468–471.

Kellokumpu-Lehtinen, P., and Pelliniemi, L. 1984. Sex ratio of human conceptuses. *Obstet. Gynecol.* **64:**220–222.

Khoury, M., Erickson, J. D., and James, L. M. 1984. Paternal effects on the human sex ratio at birth: evidence from interracial crosses. *Am. J. Hum. Genet.* **36:**1103–1111.

Kiely, M. 1987. The prevalence of mental retardation. *Epidemiol. Rev.* **9:**194–218.

Koulischer, L., and Gillerot, Y. 1980. Down's syndrome in Wallonia (south Belgium), 1971–1978: cytogenetics and incidence. *Hum. Genet.* **54:**243–250.

Krivchenia, E. 1987. Investigation of differences in Down syndrome incidences between metropolitan Atlanta and southwest Ohio, 1970–1985. Masters thesis, University of Cincinnati.

Kučera, J. 1982. Sekundarni prevence Downova syndromu v CSR 1975–1980. *Cesk. Pediatr.* **37:**404–408.

Kuroki, Y., Yamamoto, Y., Matsui, I., and Kurita, T. 1977. Down syndrome and maternal age in Japan, 1950–1973. *Clin. Genet.* **12:**43–46.

Largey, G. P., and Largey, K. A. 1971. Down's syndrome: sex difference in relation to maternal age. *Lancet* **1:**1242.

Lejeune, J., and Prieur, M. 1979. Contraceptifs oraux et trisomie 21: étude rétrospective de sept cent trente cas. *Ann. Genet.* **22:**61–66.

Lilienfeld, A. M. 1969. *Epidemiology of Mongolism.* Johns Hopkins Press, Baltimore.

Lindsjo, A. 1974. Down's syndrome in Sweden: an epidemiological study of a three-year material. *Acta Paediatr. Scand.* **63:**571–576.

Lindsten, J., Marsk, L., Berglund, K., Iselius, L., Ryman, N., Anneren, G., Kjessler, B., Mitelman, F., Nordenson, I., Wahlstrom, J., and Vejlens, L. 1981. Incidence of Down's syndrome in Sweden during the years 1968–1977, in: *Trisomy 21,* G. R. Burgio, M. Fraccaro, L. Tiepolo, and U. Wolf, eds. Springer, Berlin, pp. 195–210.

Lowry, R. B., Jones, D. C., Renwick, D. H. G., and Trimble, B. K. 1976. Down syndrome in British Columbia, 1952–73: incidence and mean maternal age. *Teratology* **14:**29–34.

Luthy, D. A., Emanuel, I., Hoehn, H., Hall, J., and Powers, E. K. 1980. Prenatal genetic diagnosis and elective abortion in women over 35: utilization and relative impact on the birth prevalence of Down syndrome in Washington State. *Am. J. Med. Genet.* **7:**375–381.

Mikkelsen, M. 1970. A Danish survey of patients with Down's syndrome born to young mothers. *Ann. N.Y. Acad. Sci.* **171:**370–378.

Mikkelsen, M. 1981. Epidemiology of trisomy 21: population, peri- and antenatal data, in: *Trisomy 21,* G. R. Burgio, M. Fraccaro, L. Tiepolo, and U. Wolf, eds. Springer, Berlin, pp. 211–226.

Mikkelsen, M. 1982. Parental origin of the extra chromosome in Down's syndrome. *J. Ment. Defic. Res.* **26:**143–151.

Mikkelsen, M. 1985. Down anomaly: new research aspects of an old well known syndrome, in: *Progress in Clinical and Biological Research,* Volume 177, K. Berg, ed. Liss, New York, pp. 293–307.

Mikkelsen, M. 1988. The incidence of Down's syndrome and progress towards its reduction. *Philos. Trans. R. Soc. London Ser. B* **319:**315–324.

Mikkelsen, M., and Aymé, S. 1987. Chromosomal findings in chorionic villi: a collaborative study, in: *Human Genetics, Proceedings of the 7th International Congress, Berlin, 1986,* F. Vogel and K. Sperling, eds. Springer, Berlin, pp. 597–606.

Mikkelsen, M., Fischer, G., Stene, J., Stene, E., and Petersen, E. 1976. Incidence study of Down's syndrome in Copenhagen, 1960–1971: with chromosome investigation. *Ann. Hum. Genet.* **40:**177–182.

Mikkelsen, M., Poulsen, H., Grinsted, J., and Lange, A. 1980. Non-disjunction in trisomy 21: study of chromosomal heteromorphisms in 110 families. *Ann. Hum. Genet.* **44:**17–28.

Mikkelsen, M., Fischer, G., Hansen, J., Pilgaard, B., and Nielsen, J. 1983. The impact of legal termination of pregnancy and prenatal diagnosis on the birth prevalence of Down syndrome in Denmark. *Ann. Hum. Genet.* **47:**123–132.

Mulcahy, M. 1983. The effect of prenatal diagnosis on the incidence of Down syndrome in western Australia. *Aust. N.Z. J. Obstet. Gynaecol.* **23:**197.

Mulcahy, M., and Reynolds, A. 1985. Demographic factors and the incidence of Down's syndrome in Ireland. *J. Ment. Defic. Res.* **29:**113–123.

Murthy, S. K., Murthy, K., Shah, V. C., and Desai, A. B. 1987. Down syndrome in Ahmedabad: karyotype analysis in 60 cases. *Indian J. Pediatr.* **54:**723–727.

Nielsen, J., Jacobsen, P., Mikkelsen, M., Niebuhr, E., and Sorensen, K. 1981. Sex ratio in Down syndrome. *Ann. Genet.* **24:**212–215.

Oster, J. 1953. *Mongolism.* Danish Science Press, Copenhagen.

Owens, J. R., Harris, F., Walker, S., McAllister, E., and West, L. 1983. The incidence of Down's syndrome over a 19-year period with special reference to maternal age. *J. Med. Genet.* **20:**90–93.

Penrose, L. S., and Smith, G. F. 1966. *Down's Anomaly.* Little, Brown, Boston.

Perry, T. B. 1971. Down's syndrome: sex difference in relation to maternal age. *Lancet* **2:**253.

Pilgaard, B., and Mikkelsen, M. 1985. Fald i incidenden af Down's syndrom i Danmark 1980–1982. *Ugeskr. Laeg.* **147:**243–245.

Priest, J. H., Fernhoff, P. M., Elsas, L. J., and Huether, C. A. 1988. Prenatal diagnosis in metropolitan Atlanta and the impact on autosomal trisomies. *Am. J. Obstet. Gynecol.* **159:**1306–1307.

Pueschel, S. M., and Rynders, J. E., eds. 1982. *Down Syndrome: Advances in Biomedicine and the Behavioral Sciences.* Ware Press, Cambridge.

Pueschel, S. M., Tingey, C., Rynders, J. E., Crocker, A. C., and Crutcher, D. M., eds. 1987. *New Perspectives on Down Syndrome.* Brooks, Baltimore.

Pulliam, L. H., and Huether, C. A. 1986. Translocation Down syndrome in Ohio 1970–1981: epidemiologic and cytogenetic factors and mutation rate estimates. *Am. J. Hum. Genet.* **39:**361–370.

Qazi, Q. H., and Lanman, J. T. 1971. Down's syndrome: sex difference in relation to maternal age. *Lancet* **2:**264.

Ros, Y., Hayez-Delatte, F., and van de Poel, H. 1971. Down's syndrome: sex difference in relation to maternal age. *Lancet* **2:**264.

Rosanoff, A. J., and Handy, L. M. 1934. Etiology of mongolism: with special reference to its occurence in twins. *Am. J. Dis. Child.* **48:**764–779.

Rudd, N. L., Dimnik, L. S., Greentree, C., Mendes-Crabb, K., and Hoar, D. I. 1988. The use of DNA probes to establish parental origin in Down syndrome. *Hum. Genet.* **78:**175–178.

Sacchi, N., Gusella, J. F., Perroni, L., Bricarelli, F. D., and Papas, T. S. 1988. Lack of evidence for association of meiotic nondisjunction with particular DNS haplotypes on chromosome 21. *Proc. Natl. Acad. Sci. USA* **85:**4794–4798.

Sharov, T. 1985. High-risk population for Down syndrome: orthodox Jews in Jerusalem. *Am. J. Ment. Defic.* **89:**559–561.

Slavin, R., Kameda, N., and Hamilton, H. 1967. A cytogenetic study of Down syndrome in Hiroshima and Nagasaki. *Jpn. J. Hum. Genet.* **12:**17–28.

Smith, G. F., and Berg, J. M. 1976. *Down's Anomaly,* 2nd ed. Churchill/Livingston, Edinburgh.

Smith, R. G., Gardner, R. W., Steinhoff, P., Chung, C. S., and Palmore, J. A. 1980. The effect of induced abortion on the incidence of Down's syndrome in Hawaii. *Fam. Plann. Perspect.* **12:**201–205.

Spencer, D. A. 1971. Down's syndrome: sex difference in relation to maternal age. *Lancet* **1:**1356.

Stein, Z., Susser, M., Warburton, D., Wittes, J., and Kline, J. 1973. Spontaneous abortion as a screening device: the effect of fetal survival on the incidence of birth defects. *Am. J. Epidemiol.* **102:**275–290.

Stein, Z., Stein, W., and Susser, M. 1986. Attrition of trisomies as a maternal screening device: an explanation of the association trisomy 21 with maternal age. *Lancet* **1:**944–946.

Stewart, G. D., Harris, P., Galt, J., and Ferguson-Smith, M. A. 1985. Cloned DNA probes regionally mapped to human chromosome 21 and their use in determining the origin of nondisjunction. *Nucleic Acid Res.* **3:**4125–4132.

Stewart, G. D., Hassold, T. J., Berg, A., Watkins, P., Tanzi, R., and Kurnit, D. M. 1988. Trisomy 21 (Down syndrome): studying nondisjunction and meiotic recombination by using cytogenetic and molecular polymorphisms that span chromosome 21. *Am. J. Hum. Genet.* **42:**227–236.

Sutherland, G. R., Clisby, S. R., Bloor, G., and Carter, R. F. 1979. Down's syndrome in South Australia. *Med. J. Aust.* **2**:58–61.

Tanaka, R. 1969. Incidence and distribution of Down's syndrome in Iwate district, Japan. *Saishin Igaku* **24**:318–321.

Teitelbaum, M. S., Mantel, N., and Stark, C. 1971. Limited dependence of the sex ratio on birth order and paternal ages. *Am. J. Hum. Genet.* **23**:271–280.

Tonomura, A., Oishi, H., Matsunaga, E., and Kurita, T. 1966. Down's syndrome: a cytogenetic and statistical survey of 127 Japanese patients. *Jpn. J. Hum. Genet.* **11**:1–16.

Uchida, I. A. 1970. Epidemiology of mongolism: the Manitoba study. *Ann. N.Y. Acad. Sci.* **171**:361–369.

Verma, R. S., and Huq, A. 1987. Sex ratio of children with trisomy 21 or Down syndrome. *Cytobios* **51**:145–148.

Visaria, P. M. 1967. Sex ratio at birth in territories with a relatively complete registration. *Eugen. Q.* **14**:132–142.

Wahrman, J., and Fried, K. 1970. The Jerusalem prospective newborn survey of mongolism. *Ann. N.Y. Acad. Sci.* **171**:341–360.

Wald, N. J., Cuckle, H. S., Densem, J. W., Nanchahal, K., Royston, P., Chard, T., Haddow, J. E., Knight, G. J., Palomaki, G. E., and Canick, J. A. 1988. Maternal serum screening for Down's syndrome in early pregnancy. *Br. Med. J.* **297**:883–887.

Walker, S., and Howard, P. J. 1986. Cytogenetic prenatal diagnosis and its relative effectiveness in the Mersey region and north Wales. *Prenat. Diagn.* **6**:13–23.

Warkany, J. 1978. Terathanasia. *Teratology* **17**:353–357.

Warkany, J. 1983. Teratology: spectrum of science, in: *Issues and Reviews in Teratology*. Volume 1, H. Kalter, ed. Plenum Press, New York, pp. 19–31.

Warren, A. C., Chakravarti, A., Wong, C., Slaugenhaupt, S. A., Halloran, S. L., Watkins, P. C., Metaxotou, C., and Antonarakis, S. 1987. Evidence for reduced recombination on the nondisjoined chromosomes 21 in Down syndrome. *Science* **237**:652–654.

Zarfas, D., and Wolf, L. 1979. Maternal age patterns and the incidence of Down's syndrome. *Am. J. Ment. Defic.* **83**:353–359.

Index